INTERPRETATION OF THREE-DIMENSIONAL SEISMIC DATA

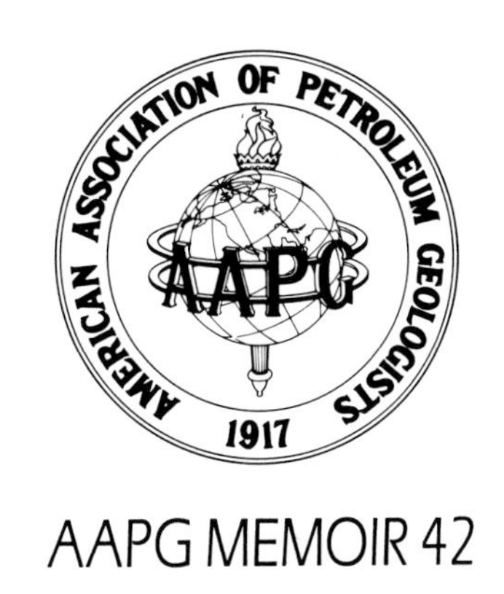

AAPG MEMOIR 42

INTERPRETATION OF THREE-DIMENSIONAL SEISMIC DATA

BY
ALISTAIR R. BROWN
GEOPHYSICAL SERVICE INC.
DALLAS, TEXAS

PUBLISHED BY
AMERICAN ASSOCIATION OF PETROLEUM GEOLOGISTS
TULSA, OKLAHOMA 74101, U.S.A.

Library of Congress Cataloging-in-Publication Data

Brown, Alistair.
Interpretation of three-dimensional seismic data.

(AAPG memoir ; 42)
Includes bibliographies and index.
1. Seismology--Methodology. 2. Seismic reflection method. 3. Petroleum--Geology--Methodology. I. Title. II. Series.
QE539.B78 1986 551.2′2 86-22341
ISBN 0-89181-318-7

Association editor: James Helwig
Science director: Ronald L. Hart
Special Projects manager: Victor V. Van Beuren
Project editor: Douglas A. White, Anne H. Thomas

PREFACE

The whole is more than the sum of the parts.
ARISTOTLE

Three-dimensional seismic data have spawned unique interpretation methodologies. This book is concerned with these methodologies but is not restricted to them. The theme is two-fold:

—How to use 3-D data in an optimum fashion, and
—How to extract the maximum amount of subsurface information from seismic data today.

I have assumed a basic understanding of seismic interpretation which in turn leans on the principles of geology and geophysics. Most readers will be seismic interpreters who want to extend their knowledge, who are freshly confronted with 3-D data, or who want to focus their attention on finer subsurface detail or reservoir properties.

Color is becoming a vital part of seismic interpretation and this is stressed by the proportion of color illustrations herein.

Alistair R. Brown
Dallas, Texas

ACKNOWLEDGMENTS

I have found the writing and organization of this book daunting, challenging and rewarding. But it certainly has not been accomplished without the help of many friends and colleagues. First, I would like to thank GSI and especially Bob Graebner for encouraging the project. Bob Sheriff, University of Houston, has been my mentor in helping me to discover what writing a book entails. Bob McBeath has been a constant help and source of technical advice; he also read all the manuscript. I am indebted to many companies who released data for publication, and also to the many individuals within those companies who provided their data and discussed its interpretation with me. In particular Roger Wright and Bill Abriel, Chevron U.S.A., New Orleans, were outstandingly helpful. Colleagues within GSI who provided significant help were Mike Curtis, Keith Burkart, Tony Gerhardtstein, Chuck Brede, Bob Howard and Jennifer Young. Last but not least, my wife, Mary, remained sane while typing and editing the manuscript on a cantankerous word processor.

CONTENTS

INTRODUCTION

History and Basic Ideas

The earth has always been three-dimensional and the petroleum reserves we seek to find or evaluate are contained in three-dimensional traps. The seismic method, however, in its attempt to image the subsurface has traditionally taken a two-dimensional approach. It was 1970 when Walton (1972) presented the concept of three-dimensional seismic surveys. In 1975, 3-D surveys were first performed on a normal contractual basis, and the following year Bone, Giles and Tegland (1976) presented the new technology to the world.

The essence of the 3-D method is areal data collection followed by the processing and interpretation of a closely-spaced data volume. Because a more detailed understanding of the subsurface emerges, 3-D surveys have been able to contribute significantly to the problems of field appraisal, development and production. It is in these post-discovery phases that many of the successes of 3-D seismic surveys have been achieved and also where their greatest economic benefits have been enjoyed. The scope of 3-D seismic for field development was first reported by Tegland (1977).

In 10 years of 3-D survey experience (1975-85) many successes and benefits have been recorded. Three particular accolades are reproduced here; others are found in the case histories of Chapter 8 and implied at many other places throughout this book.

> *"...there seems to be unanimous agreement that 3-D surveys result in clearer and more accurate pictures of geological detail and that their costs are more than repaid by the elimination of unnecessary development holes and by the increase in recoverable reserves through the discovery of isolated reservoir pools which otherwise might be missed."*
> (Sheriff and Geldart, 1983)

> *"The leverage seems excellent for 3-D seismic to pay for itself many times over in terms of reducing the eventual number of development wells."*
> (West, 1979)

> *"...the 3-D data are of significantly higher quality than the 2-D data. Furthermore, the extremely dense grid of lines makes it possible to develop a more accurate and complete structural and stratigraphic interpretation...Based on this 3-D interpretation, four successful oil wells have been drilled. These are located in parts of the field that could not previously be mapped accurately on the basis of the 2-D seismic data because of their poor quality. This eastward extension has increased the estimate of reserves such that it was possible to declare the field commercial in late 1980."*
> (Saeland and Simpson, 1982)

The fundamental objective of the 3-D seismic method is increased resolution. Resolution has both vertical and horizontal aspects and Sheriff (1985) discusses the subject qualitatively. There

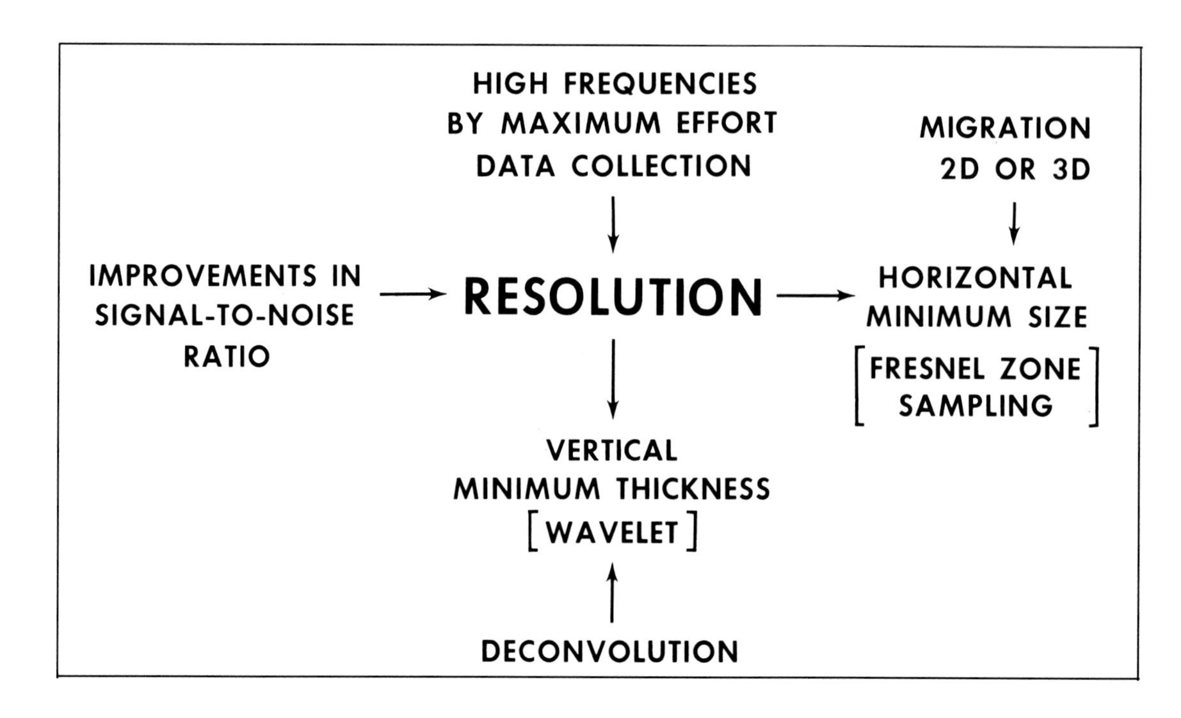

Fig. 1-1. Factors affecting horizontal and vertical seismic resolution.

are several different criteria by which the resolution of seismic data can be measured; Embree (1985) compares these and attempts to be quantitative. Figure 1-1 summarizes resolution issues. Migration is the principal technique for improving horizontal resolution, and deconvolution the principal technique for improving vertical resolution. Wavelet processing and wavelet phase will be considered from an interpretive standpoint in Chapter 2.

Migration performs three distinct functions, all of which relate to seismic resolution. The migration process (1) repositions reflections out-of-place because of dip, (2) focuses energy spread over a Fresnel zone, and (3) collapses diffraction patterns from points and edges. Seismic wavefronts travel in three dimensions and thus it is obvious that all the above are, in general, three-dimensional issues. If we treat them in two dimensions, we can only expect part of the potential improvement. In practice, 2-D lines are often located with strike and dip of major features in mind so that the effect of the third dimension can be minimized, but rarely eliminated.

The accuracy of 3-D migration depends on the velocity field, signal-to-noise ratio, migration aperture and the approach used. Assuming the errors resulting from these factors are small, the data will be much more interpretable both structurally and stratigraphically. Intersecting events will be separated, the confusion of diffraction patterns will be gone, and dipping events will be moved to their correct subsurface positions. The collapsing of energy from diffractions and the focusing of energy spread over Fresnel zones will make amplitudes more accurate and more directly interpretable in terms of reservoir properties.

Examples of 3-D Data Improvement

The interpreter of a 2-D vertical section normally assumes that the data were recorded in one vertical plane below the line traversed by the shots and receivers. The extent to which this is not so depends on the complexity of the structure perpendicular to the line. Figure 1-2 demonstrates that, in the presence of moderate structural complexity, the points at depth from which normal reflections are obtained may lie along an irregular zig-zag track. Only by migrating along *and* perpendicular to the line direction is it possible to resolve where these reflection points belong in the subsurface.

French (1974) demonstrated the value of 3-D migration very clearly in model experiments. He collected seismic data over a model containing two anticlines and a fault scarp (Figure 1-3). Thirteen lines of data were collected although only the results for Line 6 are shown. The raw data have diffraction patterns for both anticlines and the fault so the section appears very confused. The situation is greatly improved with 2-D migration and anticline number 1 (shown in green) is correctly imaged, as Line 6 passed over its crest. However, anticline number 2 (shown in yellow) should not occur on Line 6 and the fault scarp has the wrong slope. The 3-D migration has cor-

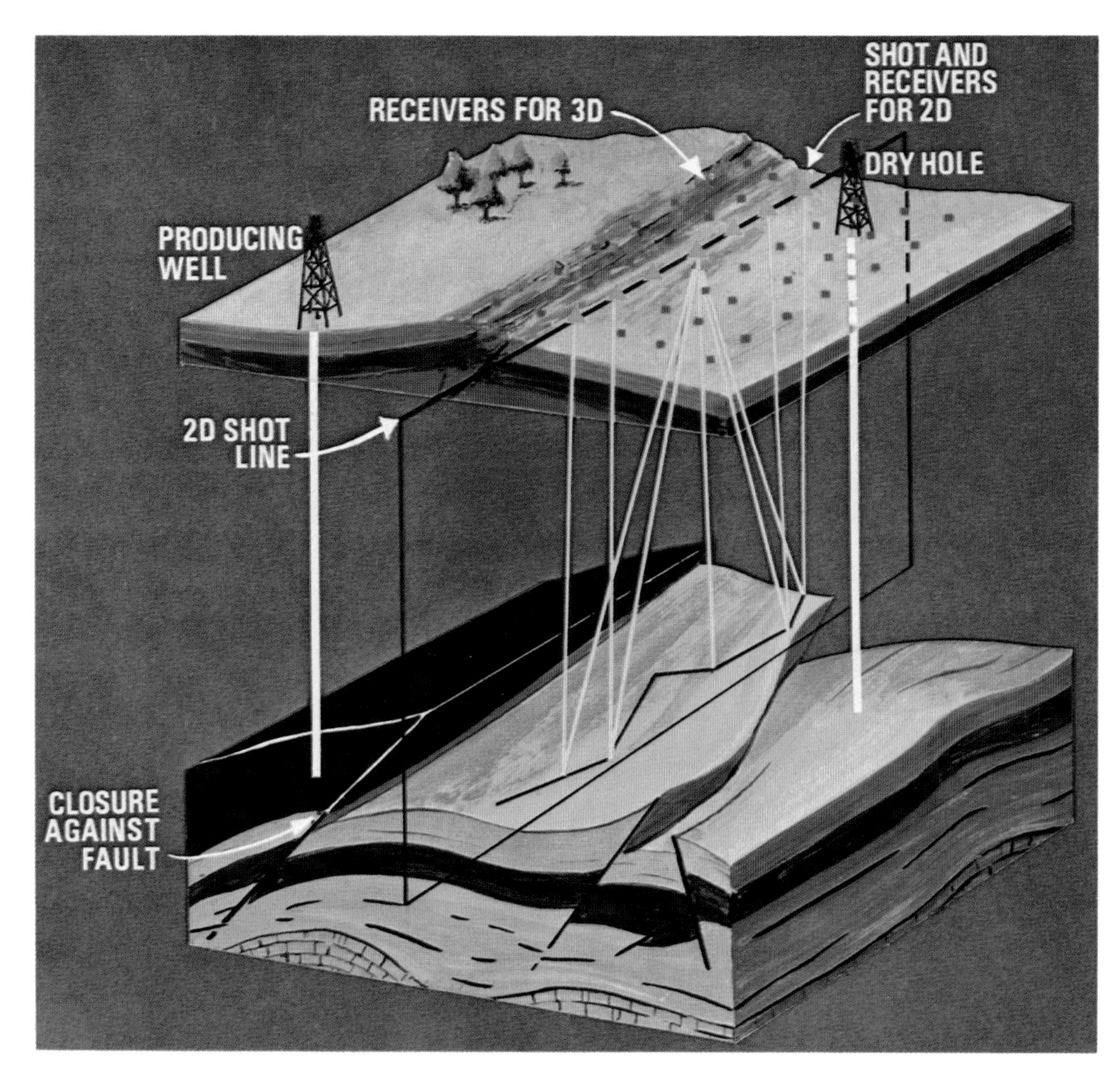

Fig. 1-2. Subsurface structure causes reflection points to lie outside the vertical plane through shots and receivers.

rectly imaged the fault scarp and moved the yellow anticline away from Line 6 to where it belongs.

Figure 1-4 demonstrates this three-dimensional event movement on real data. The same panel is presented before and after 3-D migration for six lines. Here we can observe the movement of a discrete patch of reflectivity to the left and in the direction of higher line numbers.

Figure 1-5 shows improved continuity of an unconformity reflection. The 2-D migration has collapsed most of the diffraction patterns but some confusion remains. The crossline component of the 3-D migration removes energy not in the plane of this section and clarifies the shape of the unconformity surface in detail.

Figure 1-6 shows the effect of 3-D migration in enhancing the visibility of a fluid contact reflection by removing energy not belonging in the plane of this section.

Figures 1-7 and 1-8 are portions of three lines passing through and close to a salt diapir. Line 180 shows steeply-dipping reflections at the edge of the salt mass brought into place by the 3-D migration. Line 220 shows an apparent anticline which is caused by reflections dipping up steeply toward the salt face in a plane perpendicular to that in Figure 1-8. In this prospect, 3-D migration imaged reflections underneath a salt overhang and provided valuable detail about traps located there against the salt face (Blake, Jennings, Curtis, Phillipson, 1982).

When comparing sections before and after 3-D migration to appraise its effectiveness, it is important to bear in mind the way in which reflections have moved around. In the presence of dip perpendicular to the section under scrutiny the visible data before and after 3-D migration are different. It is unreasonable to compare detailed character and deduce what 3-D migration did. It is possible to compare a section before 3-D migration with the one from the same location

Fig. 1-3. Model of two anticlines and one fault with seismic data along Line 6 showing comparative effects of 2-D and 3-D migration (from French, 1974).

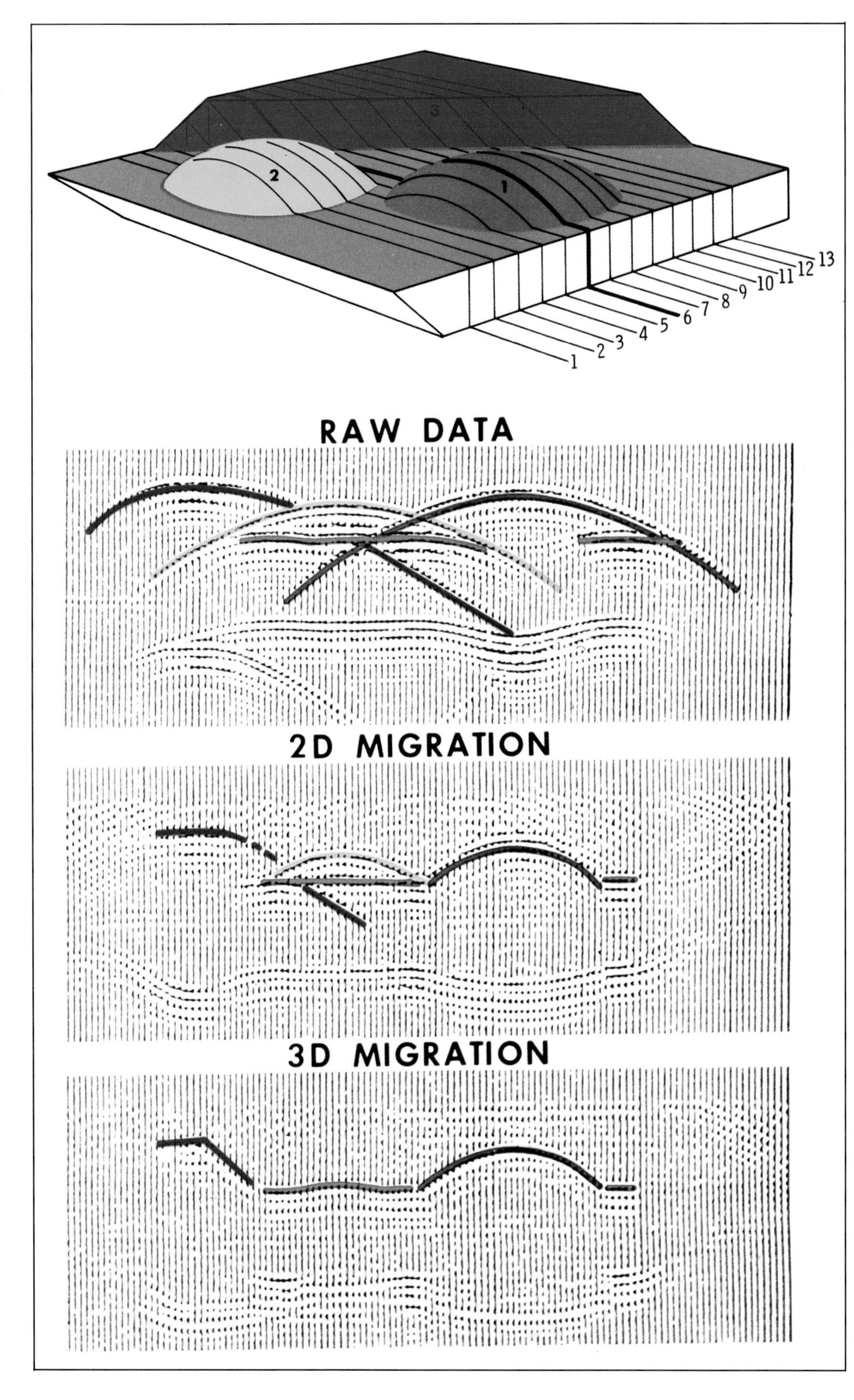

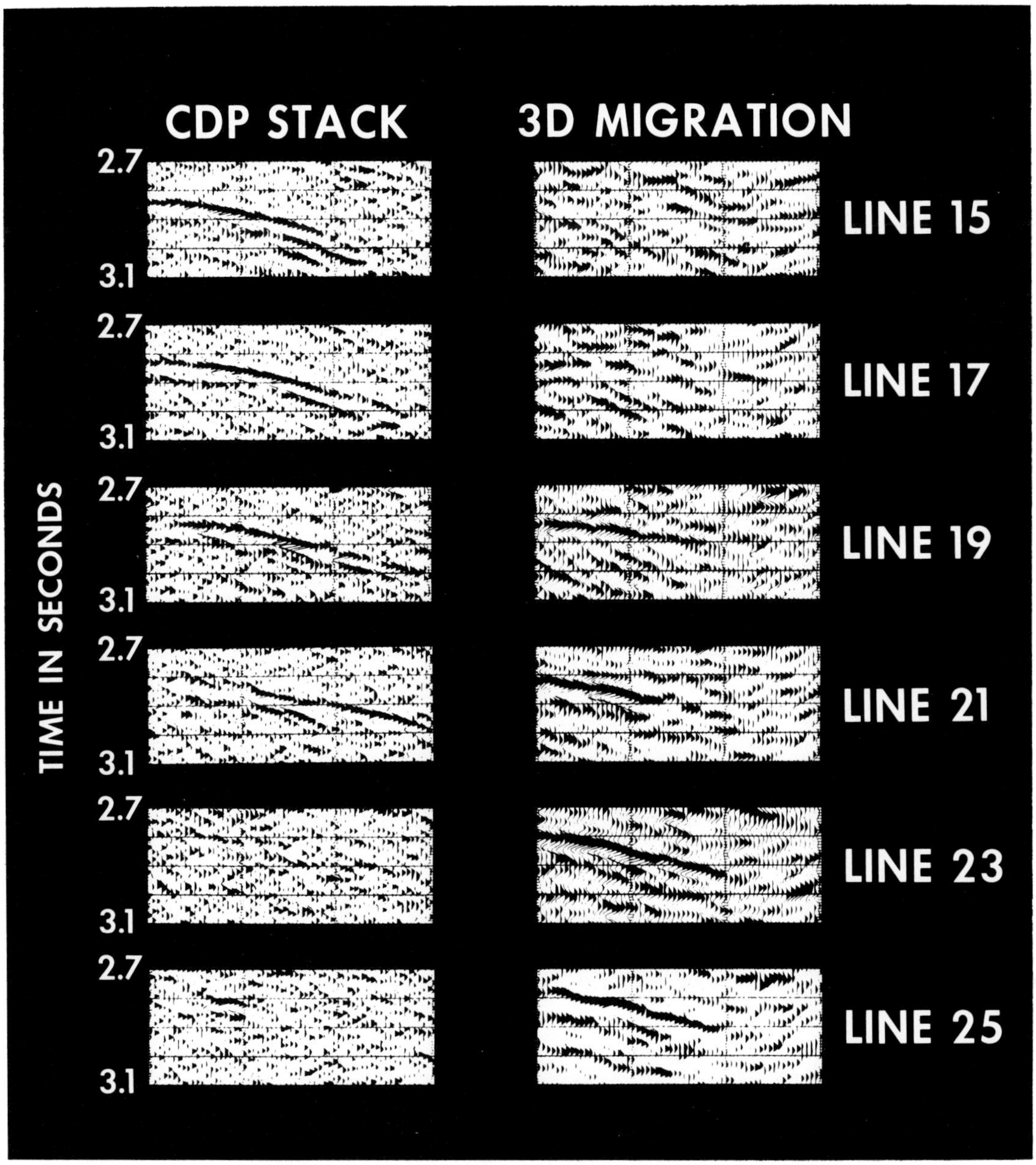

Fig. 1-4. Three-dimensional movement of a dipping reflection by 3-D migration.

after 3-D migration and find that a good quality reflection has disappeared. The migrated section is not consequently worse; the good reflection has simply moved to its correct location in the subsurface.

Sampling Requirements

The sampling theorem requires that, for preservation of information, a waveform must be sampled such that there are at least two samples per cycle for the highest frequency. Since the beginning of the digital era, we have been used to sampling a seismic trace in time. For example, 4 ms sampling is theoretically adequate for frequencies up to 125 Hz. In practice, because of system imperfections, we normally require at least three samples per cycle for the highest frequency. With this safety margin, 4 ms sampling is adequate for frequencies up to 83 Hz.

In space, the sampling theorem translates to the requirement of at least two, and preferably three, samples per shortest wavelength in every direction. In a normal 2-D survey layout this will be satisfied by the depth point spacing along lines but not by the spacing between lines. Hence the restriction that widely-spaced 2-D lines can be processed individually on a 2-D basis but not together as a 3-D volume.

If the sampling theorem is not satisfied the data are aliased. In the case of a dipping event, the

Fig. 1-5. Improved structural continuity of an unconformity reflection resulting from 2-D and 3-D migration.

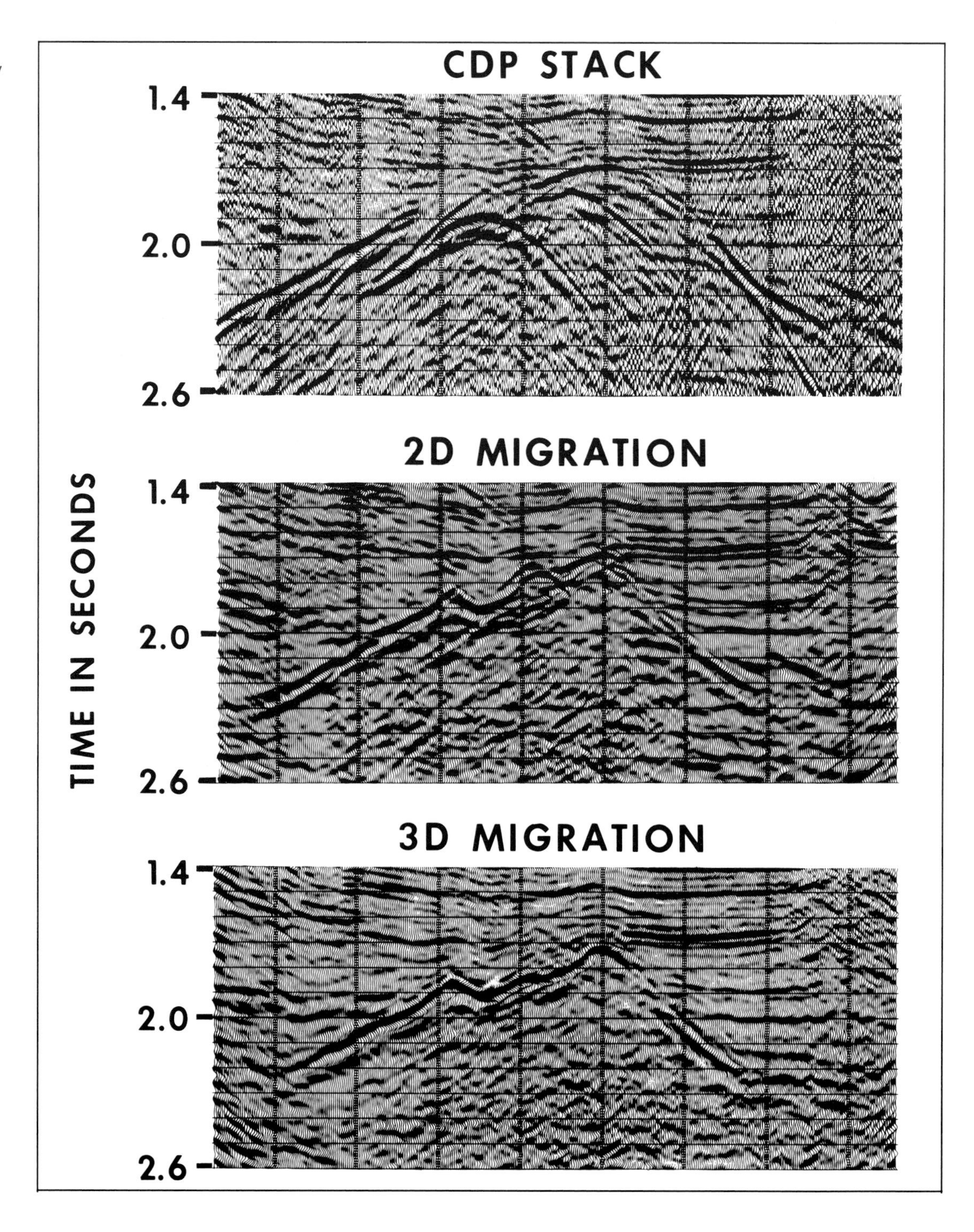

Table 1-1. Alias frequency (in hertz) as a function of subsurface spacing (in meters) and dip (in degrees) for an RMS velocity of 2500 m/s.

	SUBSURFACE SPACING				
DIP	12.5	25	50	75	100
5	574	287	143	96	72
10	288	144	72	48	36
15	193	96	48	32	24
20	146	73	37	24	18
25	118	59	30	20	15

spatial sampling of that event must be such that its principal alignment is obvious; if not, aliases occur and spurious dips result after multichannel processing. Table 1-1 shows the frequencies at which this aliasing occurs for various dips and subsurface spacings. Clearly, a 3-D survey must be designed such that aliasing during processing does not occur. Tables like the one presented can be used to establish the necessary spacing considering the dips and velocities present. In order to impose the safety margin of three samples, rather than two, per shortest wavelength, the frequency limit is normally considered to be two-thirds of each number tabulated.

Proper design of a 3-D survey is critical to its success. Sufficiently close spacing must always be the first consideration (others include adequate migration aperture and CDP fold). Figure 1-9 demonstrates a typical comparison between the subsurface sampling of a 2-D and 3-D survey. The bold dots indicate the 2-D survey depth points which satisfy the sampling theorem along

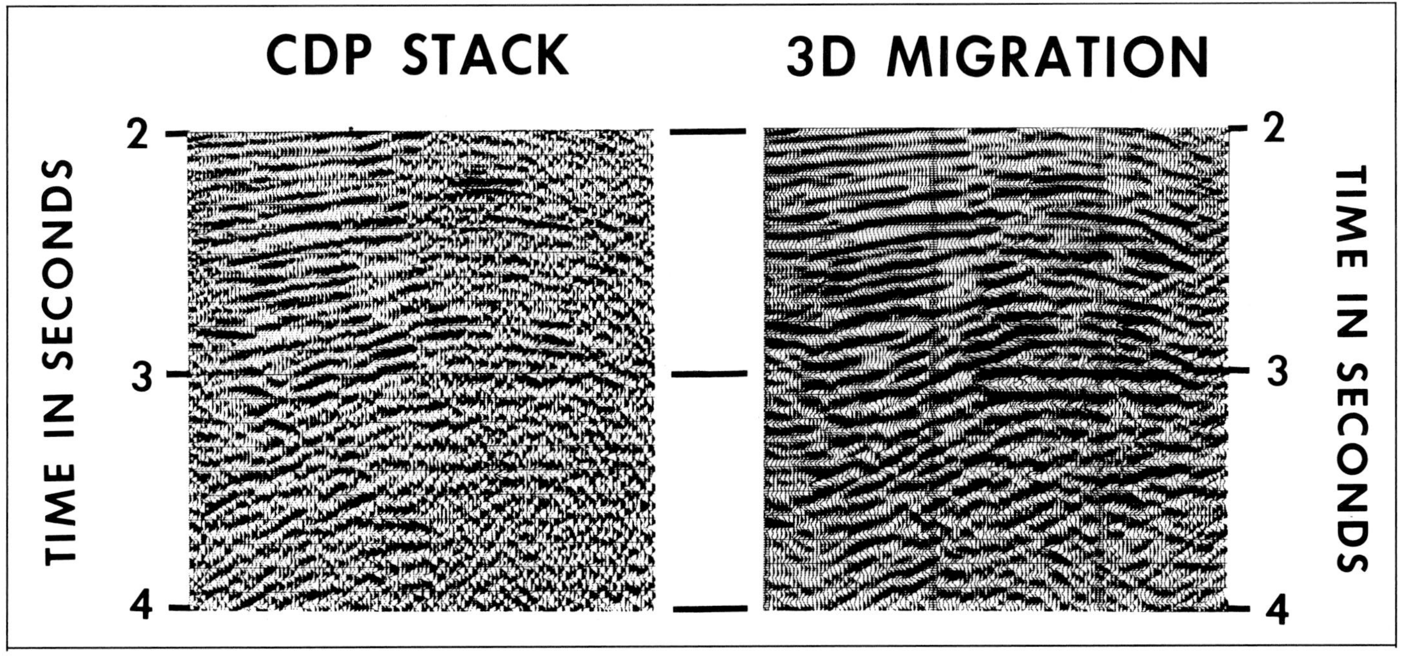

Fig. 1-6. Improved visibility of a flat spot reflection after removal of interfering events by 3-D migration.

each line. The 3-D survey requires the same spacing in both directions over the whole area. In addition to the opportunity for three-dimensional processing which the areal coverage provides, note the sampling and thus potential definition of a meandering stream channel. In practice, 3-D depth point spacing ranges between 12 and 100 m.

Volume Concept

Collection of closely-spaced seismic data over an area permits three-dimensional processing of the data as a volume. The volume concept is equally important to the seismic interpreter. With 3-D data, the interpreter is working directly with a volume rather than interpolating a volumetric interpretation from a widely-spaced grid of observations. The handling of this volume and what can be extracted from it are principal subjects of this book. One property of the volume pervades everything the 3-D interpreter does: The subsurface seismic wavefield is closely sampled in every direction, so that there is no grid loop around which the interpreter must tie, and no grid cell over which he must guess at the subsurface structure and stratigraphy. Hence, the 3-D interpreter will generate a more accurate map than his 2-D predecessor in the same area.

Figure 1-10 shows a view of a 3-D data volume through a salt dome. It demonstrates the volume concept well and the interpreter can use a display of this kind to help in appreciation of subsurface three-dimensionality. Figure 1-11 shows another cube, in this case generated interactively, which helps in the three-dimensional appreciation of a much more detailed subsurface objective. Neither of these displays, however, permits the interpreter to look *into* the volume of data.

True 3-D display of a volumetric image is a difficult problem. Nelson (1983) reviewed the applicable technologies but all fall short of what the seismic interpreter really needs. Most address very small volumes of data and also lack dynamic range. The author has personally experimented with holography and several seismic data holograms exist. However, the interpreter cannot interact with the image and the dynamic range is inadequate for most purposes.

The most useful approach to true 3-D display of a seismic data volume is the Seismodel Display Unit (Figure 1-12). Here individual vertical sections from the volume are printed onto transparent plastic plates which are positioned in accurately-engineered grooves in a metal box. A stack of these plates is then illuminated from behind and viewed from in front. The interpreter may look along a fault plane or down the axis of a structure. A plate may be removed, as shown in Figure 1-12, marked by the interpreter, and then returned to its place in the stack. The interpretation on this one section can then be seen in relationship to other adjacent sections and other plates can be marked accordingly. The principal shortcoming of the Seismodel Display Unit is that, in order to increase the transparency of each plate, the data have to be displayed with a very

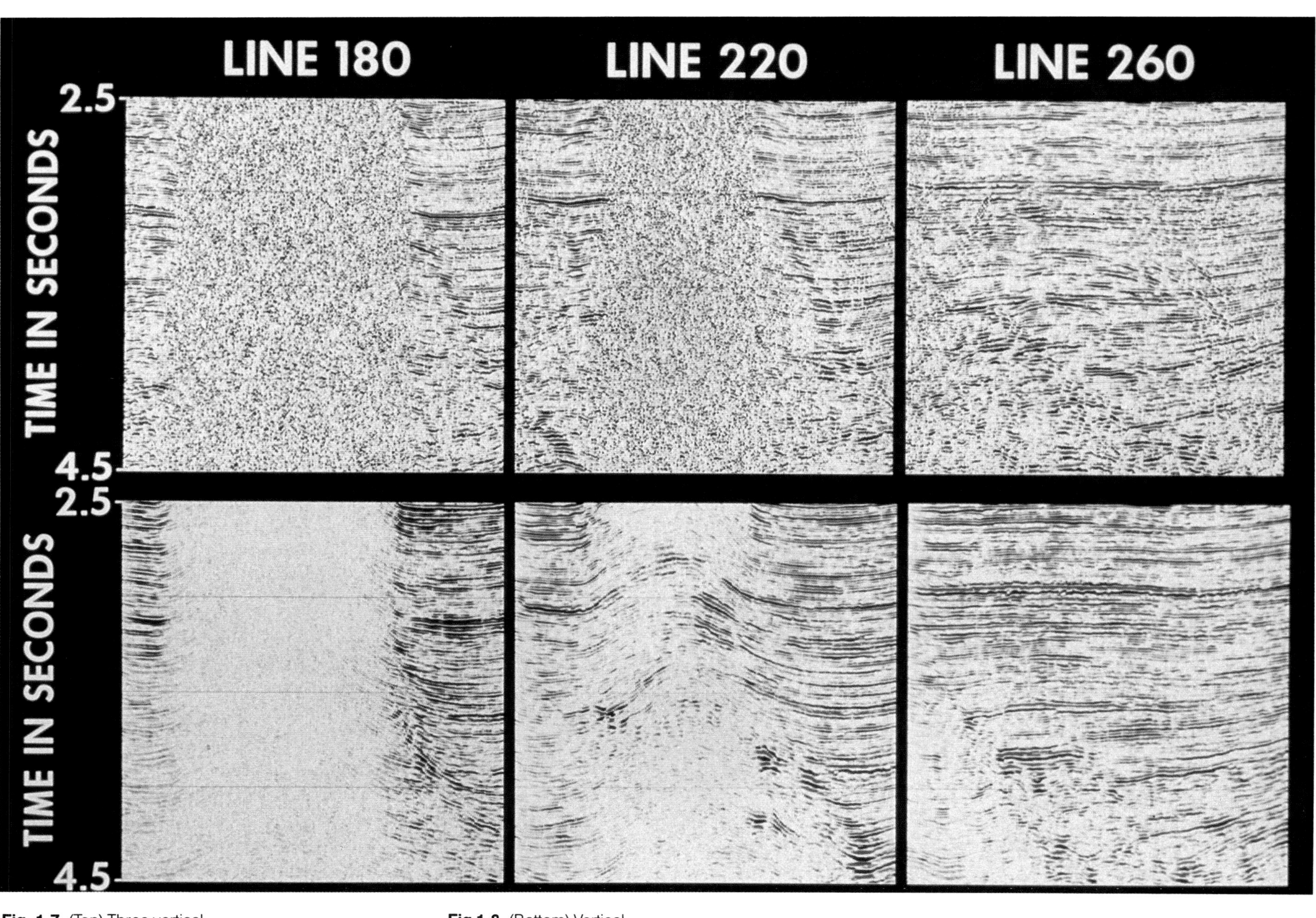

Fig. 1-7. (Top) Three vertical sections through or adjacent to a Gulf of Mexico salt dome before migration. (Courtesy Hunt Oil Company.)

Fig 1-8. (Bottom) Vertical sections along the same lines after 3-D migration showing the repositioning of several reflections adjacent to the salt face. (Courtesy Hunt Oil Company.)

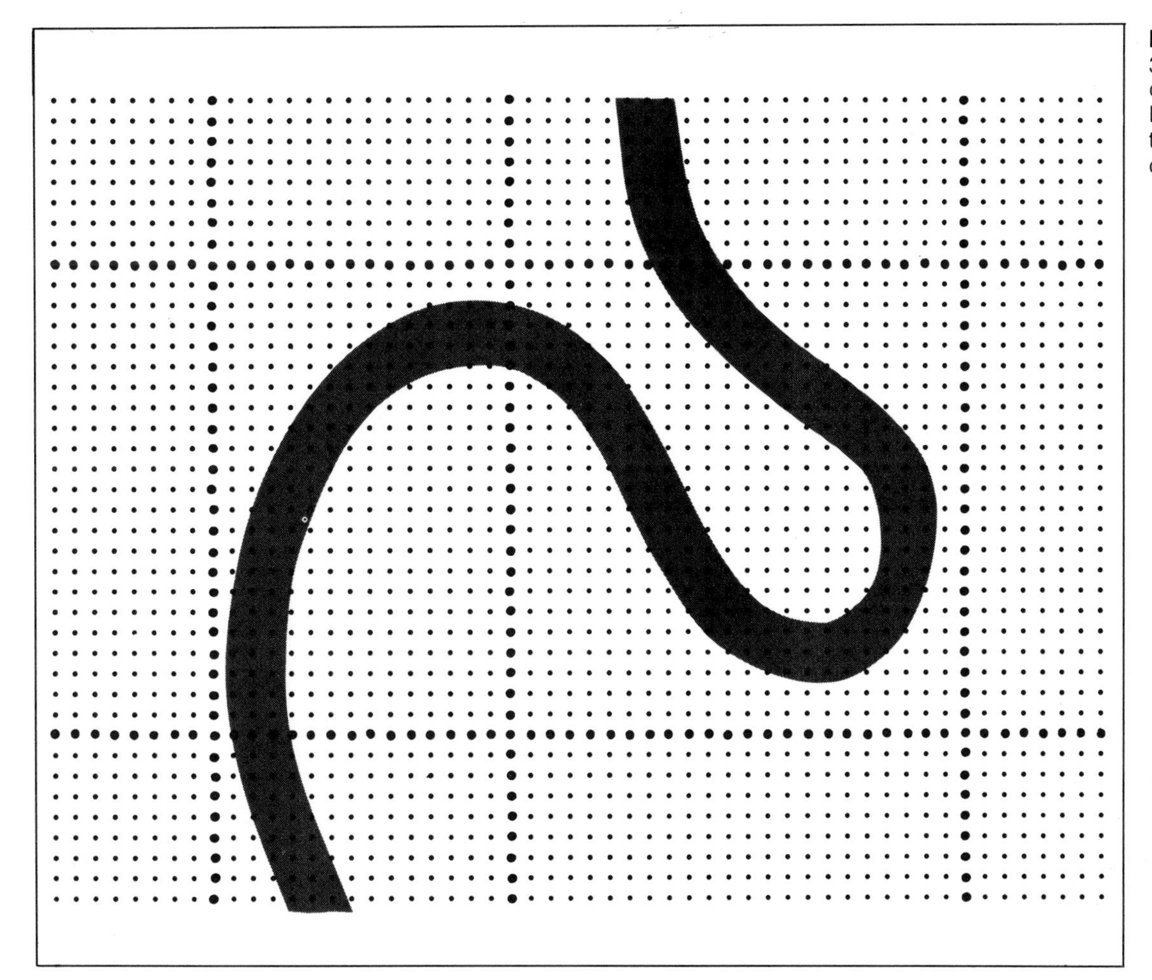

Fig. 1-9. Areal coverage of a 3-D survey compared to the coverage of a grid of five 2-D lines, and the ability of each to delineate a meandering channel.

low gain and peaks only. Hence the dynamic range of the displayed data is low, restricting its use to structural interpretation.

Slicing the Data Volume

The vast majority of 3-D interpretation is performed on slices through the data volume. There are no restrictions on the dynamic range for the display of any one slice, and therefore all the benefits of color, dual polarity, etc., can be exploited (see Chapter 2). The 3-D volume contains a regularly-spaced orthogonal array of data points defined by the acquisition geometry and probably adjusted during processing. The three principal directions of the array define three sets of orthogonal slices or sections through the data, as shown in Figure 1-13.

The vertical section in the direction of boat movement or cable set-up is called a **line** (sometimes an **inline**). The vertical section perpendicular to this is called a **crossline**. The horizontal slice is called a **horizontal section**, **Seiscrop*** **section**, or **time slice**. The terminology used for slices through 3-D data volumes has become somewhat confused. One of the objectives of this chapter is to clarify terms in common use today.

Three sets of orthogonal slices through the data volume (as defined above) are regarded as the basic equipment of the 3-D interpreter. However, many other slices through the volume are possible. A **diagonal line** may be extracted to tie two locations of interest, such as wells. A zig-zag sequence of diagonal line segments may be necessary to tie together all the wells in a prospect. In the planning stages for a production platform, a diagonal line may be extracted through the platform location along the intended azimuth of a deviated well. All these are vertical sections and are often referred to as **usertracks.**

More complicated slices are possible for special applications. A slice along or parallel to a structurally interpreted horizon, and thus parallel to bedding planes, is a **horizon slice** or **horizon**

*Trademark of Geophysical Service Inc.

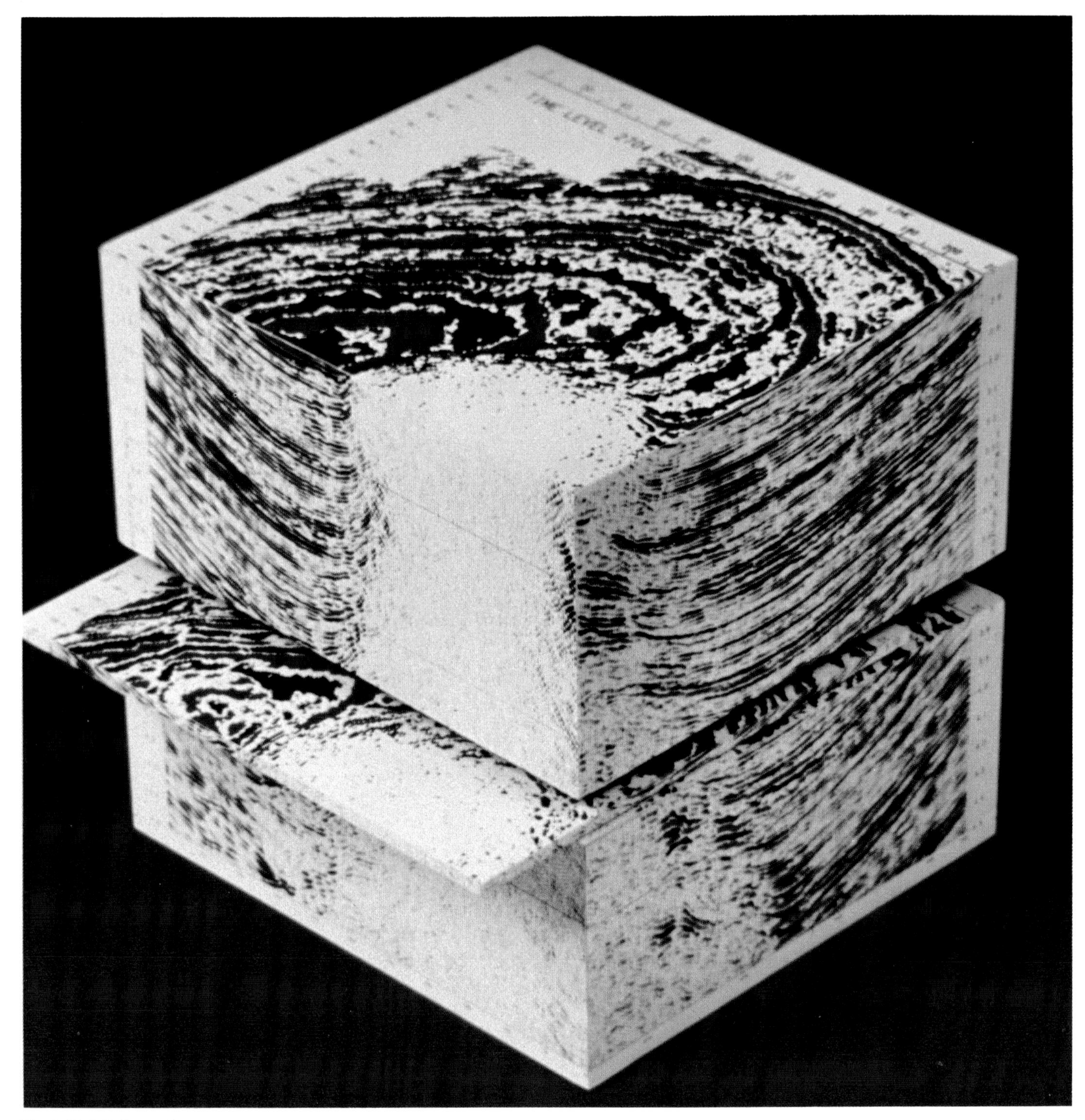

Fig. 1-10. 3-D data volume showing a Gulf of Mexico salt dome and associated rim syncline. (Courtesy Hunt Oil Company)

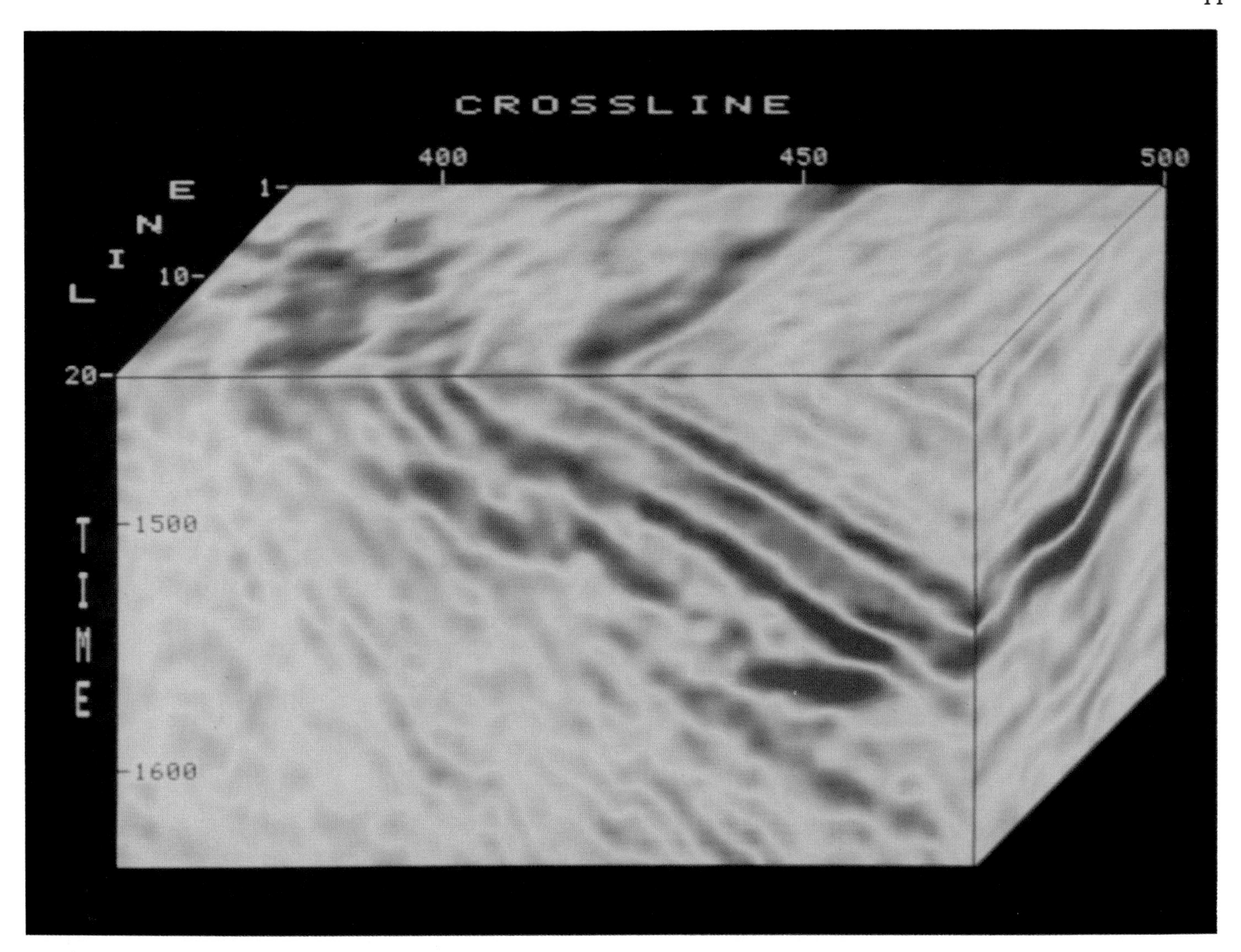

Fig. 1-11. 3-D data volume showing a bright spot from a Gulf of Mexico gas reservoir. (Courtesy Chevron U.S.A. Inc.)

Fig. 1-12. Seismodel Display Unit, one of the approaches to true 3-D display.

Seiscrop section. Slices of this kind have particular application for stratigraphic interpretation which is explored in Chapter 4. Slices can also be generated parallel to a fault face and along the surface defined by two deviated wells.

Manipulating the Slices

Because 3-D interpretation is performed with data slices and because there are a very large number of slices for a typical data volume, several innovative approaches for manipulating the data have emerged. In the early days of 3-D development a sequence of horizontal sections was displayed on film-strip and shown as a motion picture (Bone, Giles, Tegland, 1983). From this

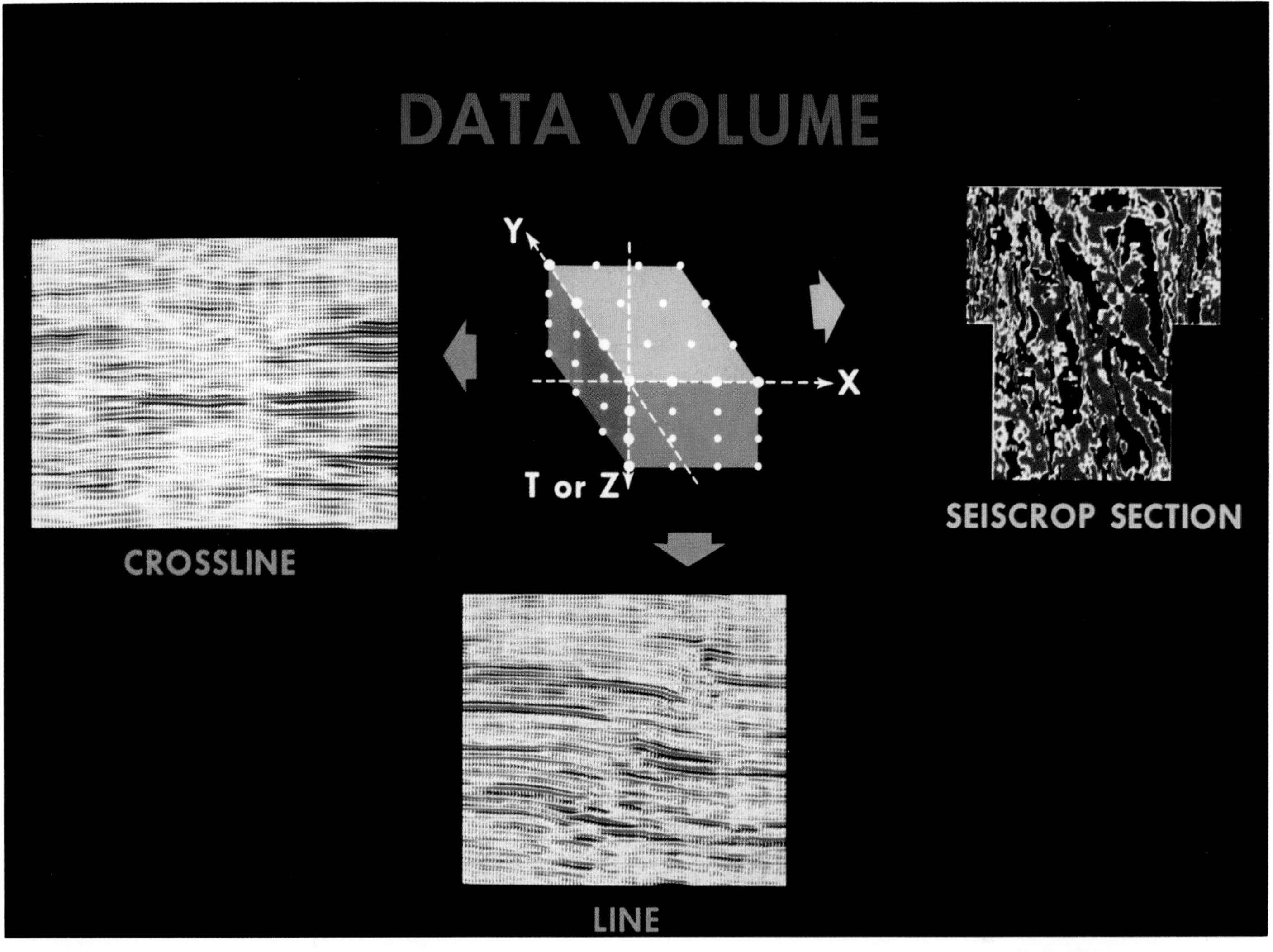

Fig. 1-13. Three sets of orthogonal slices through a data volume provide the basic equipment of the 3-D seismic interpreter.

developed the Seiscrop Interpretation Table — initially a commercially-available piece of equipment incorporating a 16mm analytical movie projector. This machine was originally developed for coaches wanting to examine closely the actions of professional athletes.

The Seiscrop Interpretation Table of today (Figure 1-14) is a custom-built device. The data, either horizontal or vertical sections, are projected from 35mm film-strip onto a large screen. The interpreter fixes a sheet of transparent paper over the screen for mapping and then adjusts the size of the data image, focus, frame advance, or movie speed by simple controls.

Today much 3-D interpretation is performed interactively (Gerhardstein and Brown, 1984). The interpreter calls the data from disk and views them on the screen of a color monitor (Figure 1-15). The large amount of regularly-organized data in a 3-D volume gives the interactive approach enormous benefits. In fact, many interactive interpretation systems addressed 3-D data first as the easier problem, and then developed 2-D interpretation capabilities later.

Most of the interpretation discussed in this book resulted from use of an interactive system, and most of the data illustrations are actual screen photographs. Furthermore, the facilities of the system contributed in several significant ways to the success of many of the projects reported here. Hence it is appropriate to review the interpretive benefits of an interactive interpretation system.

(1) **Data management** — The interpreter needs little or no paper; the selected seismic data display is presented on the screen of a color monitor and the progressive results of interpretation are returned to the digital database.

(2) **Color** — Flexible color display provides the interpreter with maximum optical dynamic range adapted to the particular problem under study.

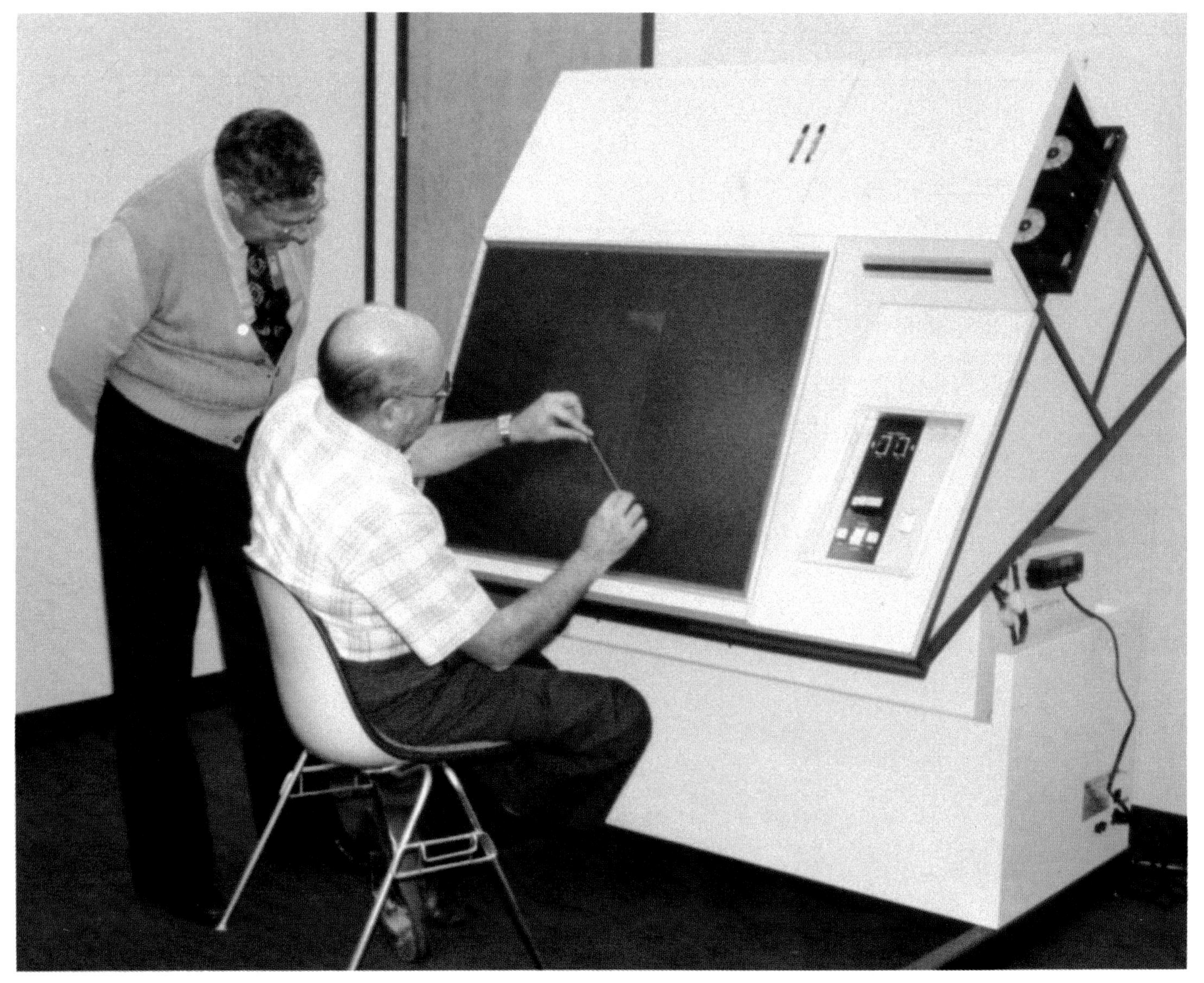

Fig. 1-14. Seiscrop Interpretation Table.

(3) **Composition of images** — Data images can be composed on the screen so that the interpreter views only those pieces of data needed, no more and no less, for the study of one particular issue. Pieces of data may be brought into juxtaposition for comparison or to help appreciate three-dimensionality.

(4) **Idea flow** — The rapid response of the system makes it easy to try new ideas. The interpreter can rapidly generate innovative map or section products in pursuit of a better interpretation.

(5) **Time** — Several automatic facilities, including tracking in two and three dimensions, posting, and contouring, save the interpreter significant amounts of time which can instead be devoted to critical interpretive decision making.

Synergism and Pragmatism in Interpretation

Seismic technology has, over the years, become increasingly complex. Whereas a party chief used to handle data collection, processing, and interpretation, experts are now increasingly restricted to each discipline. Data processing involves many highly sophisticated operations and is conducted in domains unfamiliar to the nonmathematically-minded interpreter. The ability of certain processes to transform data in adverse as well as beneficial ways is striking.

Today's seismic interpreter must understand in some detail what has been done to the data and must understand data processing well enough to ask meaningful questions of the processing staff. Today's interpreter will also benefit greatly by using high technology aids, such as an inter-

Fig. 1-15. Seismic Interactive Data Interpretation System in action.

active system. Critical to maximum effectiveness is an understanding of the advantages of color and how to work with horizontal sections, acoustic impedance sections, frequency sections, and vertical seismic profiles.

Seismic interpretation today thus involves a wide range of seismic technologies. If the results of these are studied by the interpreter in concert, significant synergism can result. However, pragmatism retains its place. The interpreter must continue to take a broad view, to integrate geology and geophysics and to make simplifying assumptions in order to get the job done. The progress of seismic interpretation depends on the continued coexistence of technological synergism and creative pragmatism.

References

Blake, B. A., J. B. Jennings, M. P. Curtis, and R. M. Phillipson, 1982, Three-dimensional seismic data reveals the finer structural details of a piercement salt dome: Offshore Technology Conference paper 4258, p. 403–406.

Bone, M. R., B. F. Giles, and E. R. Tegland, 1976, 3-D high resolution data collection, processing and display: Houston, Texas, presented at 46th Annual SEG Meeting.

———, ———, and ———, 1983, Analysis of seismic data using horizontal cross-sections: Geophysics, v. 48, p. 1172–1178.

Embree, P., 1985, Resolution and rules of thumb: Monterey, California, presented at SEG seismic field techniques workshop.

French, W. S., 1974, Two-dimensional and three-dimensional migration of model-experiment reflection profiles: Geophysics, v. 39, p. 265–277.

Gerhardstein, A. C., and A. R. Brown, 1984, Interactive interpretation of seismic data: Geophysics, v. 49, p. 353–363.

Nelson, H. R., Jr., 1983, New technologies in exploration geophysics: Houston, Texas, Gulf Publishing Company, p. 187-206.

Saeland, G. T., and G. S. Simpson, 1982, Interpretation of 3-D data in delineating a sub-unconformity trap in Block 34/10, Norwegian North Sea, *in* M. T. Halbouty, ed., The deliberate search for the subtle trap: AAPG Memoir 32, p. 217–236.

Sheriff, R. E., 1985, Aspects of seismic resolution, *in* O. R. Berg and D. Woolverton, eds., Seismic stratigraphy II: an integrated approach to hydrocarbon exploration: AAPG Memoir 39, p. 1-10.

——— and L. P. Geldart, 1983, Exploration seismology; v. 2, data-processing and interpretation: Cambridge University Press, p. 130.

Tegland, E. R., 1977, 3-D seismic techniques boost field development: Oil and Gas Journal, v. 75, no. 37, p. 79–82.

Walton, G. G., 1972, Three-dimensional seismic method: Geophysics, v. 37, p. 417–430.

West, J., 1979, Development near for Thailand field: Oil and Gas Journal, v. 77, no. 32, p. 74-76.

COLOR, CHARACTER AND ZERO-PHASENESS

Color Principles

"The total quantity of information recorded on a typical seismic line is enormous. It is virtually impossible to present all this information to the user in a comprehensible form." This quotation from Balch (1971) is even more true today than it was in 1971 and color has become an important contributor to the problem's solution. The human eye is very sensitive to color and the seismic interpreter can make use of this sensitivity in several ways. Taner and Sheriff (1977) and Lindseth (1979) were among the first to present color sections which demonstrated the additional information color can convey. Of equal importance is the increased optical dynamic range of a color section compared to its black and white variable area/wiggle trace equivalent. Both these properties are of great importance in stratigraphic interpretation.

Some understanding of color principles will help an interpreter maximize the use of color. It is helpful to visualize colors as a three-dimensional solid but there are three relevant sets of coordinates in terms of which the color solid can be expressed:

(1) the three additive primary colors — red, green, blue;
(2) the three subtractive primary colors — magenta, yellow, cyan; and,
(3) hue, saturation, density.

Figure 2-1 is a diagrammatic representation of a color cube showing the interrelationship of the above sets of coordinates. Figure 2-2 is a photograph of an actual color cube oriented to correspond to the diagram of Figure 2-1. Figure 2-3 is a photograph of the same cube from the opposite direction.

This cube was made using an Applicon color plotter, but the principles under discussion are independent of the plotting device. Any system which combines pigments employs the subtractive primaries — magenta, yellow and cyan. Figure 2-2 and 2-3 show the *absence* of any color, which is *white*, at the top and progressively increasing quantities of magenta, yellow and cyan down the upper edges of the cube. These primaries, paired in equal quantities, give the additive primaries, red, green and blue, at the three lower corners. All three subtractive primaries combined in equal quantities give black, seen at the bottom apex of the cube.

Any display system which combines light, such as a color monitor, follows the cube of Figures 2-2 and 2-3 from *bottom to top*. The *absence* of color is then *black*. Light of the three additive primary colors, red, green and blue, combine in pairs to make magenta, yellow and cyan and altogether to make white.

The cube photographs display only those colors on the surface of the cube. In fact, a much larger number of colors is inside. Down the vertical axis from white to black is the gray scale for which the **density** increases progressively (Figure 2-1). The **saturation** measures the distance from this central axis, ranging from zero on the axis to 100% on the surface of the cube. The **hue** is the rotational parameter measuring the spectral content of a color.

For the color cube illustrated in Figures 2-2 and 2-3 there are 17 levels (0-16) of each of the

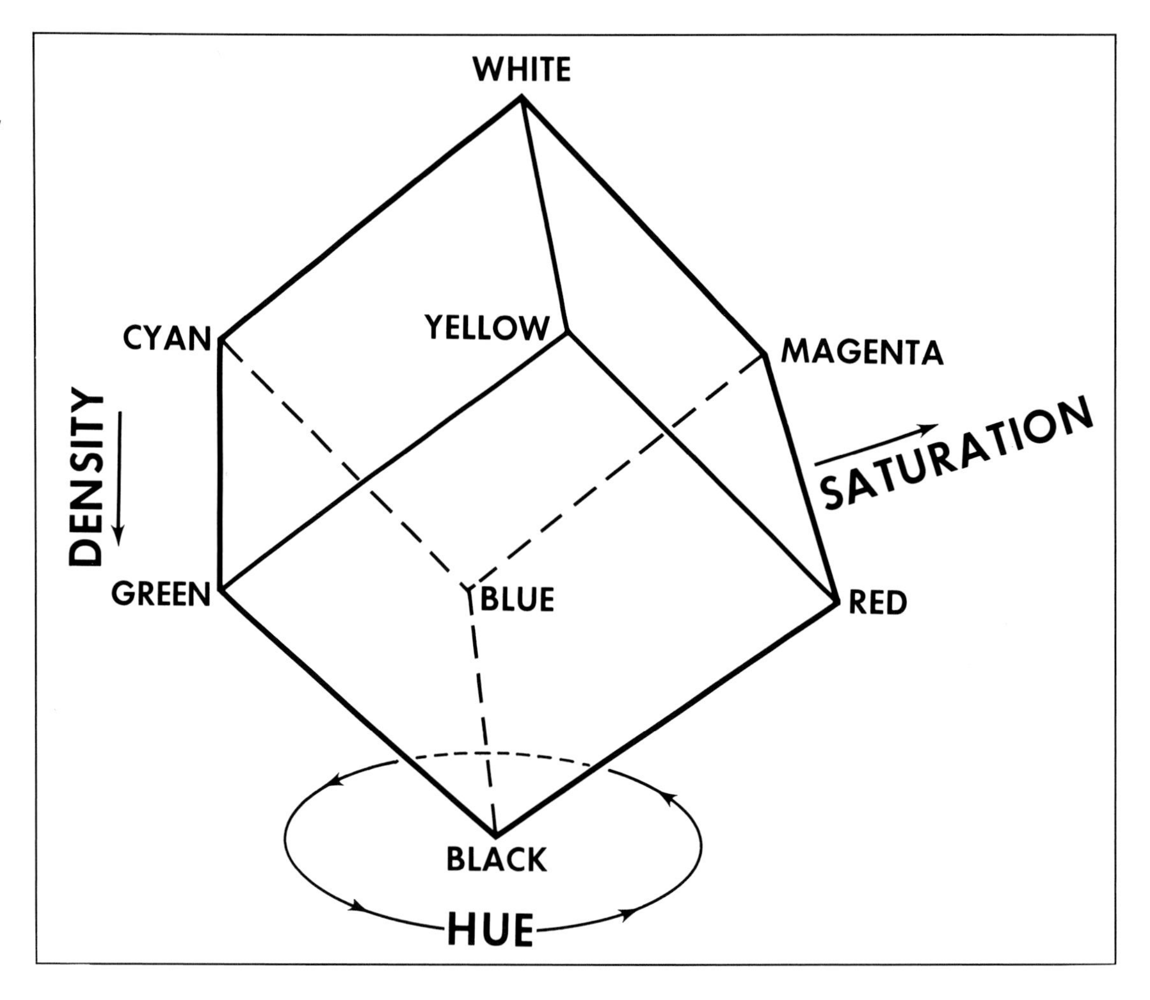

Fig. 2-1. Diagram of a color cube showing the relationship of the subtractive primary colors (magenta, yellow and cyan) to the additive primary colors (red, green and blue) to the color parameters (hue, saturation and density).

subtractive primaries — magenta, yellow and cyan. The total number of colors in the cube is thus 17 × 17 × 17 or 4,913, of which 1,538 are fully saturated colors on the surface. One way of studying the colors available inside the cube is to slice it along a chosen density level. Figure 2-4 shows density level 16, which has maximum strength magenta, yellow and cyan at the corners and gray of density 33% at the center. This display clearly demonstrates the significance of hue as the rotational parameter and saturation as the radial distance from the gray axis. The additive primaries, red, green and blue, lie on density level 32 with gray of density 67% at the center.

Figure 2-5 is a color chart used in an interactive interpretation system (Gerhardstein and Brown, 1984). It is based on the mixing of light and hence involves the additive primaries — red, green and blue. All the colors displayed in Figure 2-5 are fully saturated; that is, they lie only on the surface of the color cube. The right half of the chart is a projection of a color cube similar to that of Figures 2-2 and 2-3 when viewed from the top. The left half of the chart is a view of the same color cube from the bottom.

Interpretive Value of Color

Today's interpreter uses color in two fundamentally different ways: with a *contrasting* or with a *gradational* color scheme. A map or a section displayed in *contrasting* colors is normally accompanied by a legend, so the reader can identify the value of the displayed attribute at any point by reading the range of values associated with each color. Figure 2-6 is a structural contour map with a contour interval of 20 ms.

For an effective color display it is important that the range of values associated with each color, the number of colors used and their sequence, the contrast between adjacent colors, and the display scales are all carefully chosen. A color display must convey useful information and at the same time be aesthetically pleasing. For a map such as Figure 2-6 it is desirable to perceive equal visual contrast between adjacent colors, so that no one color boundary is more outstanding than another. A spectral sequence of colors was selected as the only really logical sequence available.

Fig. 2-2. Photograph of a color cube oriented the same as the diagram in Figure 2-1.

Figure 2-7 is a nomogram used for assessing visual color contrast. Visual contrast between two colors is, of course, somewhat subjective. Numerical color contrast is the sum of the absolute values of the differences in the amounts of the three primary colors. Zero density is white, maximum density (100%) is black, and density can have either arbitrary or percentage units between these extremes. Figure 2-7 shows that, for a particular visual color contrast, numerical contrast should be approximately proportional to average density. In other words, a larger numerical contrast is needed between darker colors.

A *gradational* color scheme is used when the interpreter is looking for trends, shapes, patterns and continuity. Figure 2-8 includes a vertical section displayed with gradational blue for positive amplitudes (peaks) and gradational red for negative amplitudes (troughs). Absolute amplitude levels are unimportant but relative levels are very important. Much stratigraphic information is implied by the lateral variations in amplitude along each reflection. The blue and red give equal visual weight to peaks and troughs. If the display gain is properly set, only a few of the highest amplitudes reach the fully saturated color and the full range of gradational shades expresses the

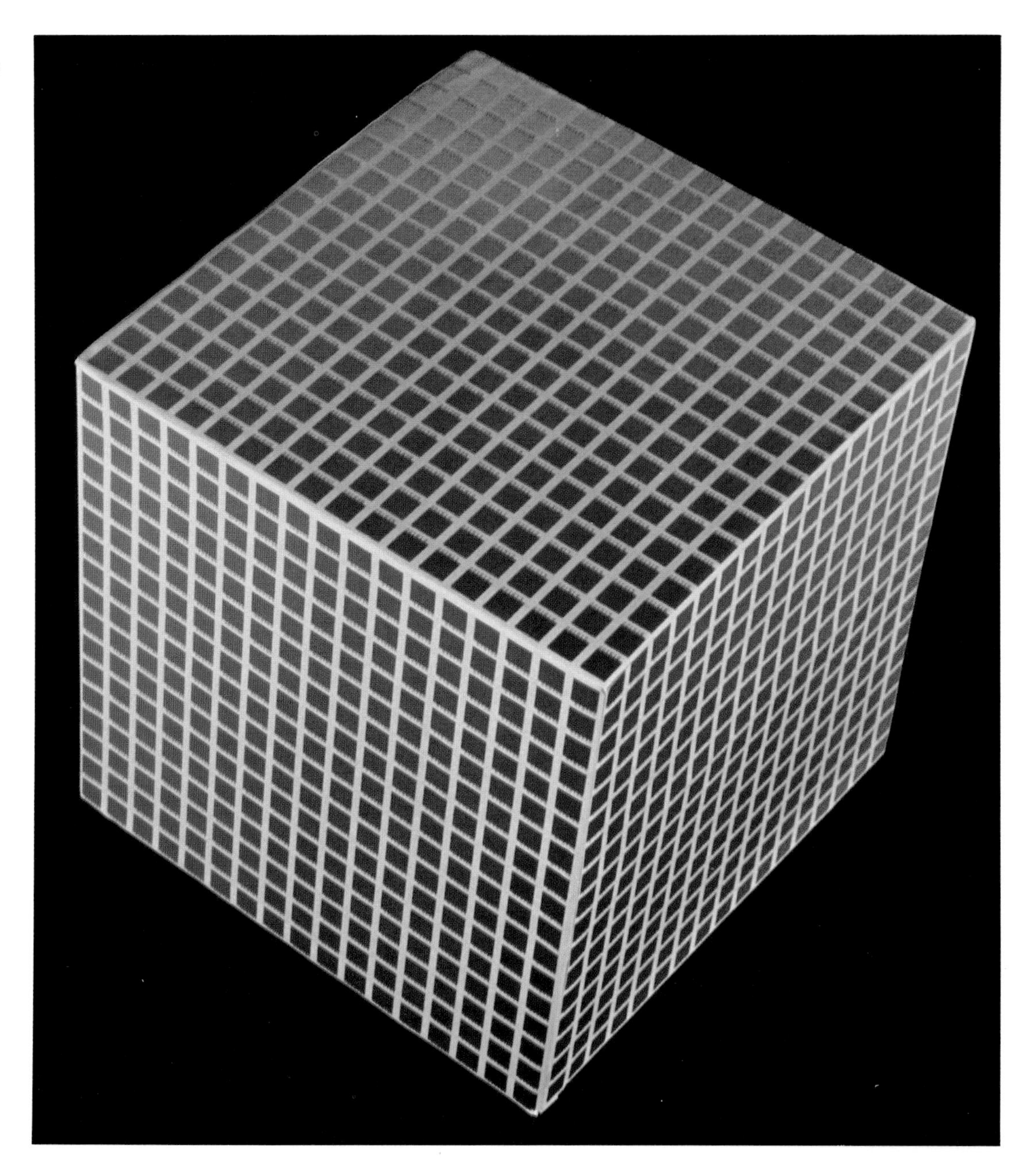

Fig. 2-3. Photograph of the same color cube as in Figure 2-2 from the back.

varying amplitudes in the data. This increased dynamic range gives the interpreter the best opportunity to judge the extent and the character of amplitude anomalies of interest.

Figure 2-8 also provides a comparison of gradational color and variable area/wiggle trace for the same piece of data. The shortcomings in the variable area/wiggle trace display relative to the color section are: (1) the visual weights of peaks and troughs are very different, which makes comparison difficult and biases the interpreter's eye towards the peaks; (2) the peaks are saturated; and (3) the troughs, where they have significant amplitudes, are not visible beneath the depth points where they belong. The red flat spot reflection is clearly visible on the color section as are the relative amplitudes of peaks and troughs. At the extreme right of the section, coincident amplitude maxima in the peak and the trough indicate a tuning phenomenon (see Chapter 6).

The need for equal visibility of peaks and troughs has long been recognized. Backus and Chen (1975) generated dual polarity variable area sections with the peaks in black and the troughs in red. Figure 2-9 is an example of this display from Galbraith and Brown (1982). Dual polarity variable area rectifies some of the shortcomings of variable area/wiggle trace but has less dynamic

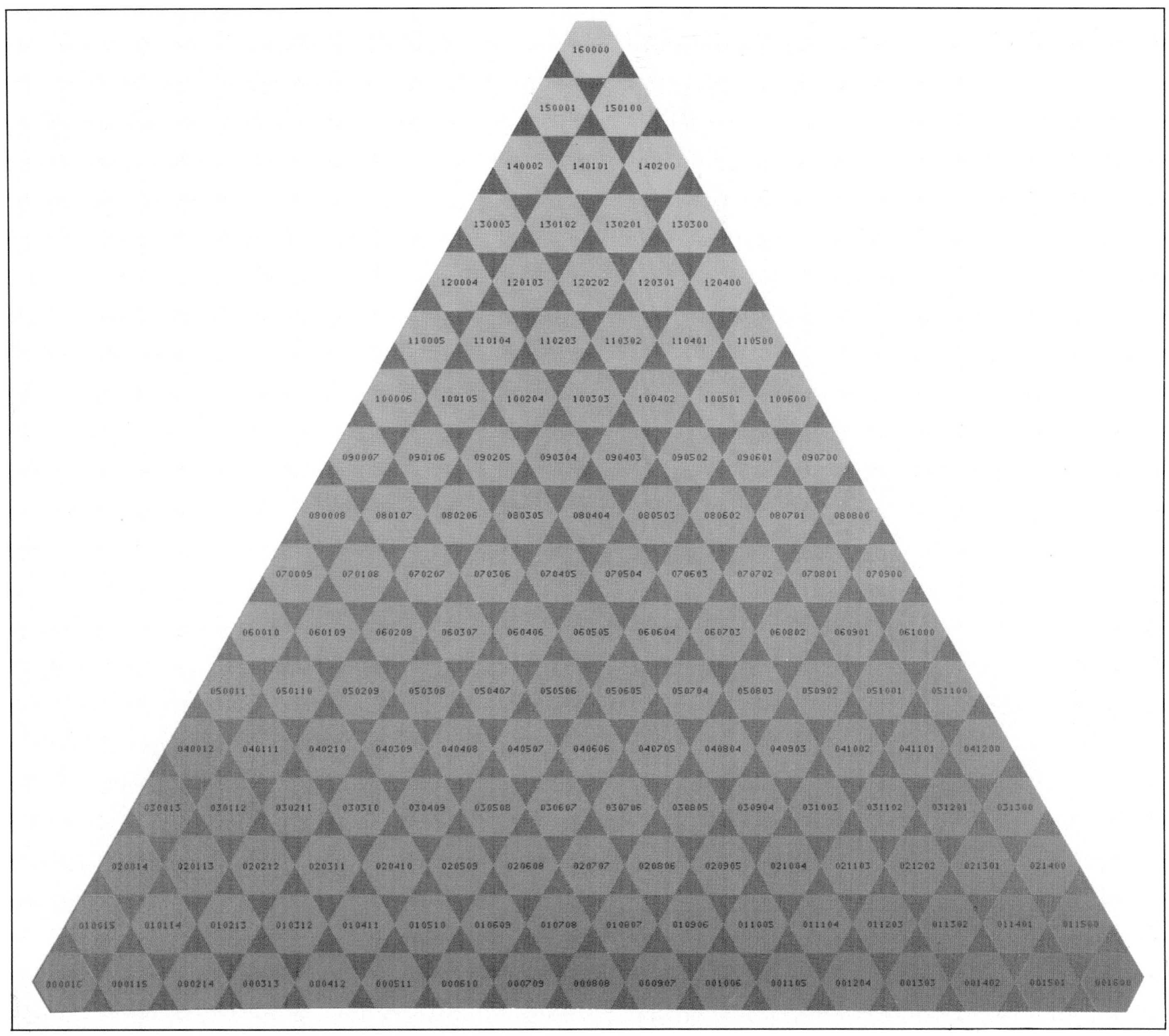

Fig. 2-4. Horizontal slice through the color cube at density level 33%, showing magenta, yellow and cyan at the corners and gray at the center.

range than gradational color. An additional benefit of dual polarity variable area relative to variable area/wiggle trace is that there effectively twice the number of event terminations at faults, making their recognition and detailed placement easier for the structural interpreter. This is apparent in Figure 2-9 and was considered an interpretive benefit by Galbraith and Brown.

A quite different display, but one that also has benefits for fault interpretation, is shown in Figure 2-10. Here a single gray scale from black for the largest peaks to white for the largest troughs enhances the visibility of low amplitude events. This also increases the number of event terminations visible at faults. The gradational blue and red, already discussed, is better for studying amplitudes but these need not be in conflict. Figure 5-16 (Chapter 5) shows gradational blue and red for the higher amplitudes and gradational gray for the lower amplitudes. This facilitates the study of bright spots in the presence of faulting.

An almost infinite number of color schemes can be applied to the same piece of data. Many of them are useful and interpreters' preferences vary. Care and thought are needed to develop a good scheme. Figure 2-11 shows four color schemes applied to the same vertical section segment. In the bottom left is the gradational blue and red, rapidly becoming an industry standard. Above this in the upper left is a scheme made with the same blue and red hues but

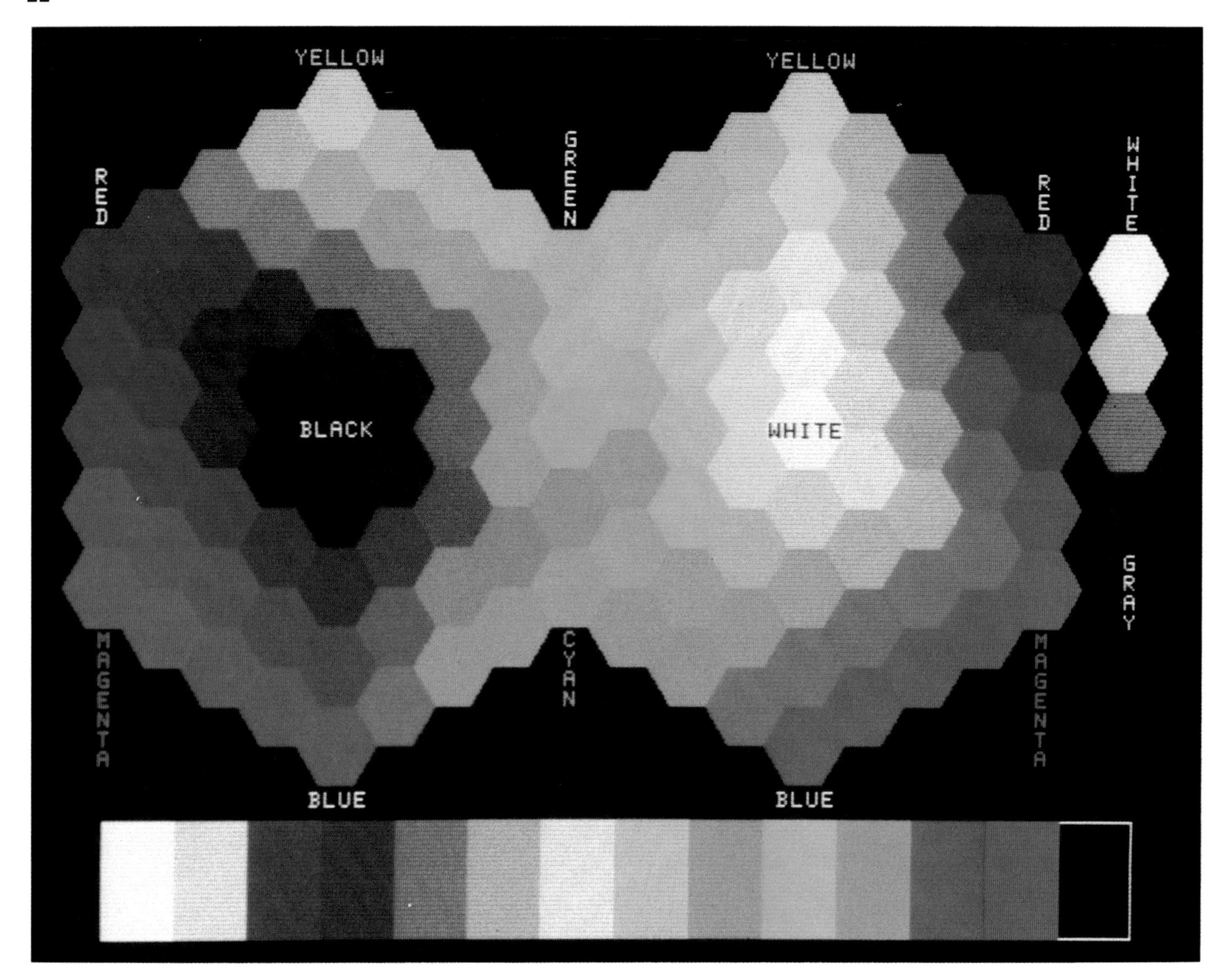

Fig. 2-5. Color selection chart from an interactive interpretation system.

many fewer intensity levels. It is sometimes useful to compare actual amplitude levels between two points. Here also the highest amplitudes have been distinguished with contrasting colors. This technique is particularly useful when it is possible to identify a particular amplitude level with the onset of a direct hydrocarbon indicator or other significant anomaly. The upper right panel of Figure 2-11 shows the color scheme preferred by another interpreter. The lower right panel illustrates a color scheme selected for a different purpose and is clearly inappropriate here.

The key to success with colors is learning how to adapt the display to suit the type of interpretation underway. Hence flexible color display has become one of the key benefits of an interactive interpretation system. Each interpreter needs to use the color schemes with which he is most comfortable, but an innovative interpreter can beneficially scan through many schemes seeking the one which conveys, to him, the most information about the issue at hand.

Neidell and Beard (1985) are vociferous in their promotion of color display. In fact, in drawing conclusions concerning point bar and channel sands, they say, "Such interpretation would not be reasonable without the support of the color display and the stratigraphic visibility which it provides." Their detailed use of color is, however, slightly different from that discussed above in that stratigraphic acoustic impedance sections are displayed with a very large number of mildly-contrasting colors.

The recognition of channels, bars and other depositional features on horizontal sections and horizon slices is becoming increasingly important for the stratigraphic interpreter. Here again

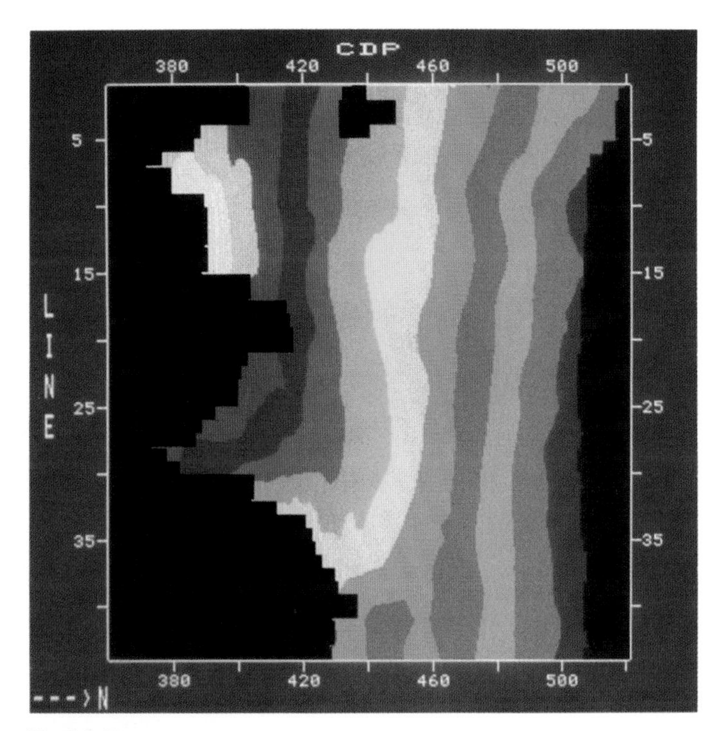

Fig. 2-6. Time structure map displayed in a contrasting spectral color scheme.

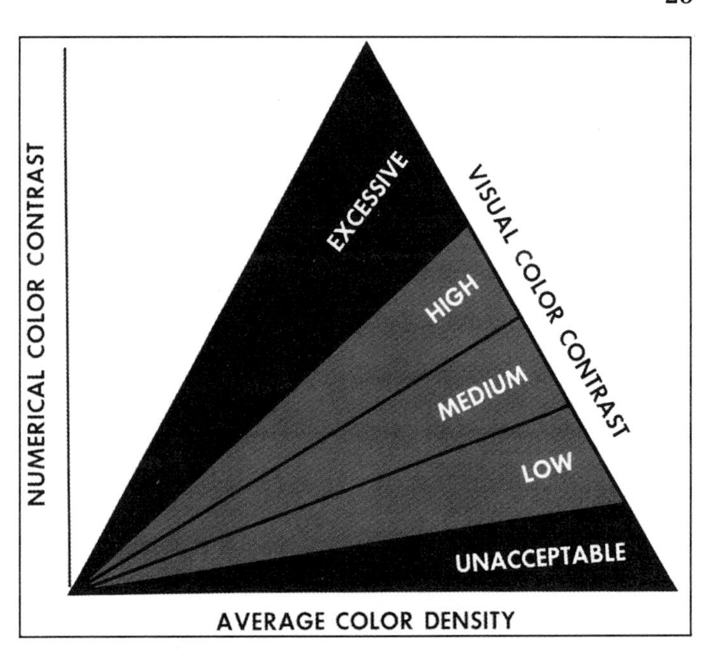

Fig. 2-7. Contrast-density nomogram used for establishing a color scheme with acceptable visual contrast between adjacent colors.

the proper use of gradational color coded to amplitude helps the detectability of these features because of the eye's ability to integrate a wide range of densities. Figures 2-12 and 2-13 illustrate an inferred channel on a horizon Seiscrop section (see Chapter 4) and the use and abuse of color for its detection. A well at about Line 55, Crossline 250, indicates that at least the lower part of the areal bright spot (Figure 2-12) is a sand-filled channel. How extensive is this channel? It seems probable that it extends to include the central zone between Lines 70 and 80 and between Crosslines 180 and 270. However, after crossing two faults, a curvilinear feature can be seen continuing to the upper right to Line 122, Crossline 330. Is this a continuation of the channel system even though the amplitude is much reduced? We do not know the answer to this question, but we have been able to observe the continuity of this extensive curvilinear feature because of the use of gradational color.

Figure 2-13 shows the same section in contrasting colors and the detectability of the inferred channel is much reduced. In fact the eye tends to be drawn to the red and pink circular maxima at Crossline 250 between Lines 45 and 60 rather than the longer arcuate high amplitude trends.

Assessment of Zero-Phaseness

Most interpreters today prefer zero-phase data. The reasons they give to support this preference include the following:

(1) the wavelet is symmetrical with the majority of the energy being concentrated in the central lobe;
(2) this wavelet shape minimizes ambiguity in associating observed waveforms with subsurface interfaces;
(3) a horizon track drawn at the center of the wavelet coincides in time with the travel time to the subsurface interface causing the reflection;
(4) the maximum amplitude occurs at the center of the waveform and thus coincides with the time horizon; and,
(5) the resolution is better than for other wavelets with the same frequency content.

Much data processing research has been devoted to wavelet processing, which can be defined as the replacement of the source wavelet, the receiver response, and the filtering effects of the

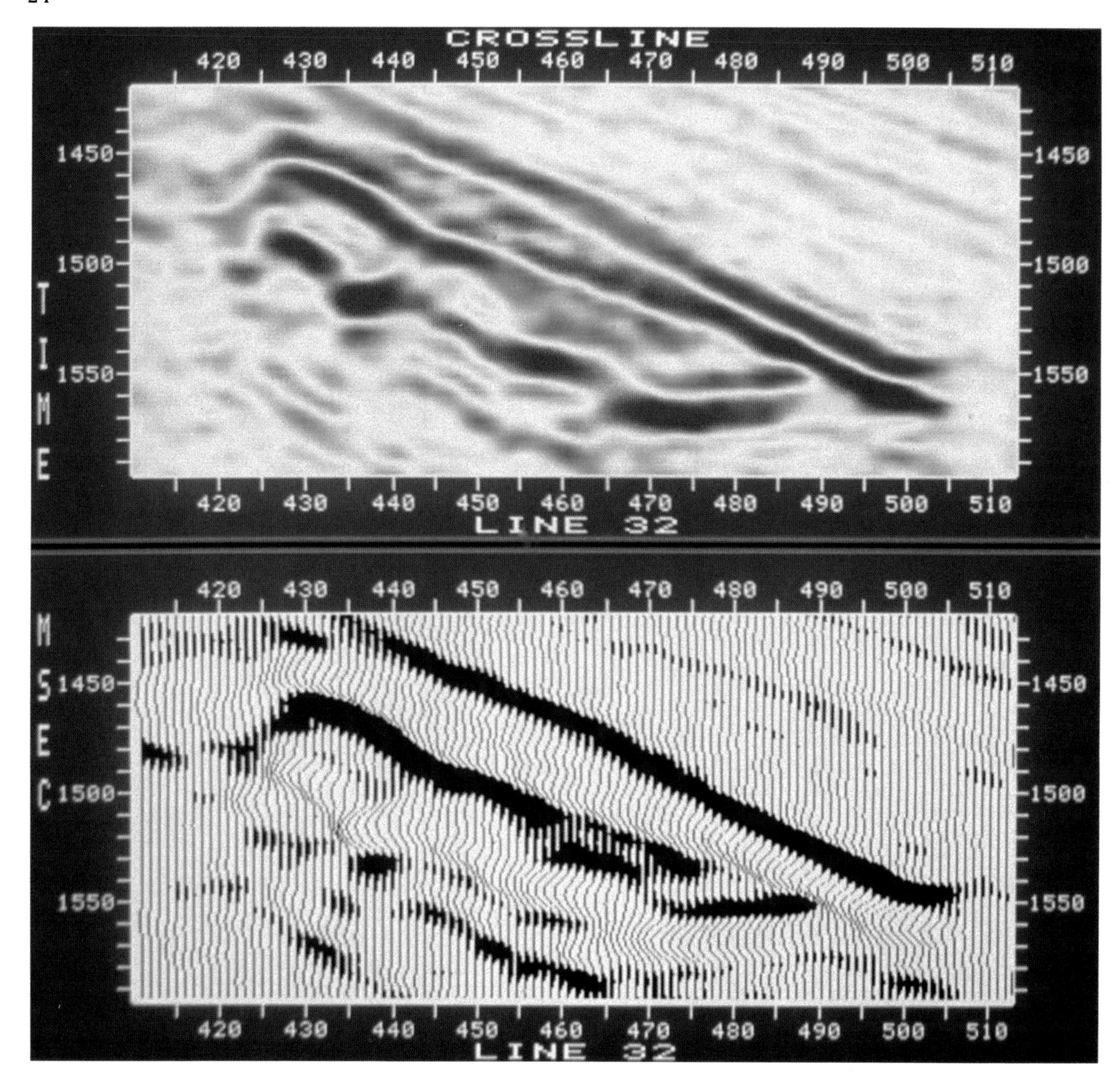

Fig. 2-8. Vertical seismic section displayed with gradational blue for peaks and gradational red for troughs compared to same section displayed in variable area/wiggle trace. (Courtesy Chevron U.S.A. Inc.)

earth by a wavelet of known and desirable characteristics. Wood (1982) outlined the principles of wavelet processing and the properties of zero-phase wavelets, and Kallweit and Wood (1982) addressed the issues of resolution. Today's interpreter, particularly one who has a stratigraphic objective, wants to be able to assess whether the data provided have been properly deconvolved to a zero-phase condition.

The interpretive assessment of zero-phaseness requires high signal-to-noise ratio reflections and maximum dynamic range color display. But first zero-phaseness will be considered on model data. Figure 2-14 shows three zero-phase wavelets and their equivalents shifted by 30, 60, and 90 degrees. The first is a Ricker wavelet, the second is derived from a bandpass filter of 2.3 octaves with gentle slopes, and the third is derived from a bandpass filter of 1.3 octaves with steep slopes. The common property of these three wavelets is that the separation of central peak

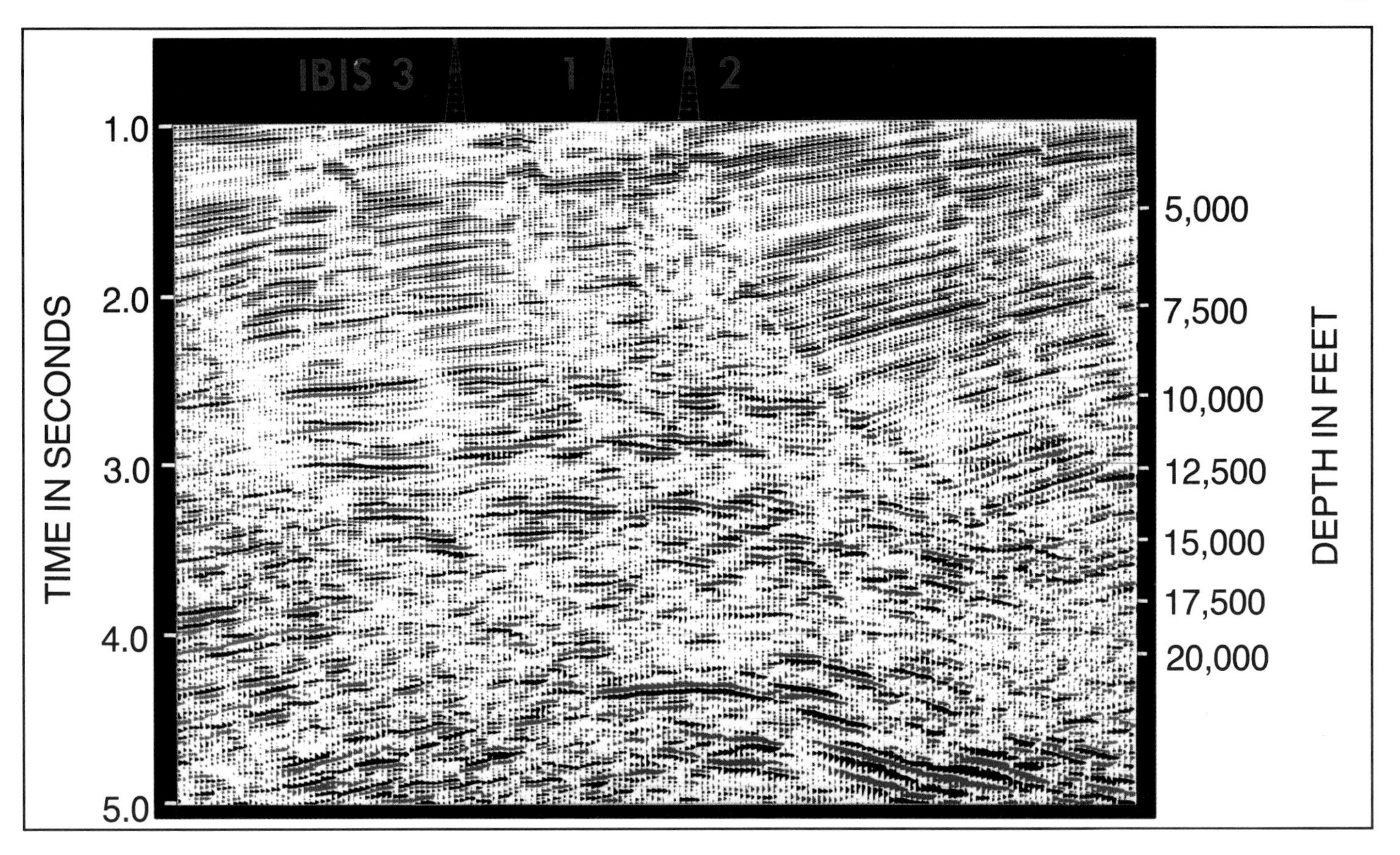

Fig. 2-9. Vertical section displayed in dual polarity variable area showing fault definition. (Courtesy Texaco Trinidad Inc.)

and first side lobe is the same for each — 16 ms. The Ricker wavelet has no side lobes beyond the first. The 2.3 octave wavelet is a good wavelet extracted from actual processed data and has low side lobes. The 1.3 octave wavelet is a poor wavelet with relatively high side lobes.

The visual assessment of zero-phaseness amounts to a visual assessment of wavelet symmetry. In these model examples 30° of distortion is visible for all the wavelets but the higher side lobe levels of the narrower band wavelet make the distortion less pronounced. For the larger distortions, for example at 60°, the central peak and the larger side lobe are more easily confused for the narrower band wavelet, so in practice it may be difficult to decide whether the peak or the trough is the principal extremum. At a distortion of 90° the time horizon lies at the zero crossing between the largest amplitude peak and trough, and these are of equal size.

Figure 2-15 is a single trace example from real data where there was a known low velocity gas sand. The top of the low velocity zone should be a peak and the base a trough (according to the polarity convention considered normal in this book). The trace labelled 0° shows peak and trough each symmetrically placed over their corresponding interfaces. The phase distortions are again fairly evident when presented in this way.

In practice, interpreters must assess zero-phaseness on a section containing many traces in case one trace is unrepresentative. We select a high amplitude reflection, which, on the basis of a simple model, can be related to a single interface. Usable reflections include hydrocarbon indicators, basement reflections and limestone markers. The interpreter can then assume that the interference of events from adjacent parallel interfaces, multiples or noise is small. Figure 2-16 illustrates a bright spot from a gas reservoir where it is assumed that the above conditions hold. In the panel labelled 0° there is one blue event from the top of the reservoir and one red event from its base, and they have approximately the same amplitude. Side lobes are low and symmetrical as far as can be determined. This is the signature expected for the zero-phase response of a gas sand.

For the 90° case in Figure 2-16 the top of the gas sand has a signature of peak-over-trough and the base one of trough-over peak. This confirms the modeling illustrated in Figure 2-14 and

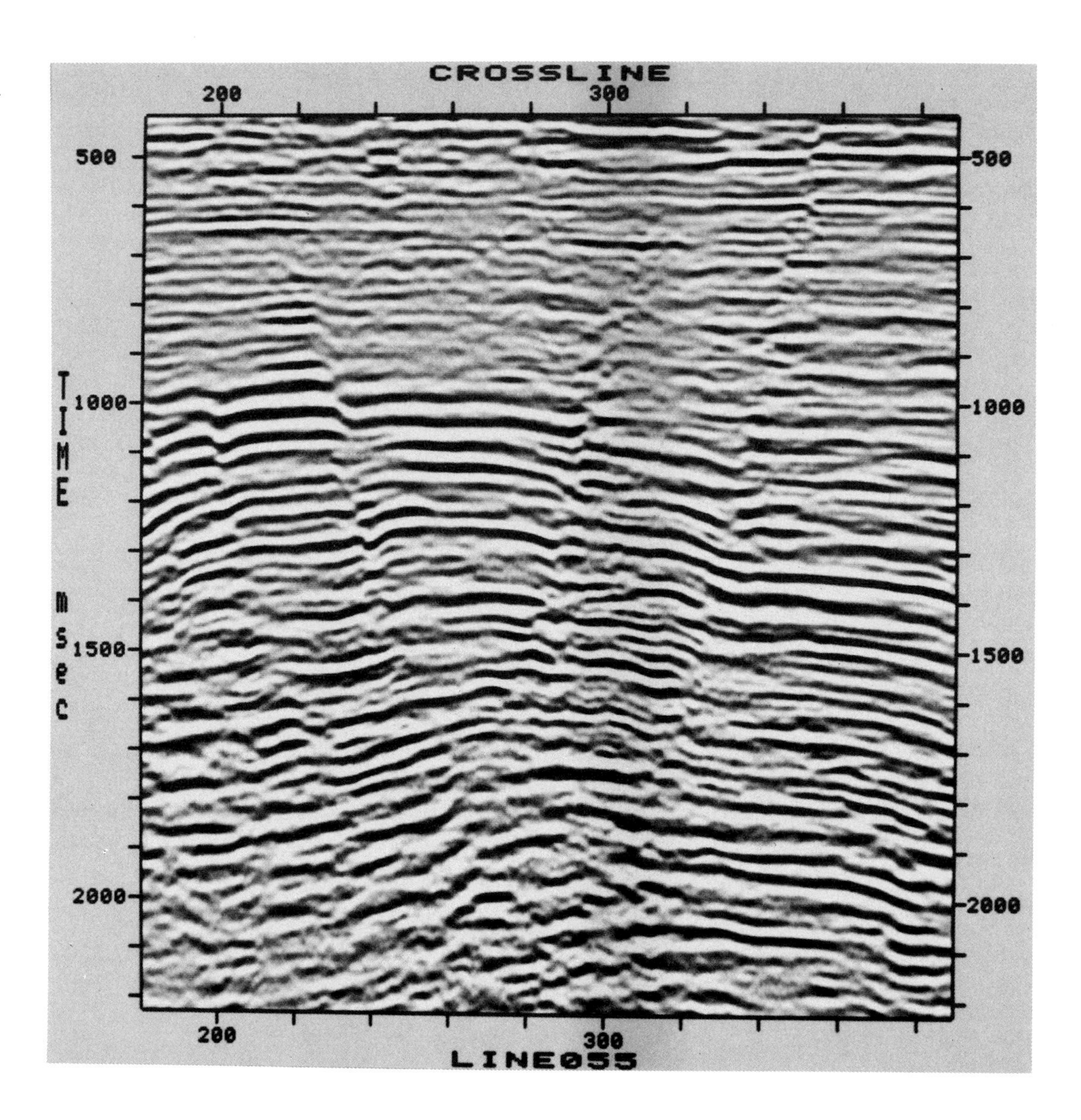

Fig. 2-10. Vertical section displayed with single gradational gray scale in order to enhance low amplitude events. (Courtesy Texas Pacific Oil Company Inc.)

certainly shows a more complex character than the zero-phase section. The intermediate levels of phase distortion show the progression from the 0° to 90° condition.

The interpreter's ability to make this kind of assessment of zero-phaseness depends critically on the display used. Figure 2-17 presents the same data panel in the same four phase conditions for three different modes of display. Variable area/wiggle trace demonstrates how the visual imbalance between peaks and troughs makes the assessment of relative amplitudes extremely difficult. Dual polarity variable area has corrected the visual imbalance but demonstrates the limited dynamic range of variable area as the bright events are all saturated. Gradational color demonstrates the visual balance between peaks and troughs and also the improved dynamic range. Relative amplitudes of peaks, troughs and side lobes can now be assessed with maximum available clarity for fairly high trace density. One disadvantage, however, of gradational color display is the stringency imposed on the reproduction process. The illustration that you, the reader, are studying is of reduced quality compared to the screen image of the color monitor on which the original assessment was made.

If the phase of the data is unknown and cannot be assessed, instantaneous amplitude (also known as envelope amplitude or reflection strength; Taner and Sheriff, 1977) provides a display in which amplitude can be studied independent of phase. Figure 2-18 shows identical instantaneous amplitude sections corresponding to the four regular amplitude sections with different phases.

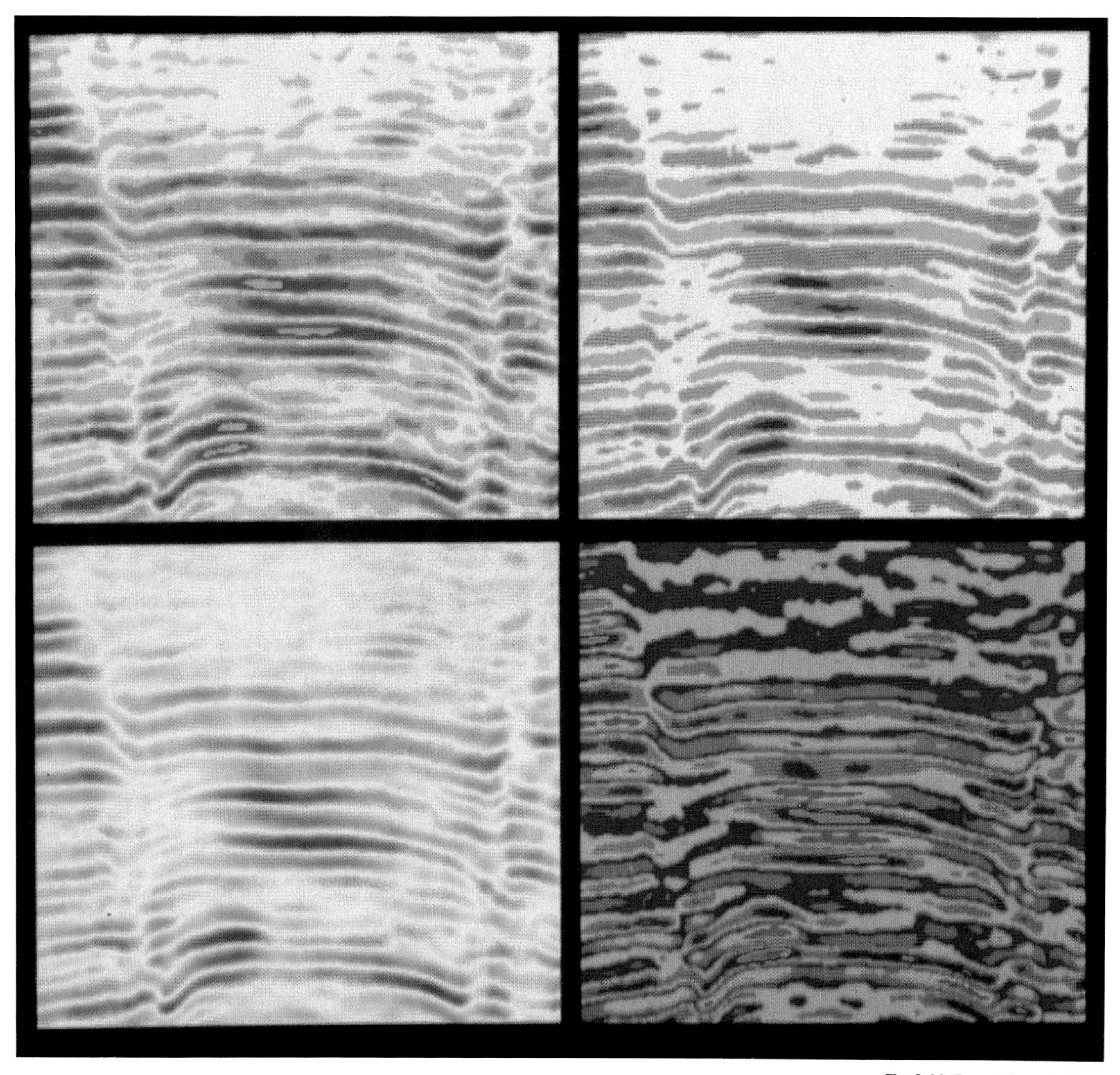

Fig. 2-11. Four different color schemes applied to the same vertical section segment. (Courtesy Texas Pacific Oil Company Inc.)

Any high amplitude reflection which can be assumed to originate from a single interface is usable for assessing zero-phaseness. A fluid contact reflection, or flat spot, is normally an excellent candidate. If the structural horizons have moderate dip and the reservoir is fairly thick, the flat spot reflection will be well resolved and structurally unconformable, and hence will give the interpreter maximum visibility of the phase of his data. The flat spot in Figure 2-19 shows a clear symmetrical trough with low and variable side lobes indicating that the data are close to zero phase.

Figure 2-20 shows an outstanding basement reflection which is probably also from a single subsurface interface. The waveform of the reflection is clear, almost symmetrical, and spatially consistent. This indicates that the data are close to zero phase, at least around the time of 3 seconds. (Please note, however, that this section has the opposite polarity convention to that considered normal in this book.)

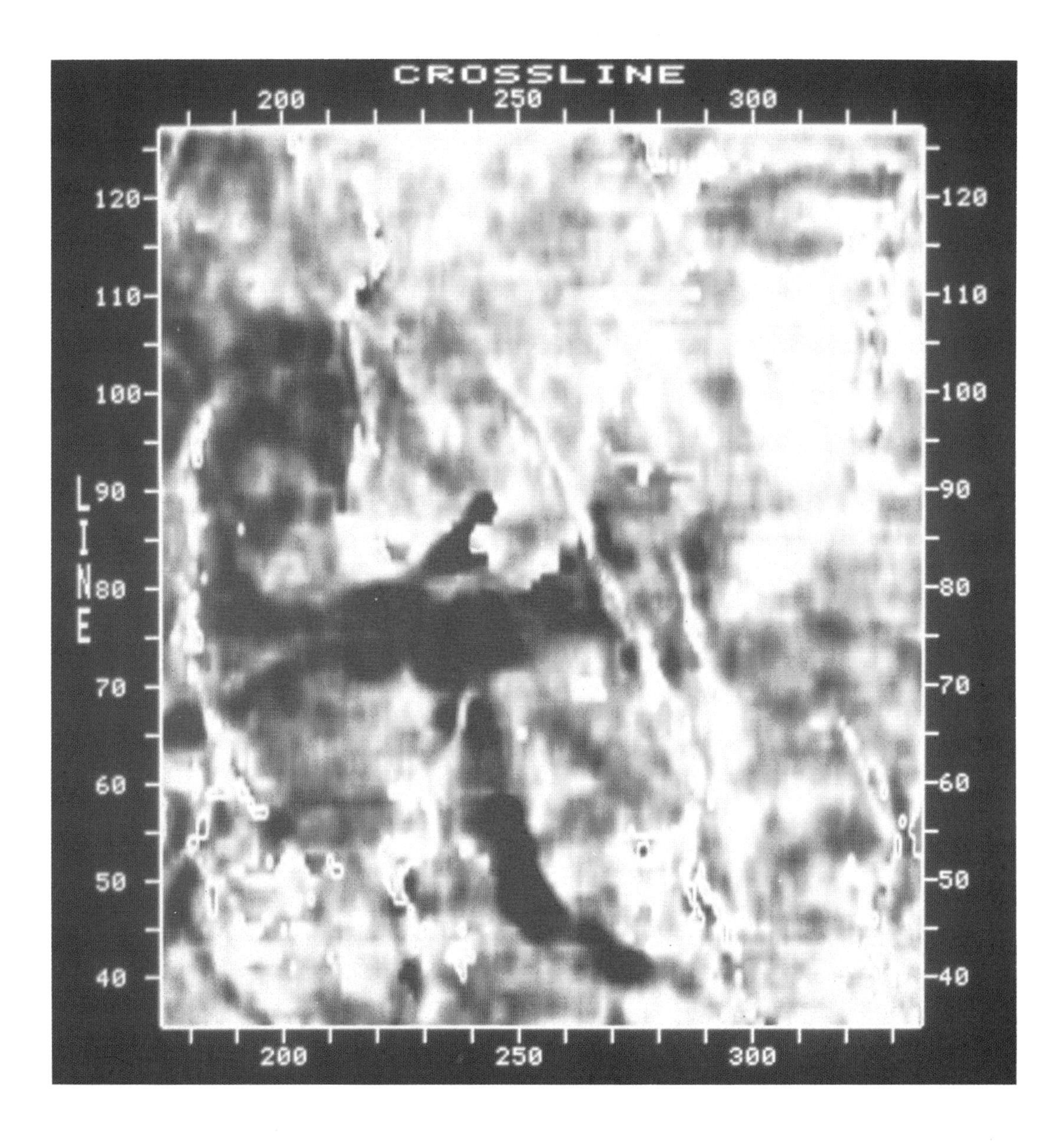

Fig. 2-12. Horizon Seiscrop section showing an inferred channel system displayed with a gradational color scheme. (Courtesy Texas Pacific Oil Company Inc.)

Psychological Impact of Color

Studies on the psychological impact of color have shown that hues of yellow, orange and red are advancing and attracting, while hues of green and blue are cooler and receding. The interpreter can take advantage of this in communicating his results. It would seem logical to display the structural highs, the isopach thicks and the bright spots in advancing colors in order to promote their prospectivity. Figure 2-6 is a structure map which demonstrates this point.

Figures 2-21, 2-22 and 2-23 are the same horizon slice displaying reflection amplitude over a Gulf of Mexico reservoir, but presented with three different color schemes. In Figure 2-21 these data are represented in a green gradational scheme to accentuate the lineations due to faulting. The gradational colors accentuate these lineations by using the full dynamic range of color density and allow the eye to integrate all of the data quickly.

Figure 2-22 shows the same data displayed with a gradational color scheme using a wider range of hues. Now the relative strength of the amplitudes has much more impact on the eye; the advancing reds and yellows appear much more interesting than the cooler greens and blues. By using this scheme, the large anomaly near the top of the section draws considerable attention. A successful well was targeted and drilled, based on this display.

Yet another display of the same data (Figure 2-23) shows that a large area of high amplitude may be considered prospective. Here the low amplitude zones have been colored with fairly

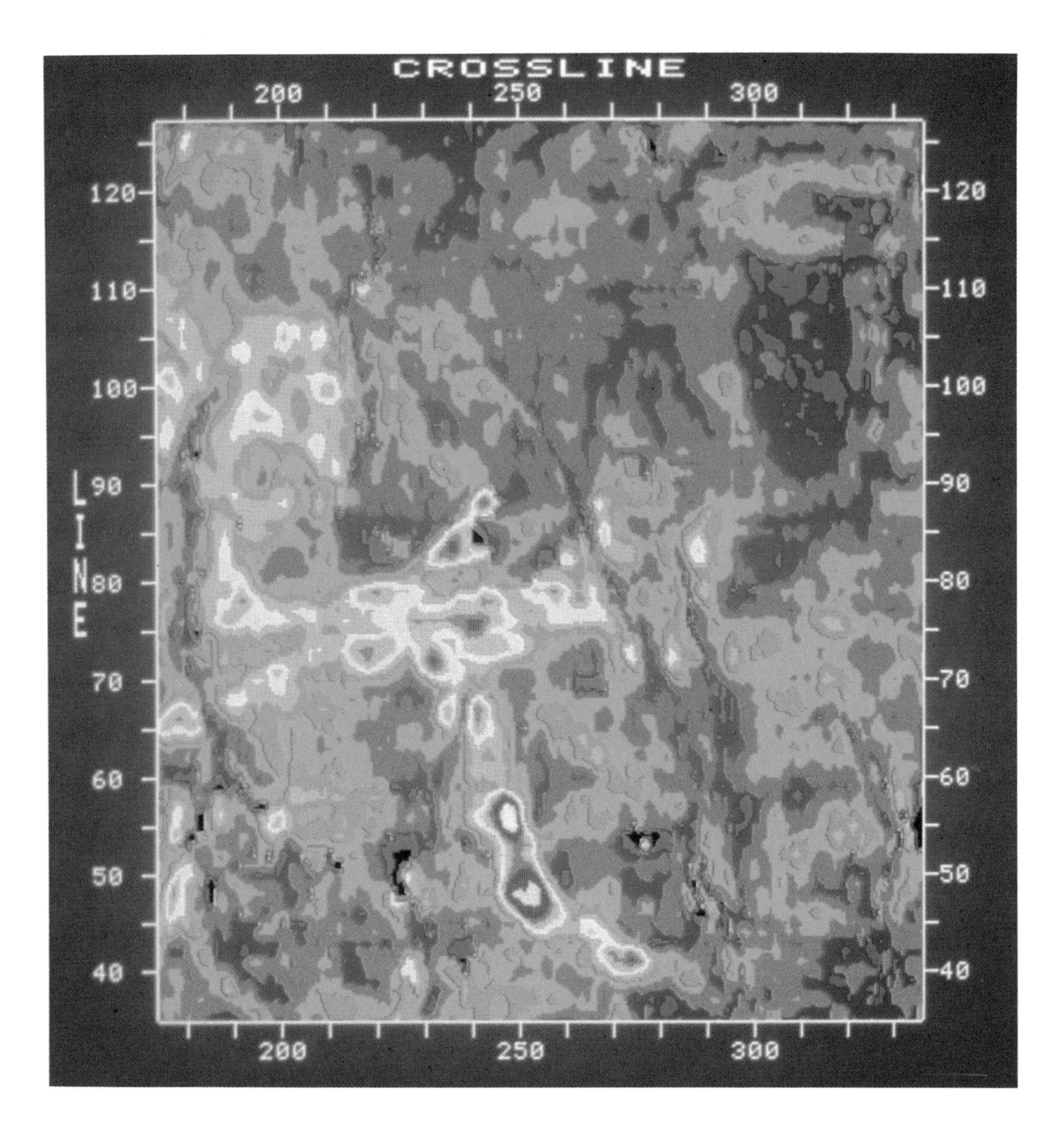

Fig. 2-13. Same horizon Seiscrop section as in Figure 2-12 displayed with a contrasting color scheme, which reduces visibility of the channel system. (Courtesy Texas Pacific Oil Company Inc.)

neutral grays. Further drilling potential can be considered on the basis of this display if amplitude strength is the key to developing this reservoir.

Thus one horizon slice was used for three different purposes by employing three different color schemes. The first drew attention to the faulting, the second to a particular anomaly, and the third to total drilling potential. Separate features of the data were enhanced differently by the different uses of color.

References

Backus, M. M., and R. L. Chen, 1975, Flat spot exploration: Geophysical Prospecting, v. 23, p. 533–577.

Balch, A. H., 1971, Color sonograms; a new dimension in seismic data interpretation: Geophysics, v. 36, p. 1074–1098.

Galbraith, R. M., and A. R. Brown, 1982, Field appraisal with three-dimensional seismic surveys offshore Trinidad: Geophysics, v. 47, p. 177–195.

Gerhardstein, A. C., and A. R. Brown, 1984, Interactive interpretation of seismic data: Geophysics, v. 49, p. 353–363.

Kallweit, R. S., and L. C. Wood, 1982, The limits of resolution of zero-phase wavelets: Geophysics, v. 47, p. 1035–1046.

Fig. 2-14. Effect of phase shifting constant phase wavelets.

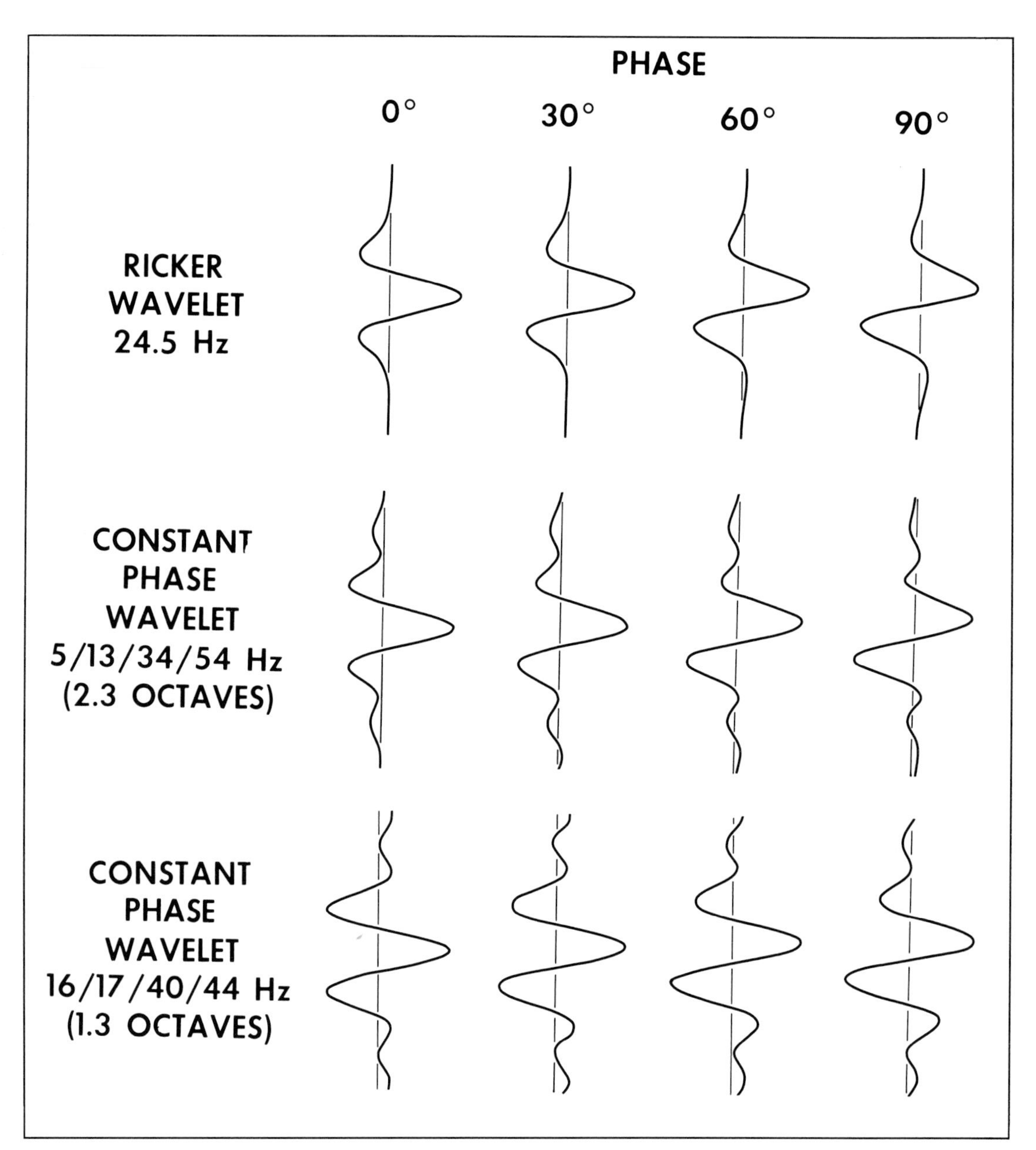

Lindseth, R. O., 1979, Synthetic sonic logs–a process for stratigraphic interpretation: Geophysics, v. 44, p. 3–26.

Neidell, N. S., and J. H. Beard, 1985, Seismic visibility of stratigraphic objectives: Society of Petroleum Engineers Paper 14175.

Taner, M. T., and R. E. Sheriff, 1977, Application of amplitude, frequency and other attributes to stratigraphic and hydrocarbon determination, *in* C. E. Payton, ed., Seismic stratigraphy–applications to hydrocarbon exploration: AAPG Memoir 26, p. 301–327.

Wood, L. C., 1982, Imaging the subsurface, *in* K. C. Jain, and R. J. P. deFigueiredo, eds., Concepts and techniques in oil and gas exploration: Society of Exploration Geophysicists Special Publication, p. 45–90.

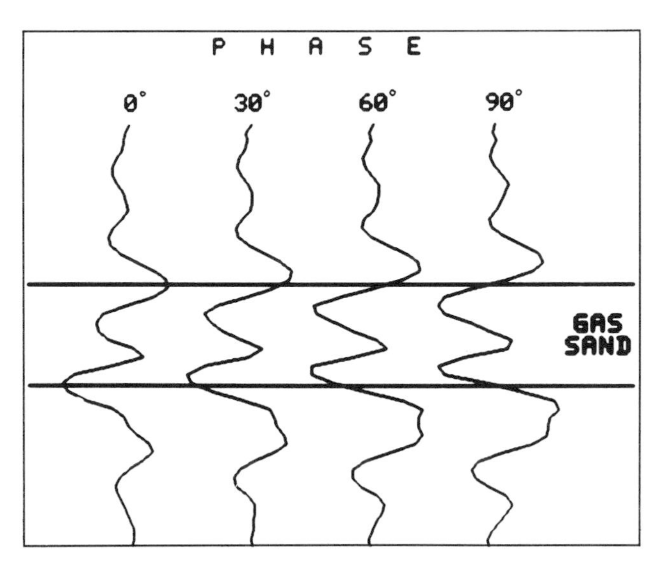

Fig. 2-15. Effect of phase shifting a real data trace showing reflections from the top and base of a gas sand. (Courtesy Chevron U.S.A. Inc.)

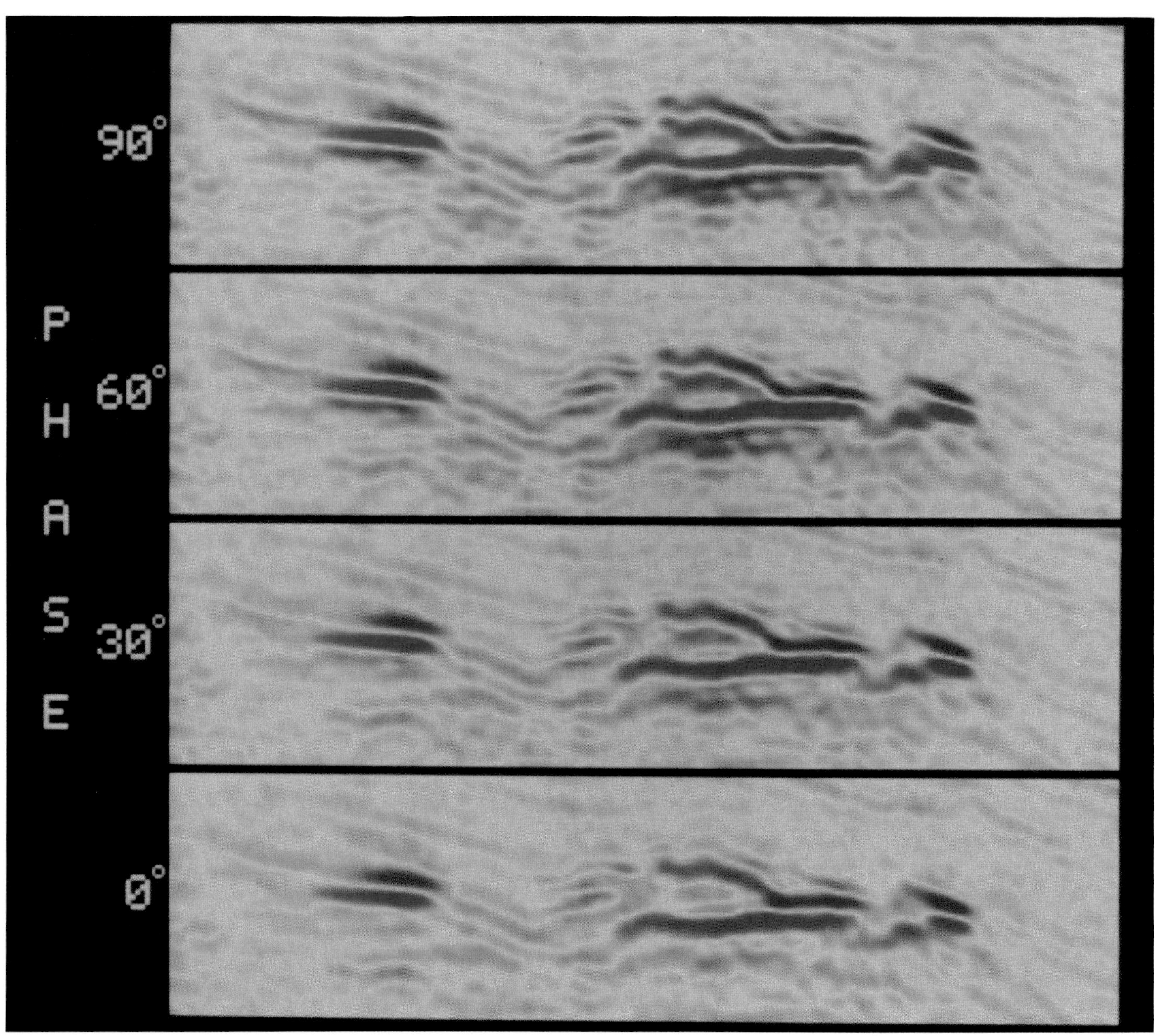

Fig. 2-16. Bright reflections from the top and base of a gas sand with constant phase shifts applied. (Courtesy Chevron U.S.A. Inc.)

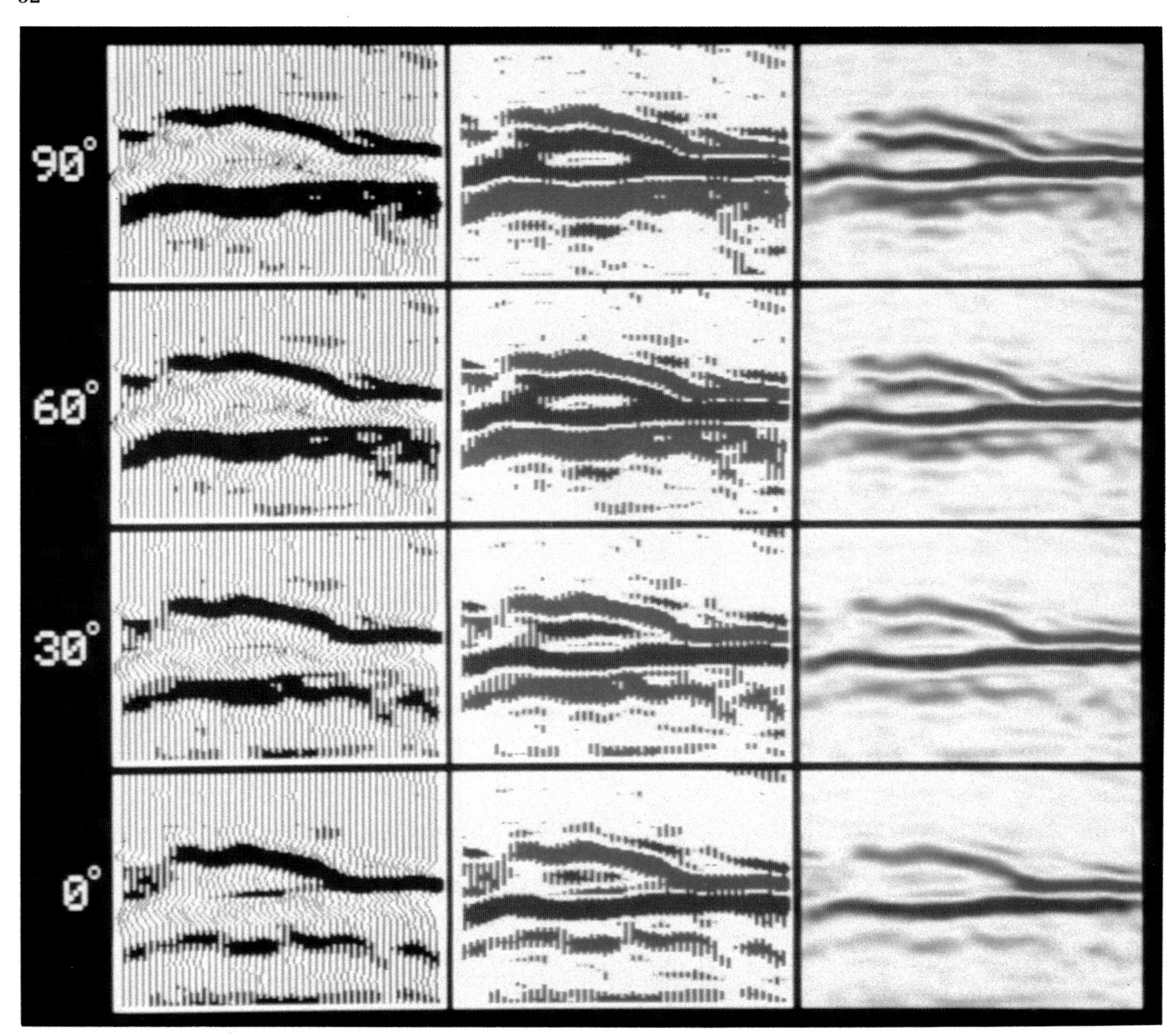

Fig. 2-17. Comparison between variable area/wiggle trace, dual polarity variable area and gradational color for the interpretive assessment of zero-phaseness. (Courtesy Chevron U.S.A. Inc.)

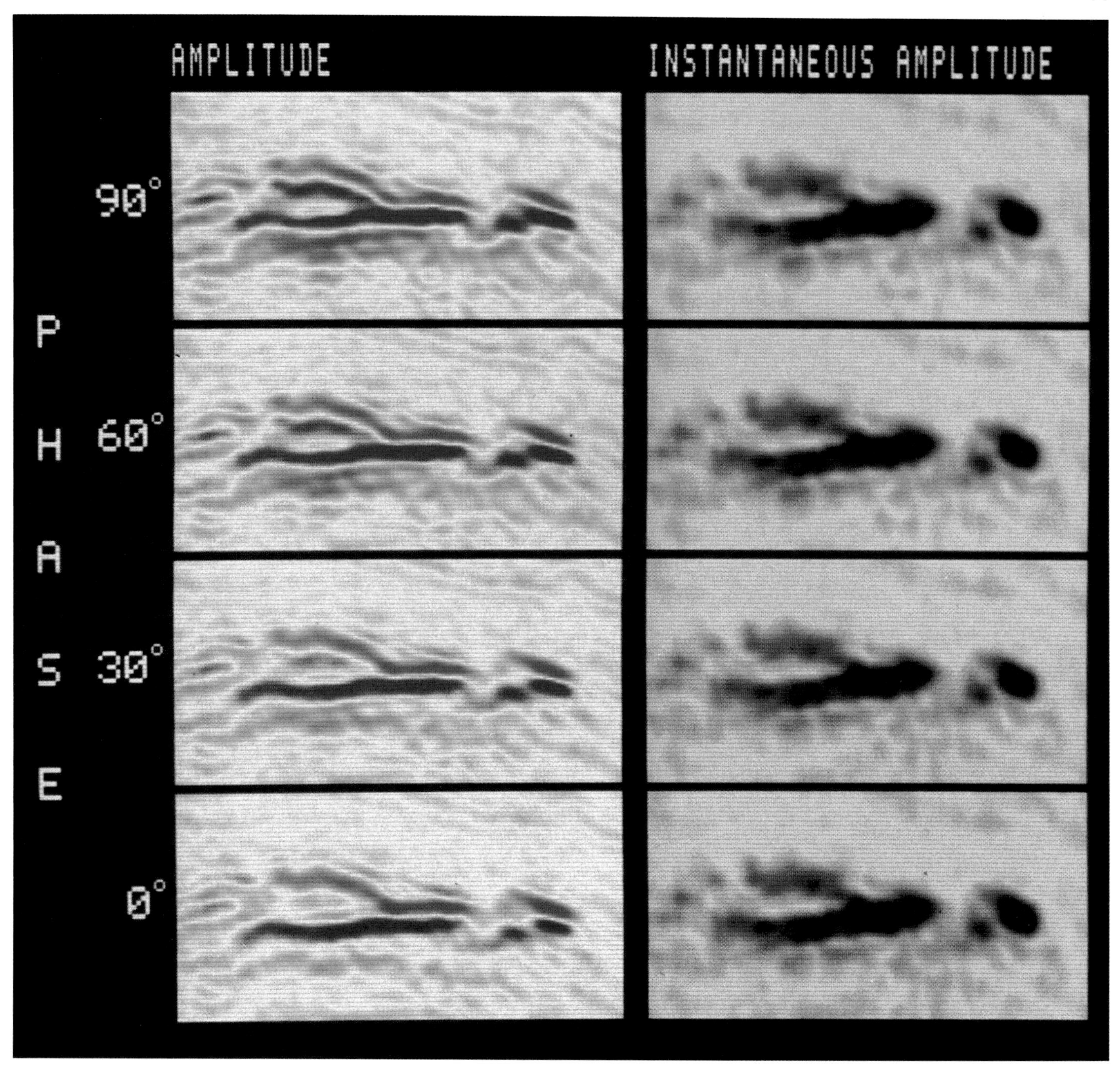

Fig. 2-18. Use of instantaneous amplitude to obscure the effects of phase distortion. (Courtesy Chevron U.S.A. Inc.)

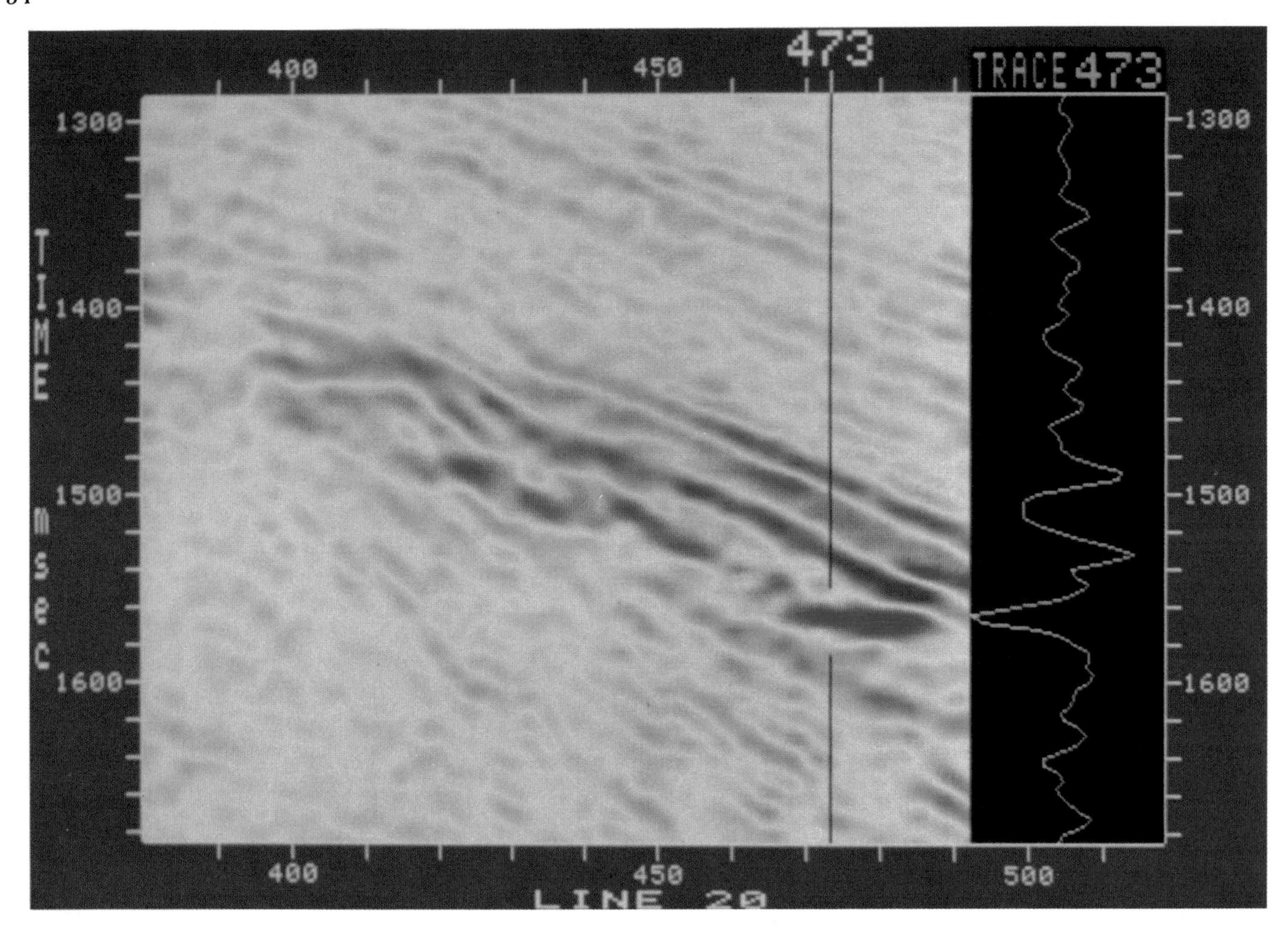

Fig. 2-19. Flat spot reflection displaying zero-phaseness, visible in gradational red for many traces and in wiggle format for one trace. (Courtesy Chevron U.S.A. Inc.)

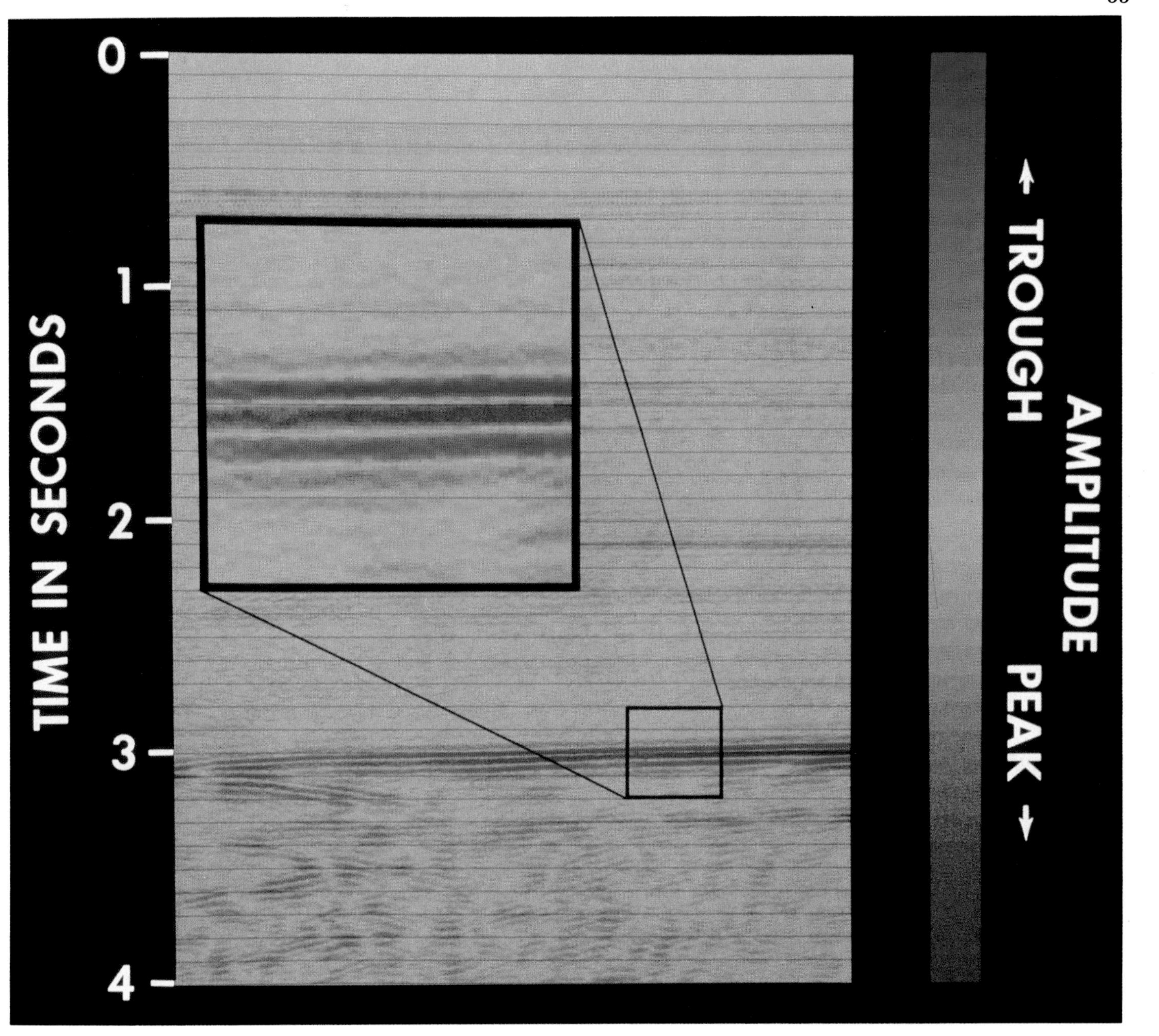

Fig. 2-20. Basement reflection displaying zero-phaseness.

Fig. 2-21. Horizon Seiscrop section displayed in gradational green to accentuate lineations due to faulting. (Courtesy Chevron U.S.A. Inc.)

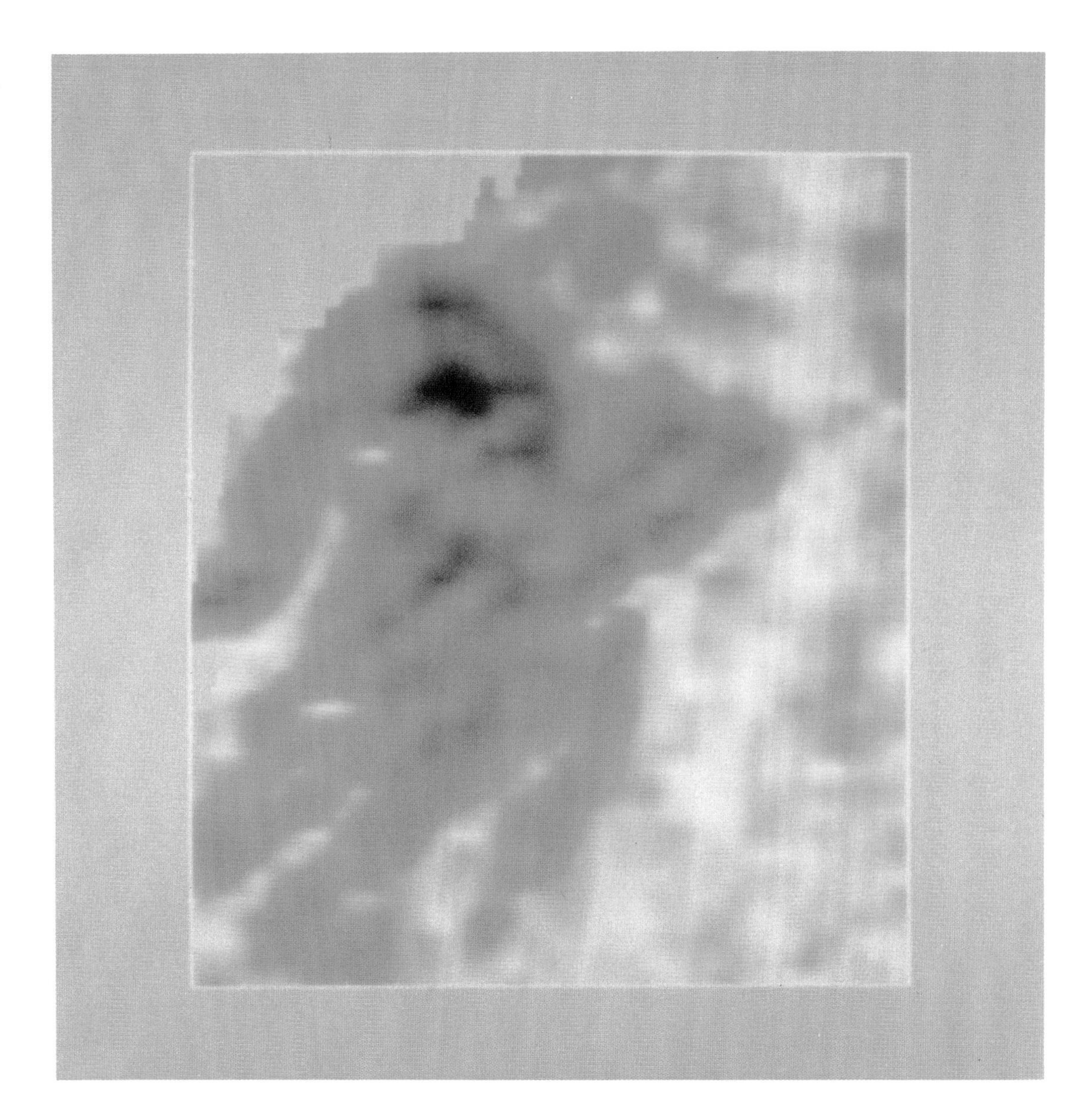

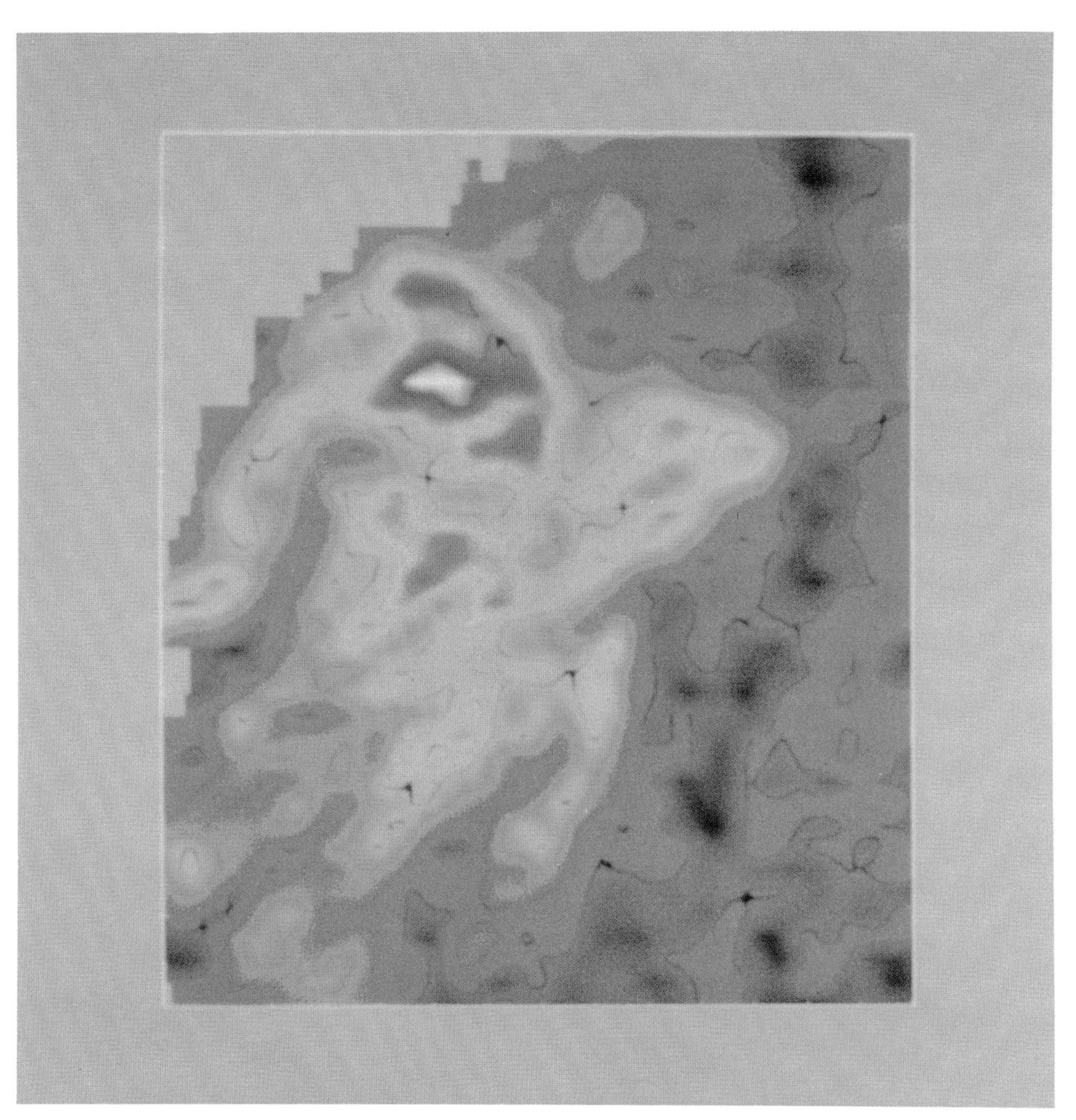

Fig. 2-22. Same horizon Seiscrop section as in Figure 2-21 displayed in a wider range of hues to draw attention to the high amplitudes using advancing colors. (Courtesy Chevron U.S.A. Inc.)

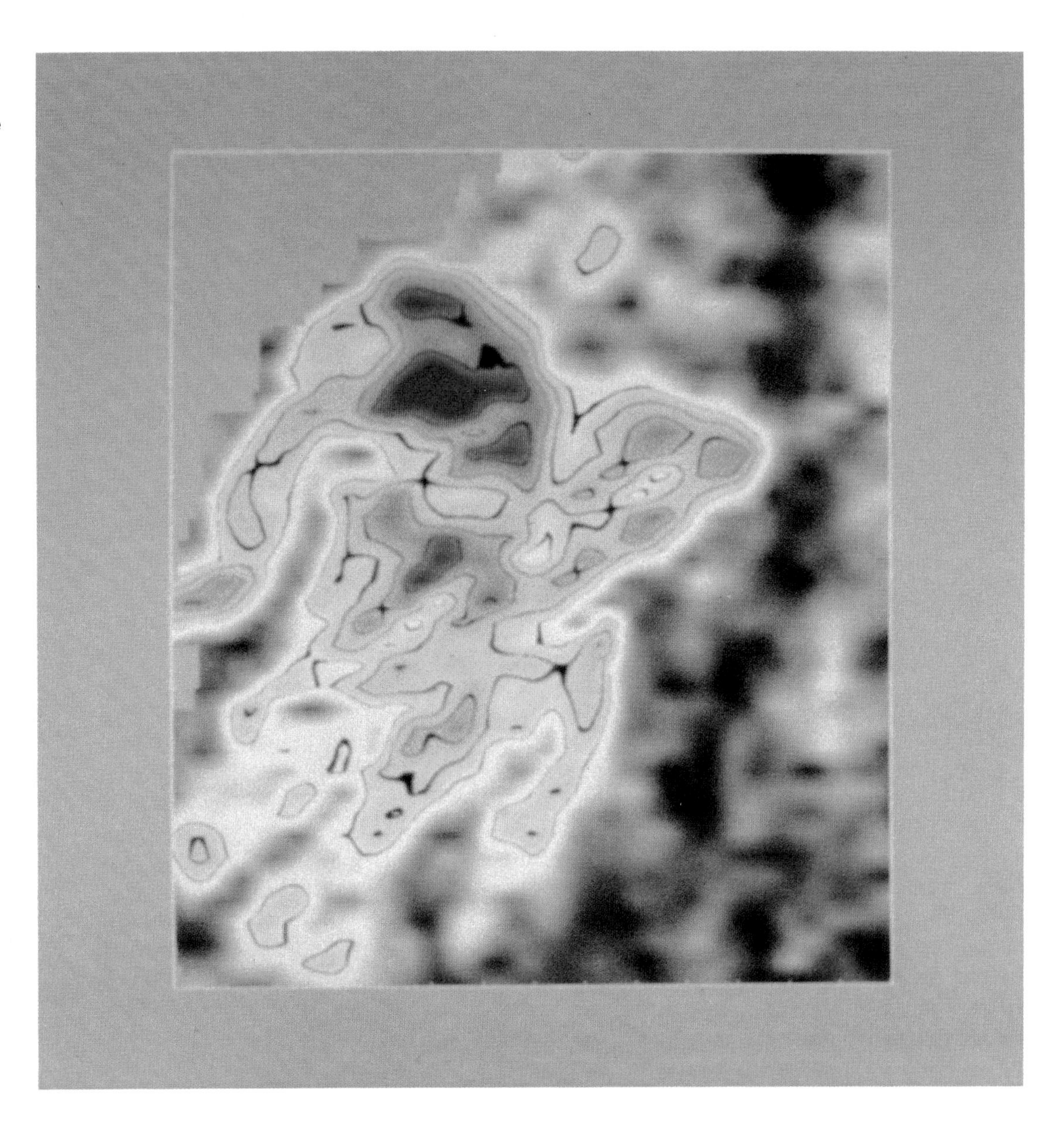

Fig. 2-23. Same horizon Seiscrop section as in Figure 2-21 displayed in reds, yellows and grays to accentuate total drilling potential. (Courtesy Chevron U.S.A. Inc.)

STRUCTURE FROM HORIZONTAL SECTIONS

Direct Contouring

The 3-D seismic interpreter works with a volume of data. Normally this is done by studying some of each of the three orthogonal slices through the volume. This chapter explores the unique contribution of the horizontal section to structural interpretation. The interpreter of structure needs to be able to judge when to use horizontal sections and when to use vertical ones in the course of an overall interpretive project.

Figure 3-1 demonstrates the conceptual relationship between a volume of subsurface rock and a volume of seismic data. Consider the diagram first to represent subsurface rocks and the gray surface to be a bedding plane. The two visible vertical faces of the rectangular solid show the two dip components of the plane; the horizontal face shows the strike of the plane. Now consider the rectangular solid of Figure 3-1 to be the equivalent volume of seismic data. The gray plane is now a dipping reflector and its intersections with the three orthogonal faces of the solid show the two components of dip and the strike as before. Hence the attitude of a reflection on a horizontal section indicates directly the strike of the reflecting surface.

Contours follow strike and indicate a particular level in time or depth. When an interpreter picks a reflection on a horizontal section, it is directly a contour on some horizon at the time (or depth) at which the horizontal section was sliced through the data volume.

Figure 3-2 shows three horizontal sections, four milliseconds apart. By following the semicircular black event (peak) from level to level and drawing contours at an appropriate interval, the structural contour map at the bottom of Figure 3-2 was generated. Note the similarity in shape between the sections and the map for the anticlinal structure and the strike east of the faults. In the central panel the peaks from 1352 ms are printed in black and the peaks from 1360 ms in blue/green. This clearly demonstrates the way in which the events have moved with depth.

Figures 3-3 and 3-4 provide one vertical section and several horizontal sections from which the relationship between the two perspectives can be appreciated. Line P (Figure 3-3) runs north-south through the middle of the prospect with south at the right. The time interval 2632-2656 ms shows some continuous reflections. Proceeding from south to north (right to left, Figure 3-3; bottom to top, Figure 3-4) the structure is first a broad closed anticline, then a shoulder, then a smaller anticline.

Figures 3-4 and 3-5 demonstrate a simple exercise in direct contouring from a suite of horizontal sections. The red event (trough) expanding in size from left to right in Figure 3-4 has been progressively circumscribed from left to right in Figure 3-5. The last frame is a raw contour map of this horizon. Clearly it needs interpretive smoothing based on the interpreter's geologic concepts of the area, but this first structural representation has been made quickly and efficiently without the traditional intermediate tasks of timing, posting and contouring.

Figure 3-6 shows 24 horizontal sections covering an area of about 5 sq mi (13 sq km). These can be used as a structural interpretation exercise. Obtain a small piece of transparent paper and register it over the rectangular area. Begin with the upper left frame and find the red event in its lower right corner. Mark this event by following its maximum amplitude and then mark its

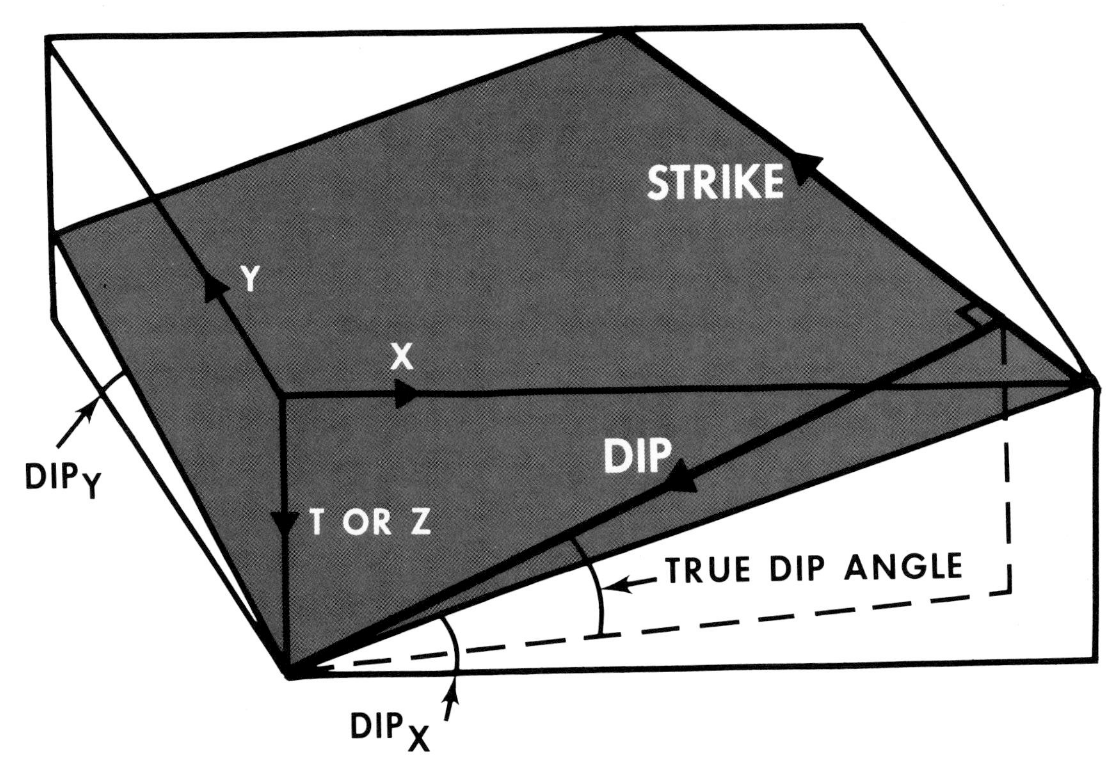

Fig. 3-1. Relation between dip and strike of a seismic reflector within a data volume.

changed position from frame to frame until you reach 2160 ms. Your resultant contour map should show that the dip is generally northwest and that the strike swings about 40° toward the north over the structural range of the map. You will probably detect a fault toward the west of the area as well. If you study the arcuate events west of the fault, you will recognize a small anticline closing against the fault and a small syncline south of it. There is no way to establish the correlation across the fault.

An event on a horizontal section is generally broader than on a vertical section as dips are usually less than 45°. Figure 3-7 shows the effect of dip and frequency on the width of an event on a horizontal section. A gently dipping event is very broad and a steeply dipping event is much narrower. Increasing dip and increasing frequency both make horizontal section events narrower.

Fault Recognition and Mapping

When an interpreter works with 3-D data after having previously mapped from 2-D data over the same prospect, the most striking difference between maps is commonly the increased fault detail in the 3-D map. Figures 3-8 and 3-9 provide a typical comparison.

We expect to detect faults from alignments of event terminations. Figure 3-10 shows a vertical section from the 3-D data which provided the map of Figure 3-9. The event terminations clearly show several faults. The horizontal section of Figure 3-11 is from the same data volume and, in contrast, does not show clear event terminations. Figure 3-12 shows four horizontal sections from a different prospect but one in a similar tertiary clastic environment. Here event terminations clearly indicate the positions of three major faults on each of the four sections.

Why are event terminations visible at the faults in Figure 3-12 but not in Figure 3-11? The answer lies simply in the relationship between structural strike and fault strike. Any horizontal section alignment indicates the strike of the feature. If there is a significant angle between structural strike and fault strike, the events will terminate. If structural strike and fault strike are parallel, or almost so, the events will not terminate but will parallel the faults. Comparison of Figures 3-11 and 3-9 demonstrates that situation.

Because an alignment of event terminations on a horizontal section indicates the strike of a fault, the picking of a fault on a horizontal section provides a contour on the fault plane. Thus picking a fault on a succession of suitably spaced horizontal sections constitutes an easy approach to fault plane mapping. The faults evident in Figure 3-12 have been mapped in this way.

Continued on page 48

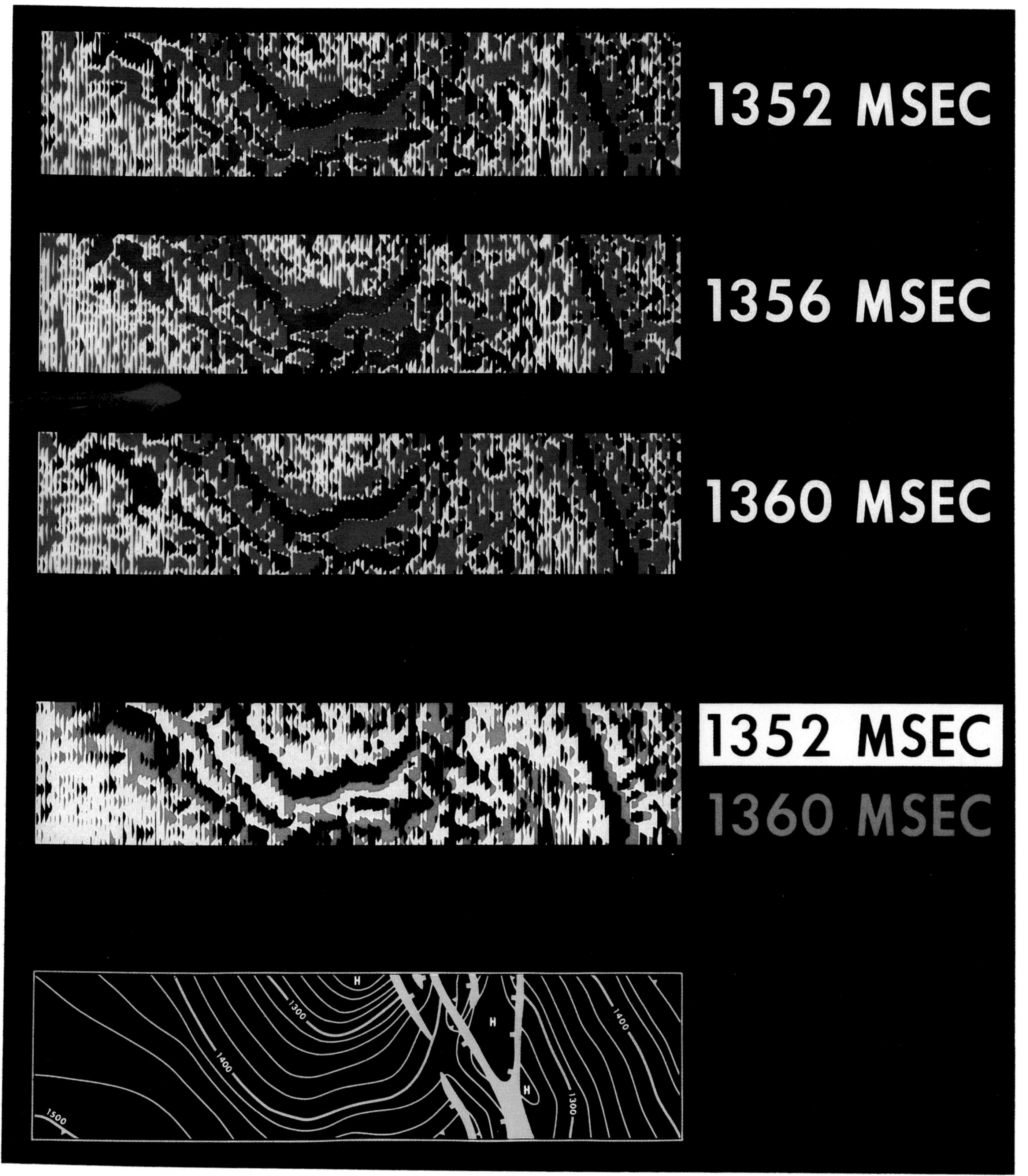

Fig. 3-2. Dual polarity Seiscrop sections from offshore Holland; two-level single polarity Seiscrop section, showing movement of events from 1352 ms to 1360 ms; interpreted contour map on horizon seen as strongest event on Seiscrop sections.

Fig. 3-3. North-south vertical section from Peru through same data volume sliced in Figure 3-4. (Courtesy Occidental Exploration and Production Company.)

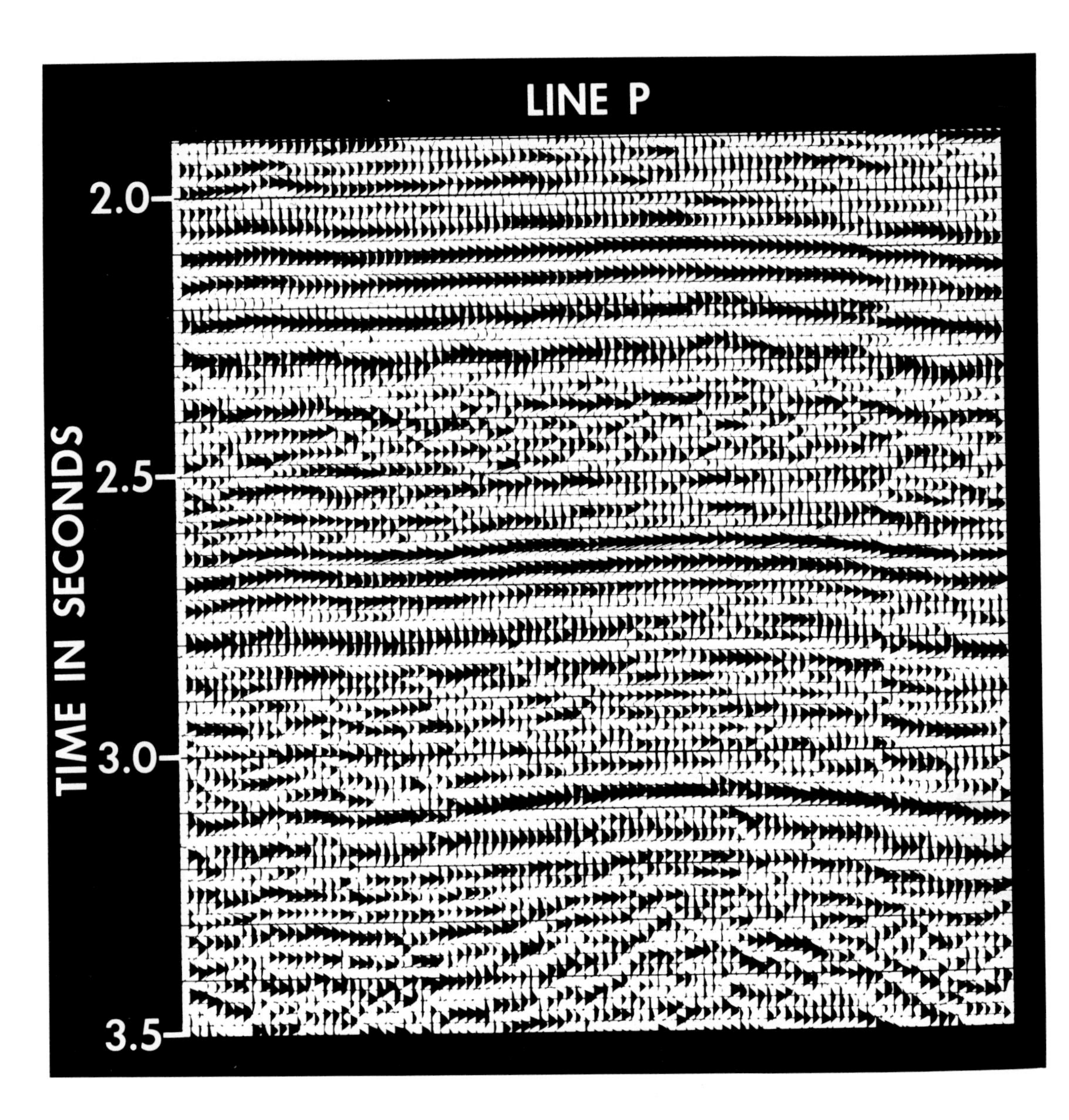

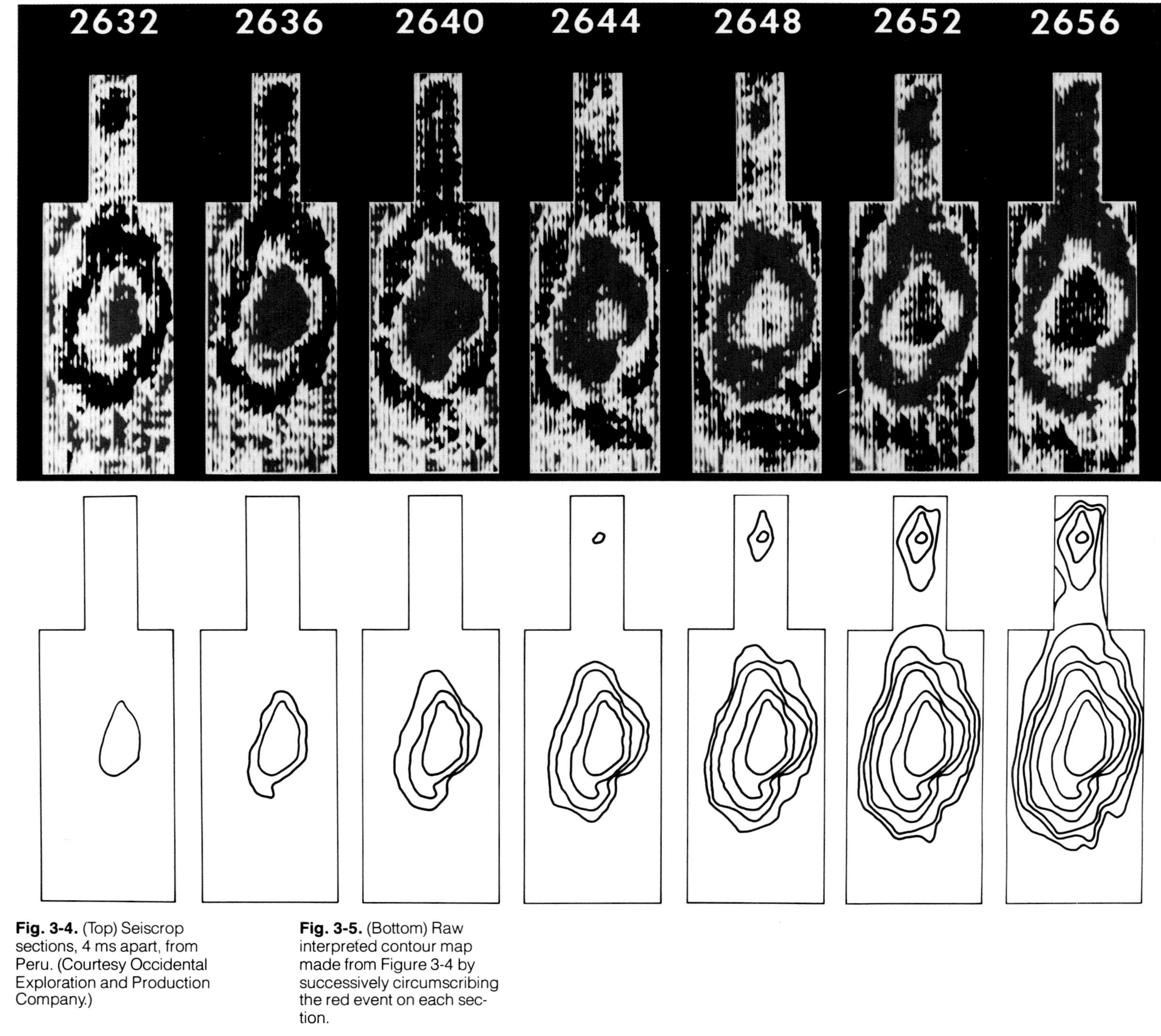

Fig. 3-4. (Top) Seiscrop sections, 4 ms apart, from Peru. (Courtesy Occidental Exploration and Production Company.)

Fig. 3-5. (Bottom) Raw interpreted contour map made from Figure 3-4 by successively circumscribing the red event on each section.

Fig. 3-6. Seiscrop sections, 8 ms apart, from offshore Trinidad. (Courtesy Texaco Trinidad Inc.)

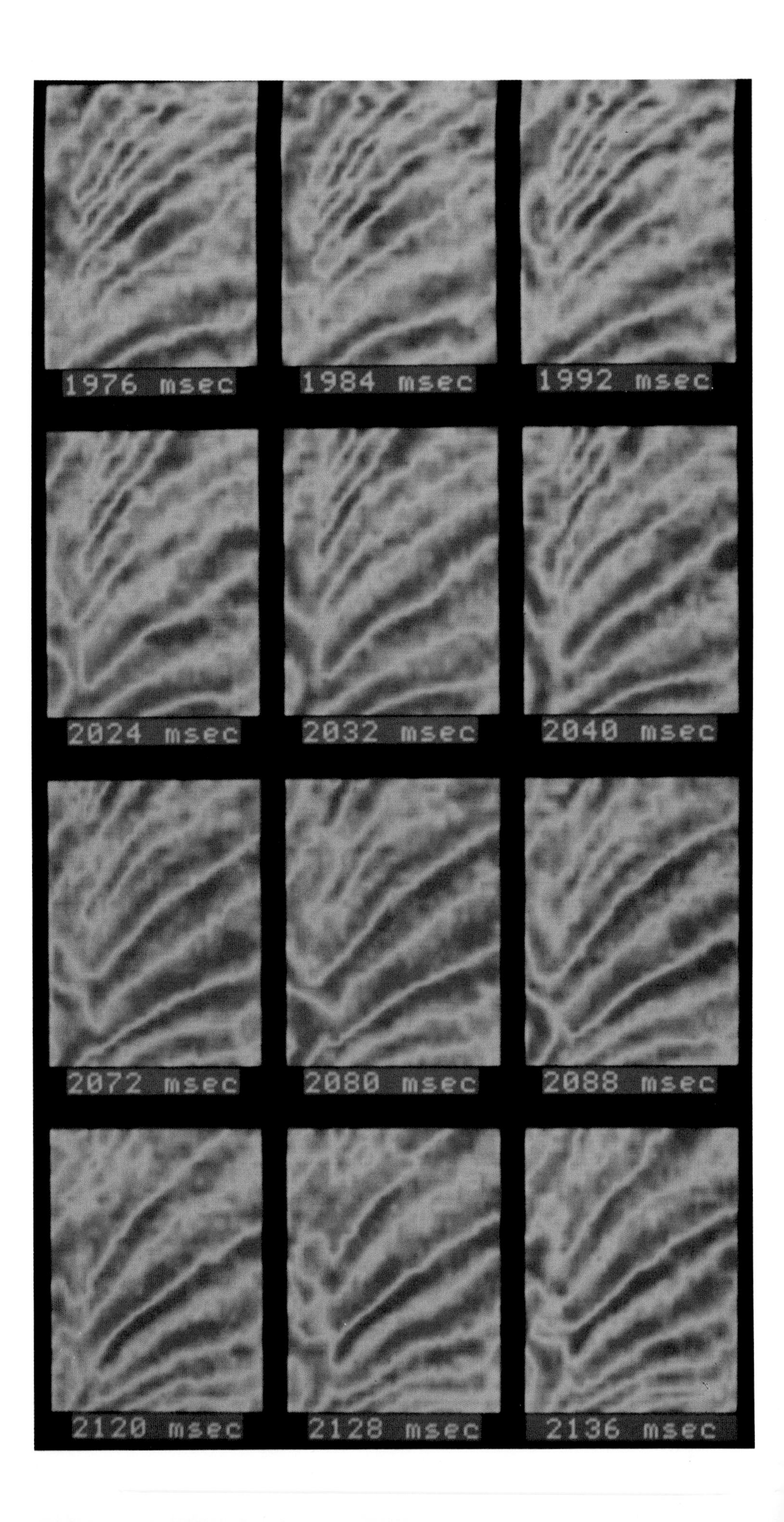

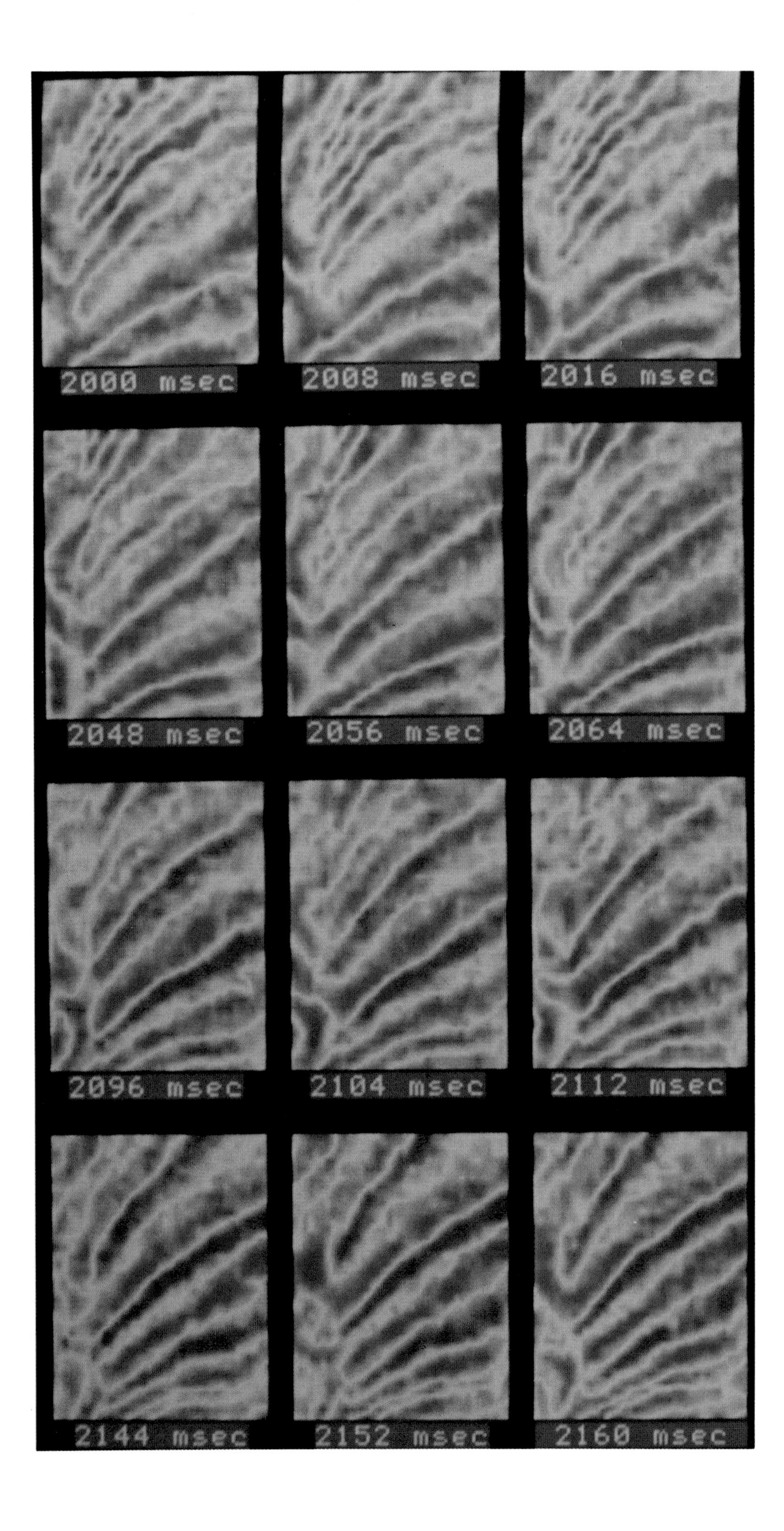
2000 msec
2008 msec
2016 msec
2048 msec
2056 msec
2064 msec
2096 msec
2104 msec
2112 msec
2144 msec
2152 msec
2160 msec

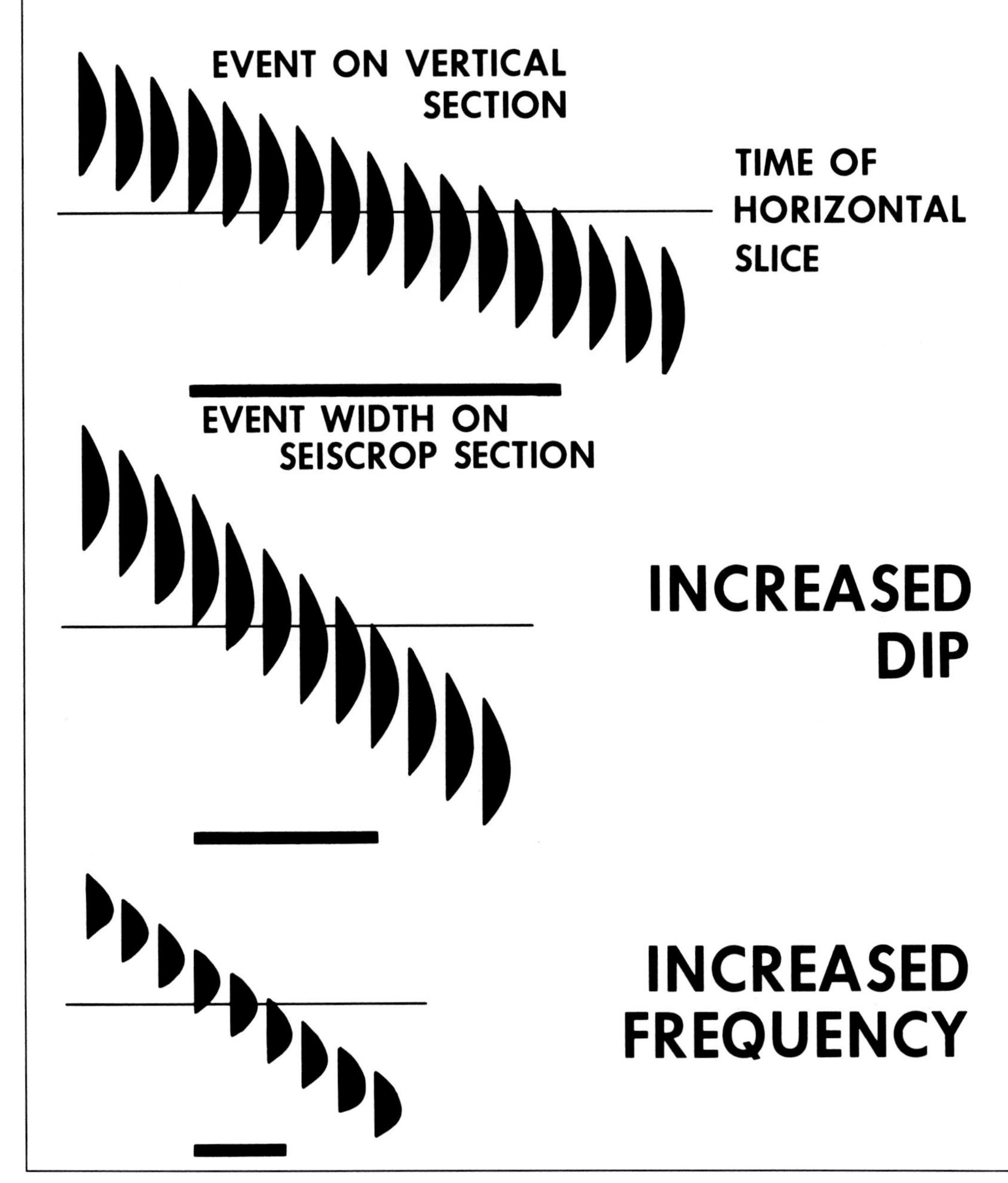

Fig. 3-7. The width of an event on a Seiscrop section decreases with increased dip and also with increased frequency.

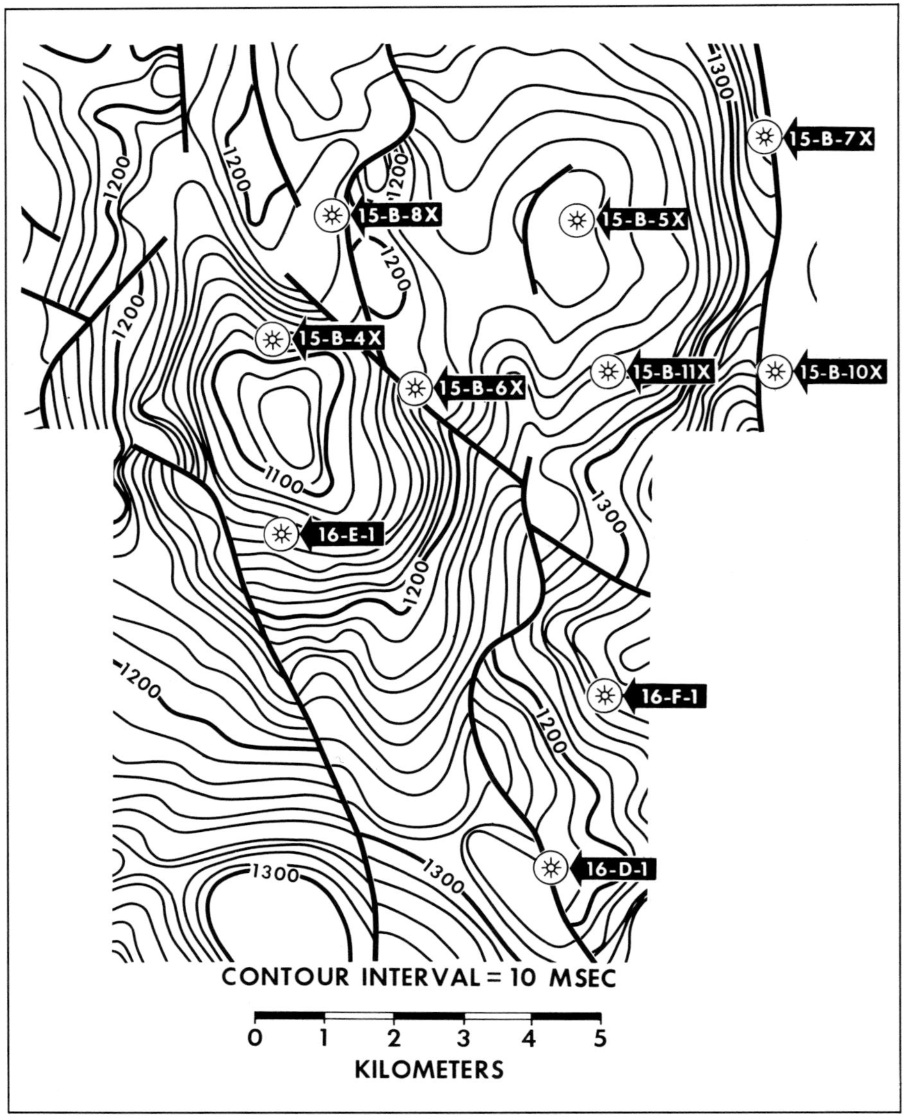

Fig. 3-8. Structural contour map derived from 2-D data from the Gulf of Thailand. (Courtesy Texas Pacific Oil Company Inc.)

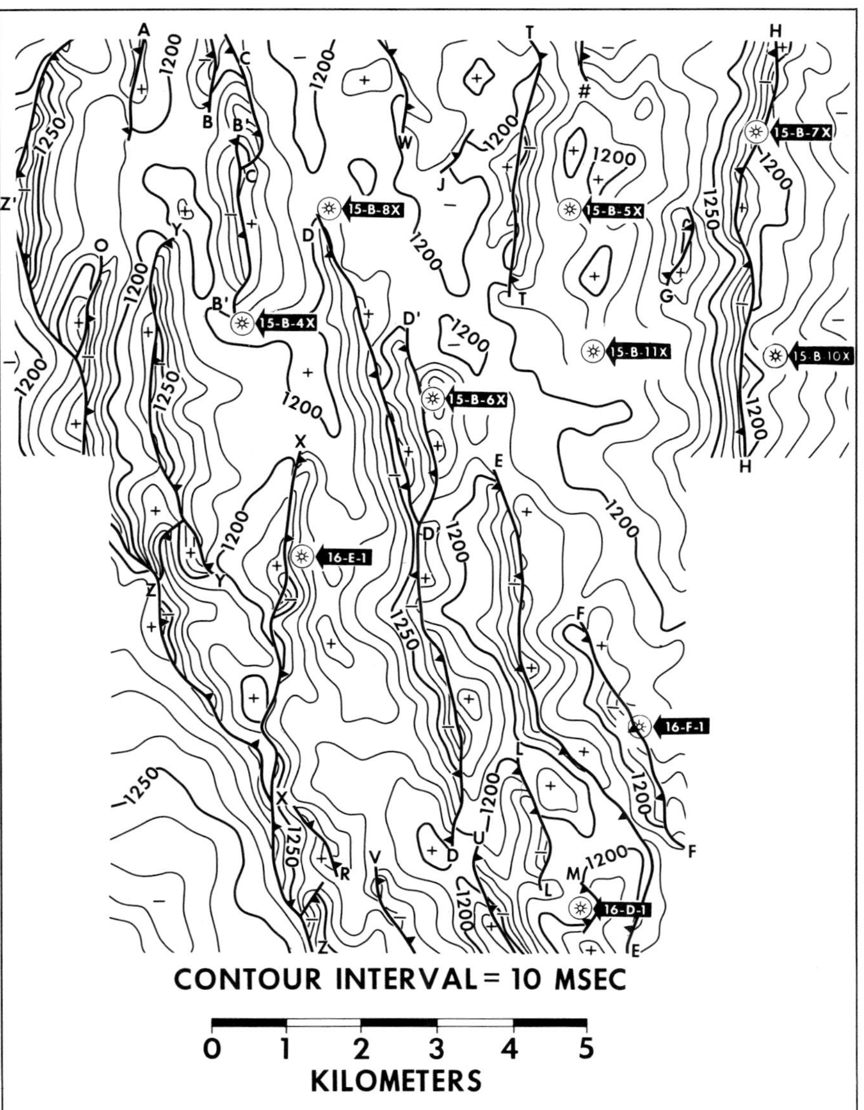

Fig. 3-9. Structural contour map derived from 3-D data from the Gulf of Thailand for the same horizon mapped in Figure 3-8. (Courtesy Texas Pacific Oil Company Inc.)

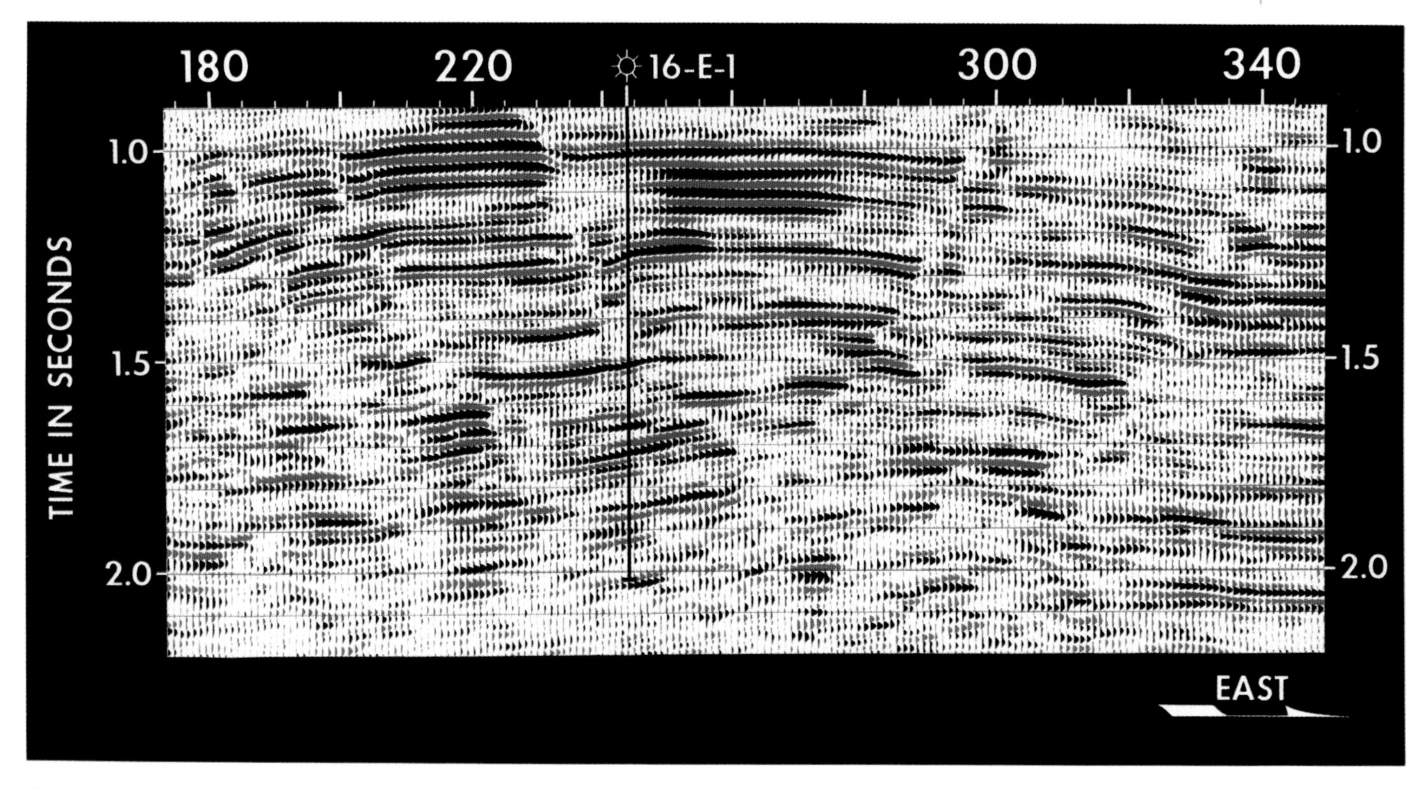

Fig. 3-10. Line 55 from Gulf of Thailand 3-D data. (Courtesy Texas Pacific Oil Company Inc.)

In the lower right corner of the horizontal section at 2260 ms (Figure 3-12) two fault blocks show events of quite different widths. This is the effect of dip which was explained by Figure 3-7. We also see a similar effect of dip in Figure 3-11 where the faults are mostly traced by narrow sinuous events striking approximately north-south.

Composite Displays

The horizontal section of Figure 3-13 shows a rim syncline surrounding a salt diapir. The narrow events around the salt indicate the steep dips near the intrusion.

Figures 3-14 and 3-15 show a deeper horizontal section from the same volume without and with interpretation. The horizon of interest, marked in green on Figure 3-15, is intersected twice, once on either side of the rim syncline. The faulting at this level, marked in yellow, is complex but can be seen fairly well on this one horizontal section. From pre-existing 2-D data in the area only one of these faults had been identified (Blake, Jennings, Curtis, Phillipson, 1982).

The interpreter of 3-D data is not restricted to single slice displays. Because the work is done with a data volume, composite displays can be helpful in appreciating three-dimensionality and also in concentrating attention on the precise pieces of data that provide insight into the problem at hand.

Figure 3-16 is a composite of horizontal and vertical sections spliced together along their line of intersection. The vertical section shows that the circular structure is a syncline. The horizontal section pinpoints the position of its lowest point. The fault on the left of this structure can be followed across the horizontal section. Figure 3-17 provides a different view of the structure. The same horizontal section is here spliced to the portion of the vertical section above in the volume.

It is possible to make cube displays showing, simultaneously, three orthogonal slices through the volume (Figures 1-10 and 1-11). These can certainly aid in the appreciation of three-dimensionality but have limited application in the mainstream of the interpretation process, because two of the faces of any cube displayed on a monitor or piece of paper will always be distorted.

Figures 3-18 and 3-19 illustrate the study of a trio of normal faults. In Figure 3-18 one horizon has been tracked indicating the interpreted correlation across the faults. At the bottom of this figure a portion of the data from each of the four fault blocks is enlarged and again carries the interpreted track. Each block has been adjusted vertically to bring the track segments into conti-

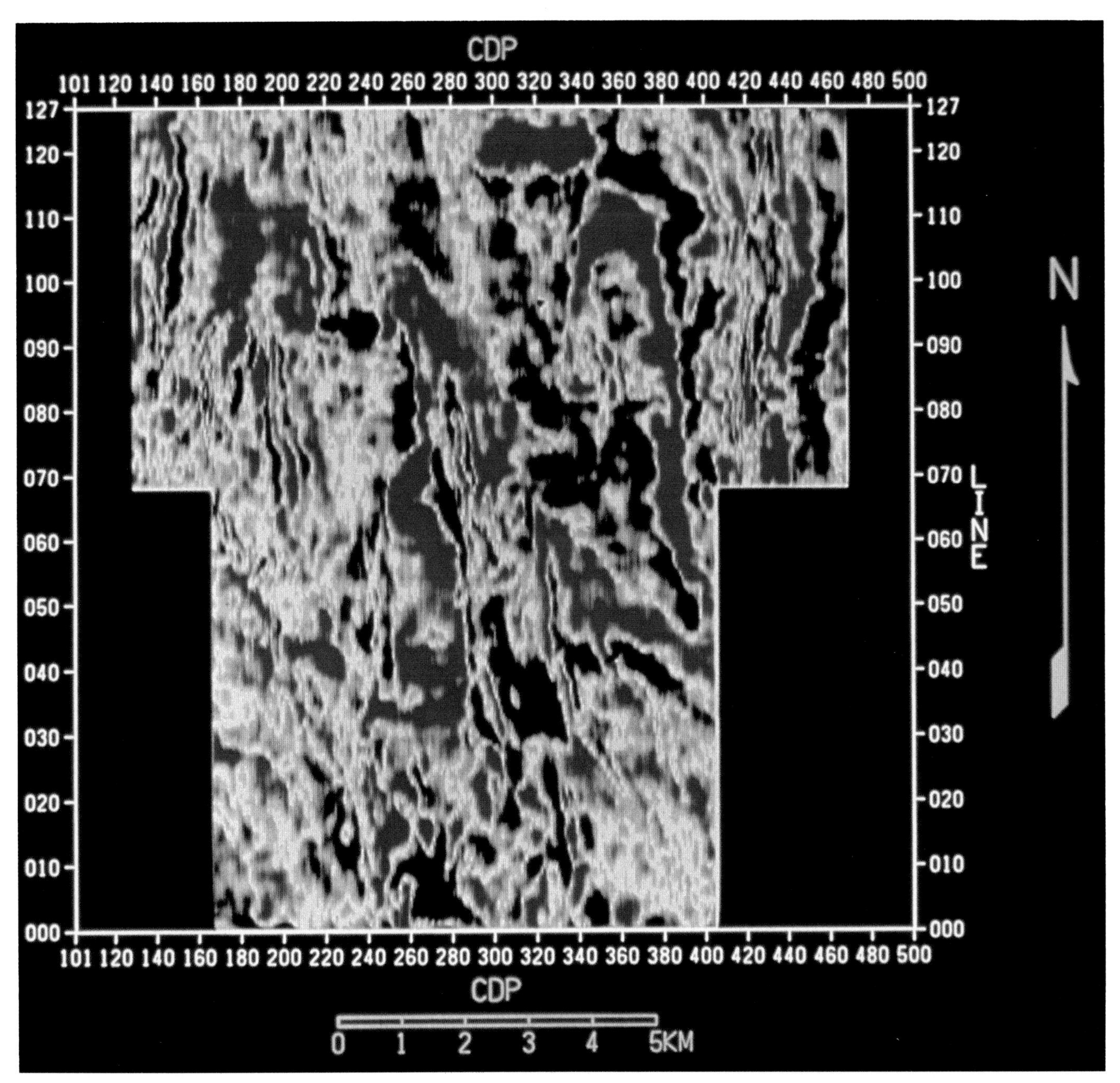

Fig. 3-11. Seiscrop section at 1388 ms from Gulf of Thailand. (Courtesy Texas Pacific Oil Company Inc.)

Fig. 3-12. Seiscrop sections from offshore Trinidad. Event terminations indicate faulting. (Courtesy Texaco Trinidad Inc.)

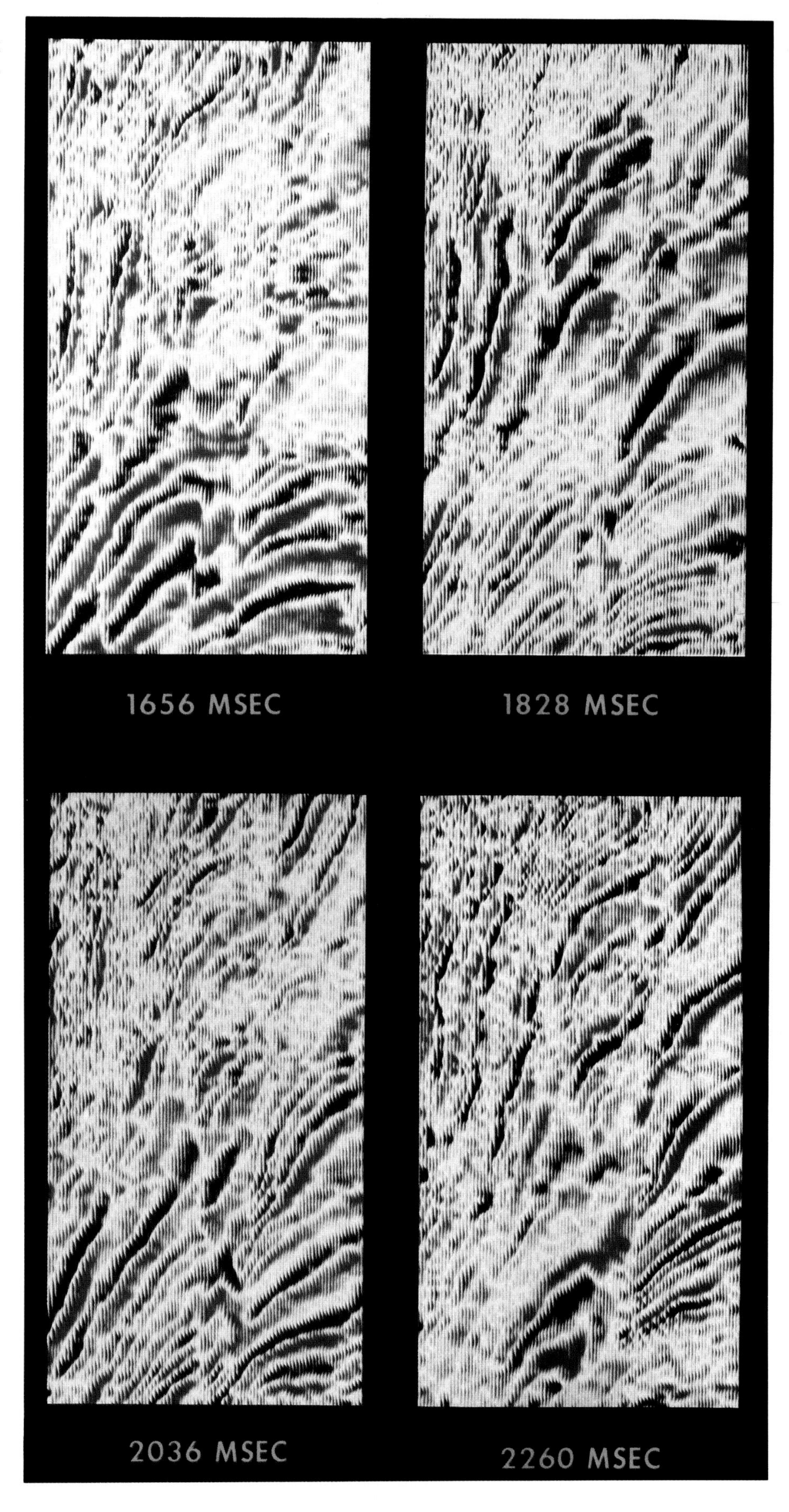

nuity so that the correlation between these blocks of data can be assessed easily. Note how this display accentuates the apparent growth on the center fault of the three. In Figure 3-19 the composite horizontal and vertical section display permits the study of the horizontal extension ol each of these three faults.

Interpretation Procedures

The interpreter of 3-D data has a real opportunity to generate accurate subsurface structure maps but to do so a large amount of data must be studied. The Seiscrop Interpretation Table and interactive workstation, discussed in Chapter 1, are both devices to help the 3-D interpreter manage this large volume of data.

Figure 3-20 charts a possible procedure for 3-D structural interpretation using the Seiscrop Interpretation Table. The Seiscrop, or horizontal section, movie and the vertical section movie viewed as motion pictures provide an effective way of overviewing the structure before mapping begins. The viewer immediately gets a general impression of the fault directions, the structural culminations, general strike and dip directions, and variations in data quality. This is the only occasion on which a Seiscrop movie is used literally as a movie.

The next stage will probably be to pick faults and make a preliminary interpretation on a selected set of vertical sections in the line and crossline directions, for example on a one kilometer grid. This will provide the approximate extent of the first fault block in which mapping will begin. Normally the interpreter will identify the horizon to be followed at a well.

Using the selected set of vertical sections the approximate fault locations are marked on the base map on the screen of the table. The event to be mapped is then identified on one horizontal section and followed up and down within the first fault block, drawing contours from the horizontal sections at the desired interval. The faults surrounding the first fault block are marked in detail at the same time. Several iterations through the sections covering the structural relief of the horizon in this first fault block may be necessary before the interpreter is satisfied with the contours drawn.

Selected vertical sections are revisited to establish the correlation into the next fault block and the procedure then repeats in that fault block. The interpreter thus works from fault block to fault block until the prospect is covered; alternatively the same horizon may be carried in two or more blocks at the same time.

When the interpreter encounters a problem in understanding the data at a particular location, reference to vertical sections through that point in line, crossline, and other directions is made. Selected azimuth lines may be specially extracted from the data volume for the purpose. Once the problem is resolved, the interpreter should be able to return to the horizontal sections to continue contouring.

Throughout the structural interpretation the interpreter should be watching for amplitude anomalies and lineations that may have stratigraphic significance. Horizontal sections have the unique ability of bringing to the interpreter's attention subtle features which later prove significant. This is discussed at length later, especially in Chapter 4.

Figure 3-21 charts a possible procedure for 3-D interpretation using an interactive workstation. The interactive capabilities required to follow this procedure include:

(1) automatic and manual tracking of horizons on vertical and horizontal sections;
(2) automatic spatial horizon tracking and editing through a 3-D data volume;
(3) correlation of vertical sections with well data;
(4) storing and manipulation of seismic amplitudes;
(5) manipulation of maps; and,
(6) flexible use of color.

This approach incorporates many of the notions from the previous procedure but utilizes the greatly extended capabilities. The procedure of Figure 3-21 also addresses several areas of stratigraphic and reservoir interpretation which will be discussed in later chapters.

Advantages and Disadvantages of Different Displays

With increasingly successful amplitude preservation in seismic processing, interpreters are increasingly suffering from the limited optical dynamic range of conventional seismic displays. Too common are the variable area sections where some events of interest are heavily saturated and others have barely enough trace deflection to be visible. This applies to all displays, vertical and horizontal, made with variable area techniques. Horizontal sections, historically, were first

Fig. 3-13. Seiscrop section at 3252 ms from Eugene Island area of Gulf of Mexico showing interpreted shape of salt plug. (Courtesy Hunt Oil Company.)

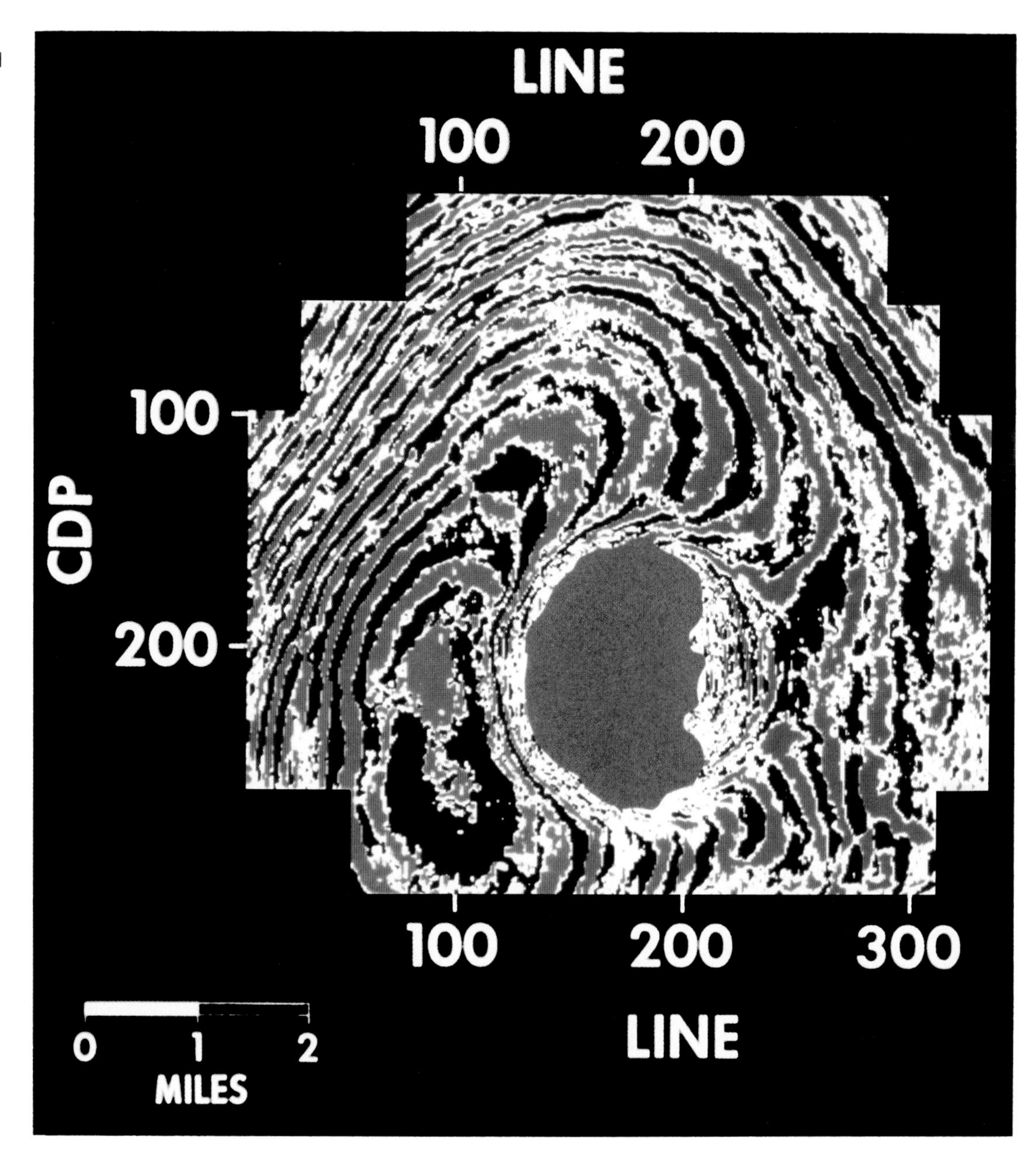

made with variable area using one polarity only, normally peaks. This soon evolved into dual polarity variable area giving equal weight to peaks and troughs (see Chapter 2). This is exemplified by the upper row of sections in Figure 3-22 and explained in detail by the diagram of Figure 3-23.

Dual polarity variable area provides five clearly discernible amplitude levels. The highest amplitude peaks are saturated and appear as continuous black areas; the medium amplitude peaks do not coalesce and appear as discontinuous black areas which look gray; the lowest amplitudes are below the variable area bias level and appear white; the medium amplitude troughs appear pink; and the highest amplitude troughs are continuous red areas. In structural mapping the interpreter must pick a consistent point on the seismic waveform, although it doesn't normally matter which is chosen. Most commonly, a red event (trough) is followed and picked on the edge of the pink, that is close to, but not quite at, the zero crossing. This point is fairly consistent in phase as the amplitude of the event changes with position over the prospect. It is more consistent than the edge of the red, the other simple option with this mode of display.

If the detail in the seismic waveform provided by dual polarity variable area is inadequate,

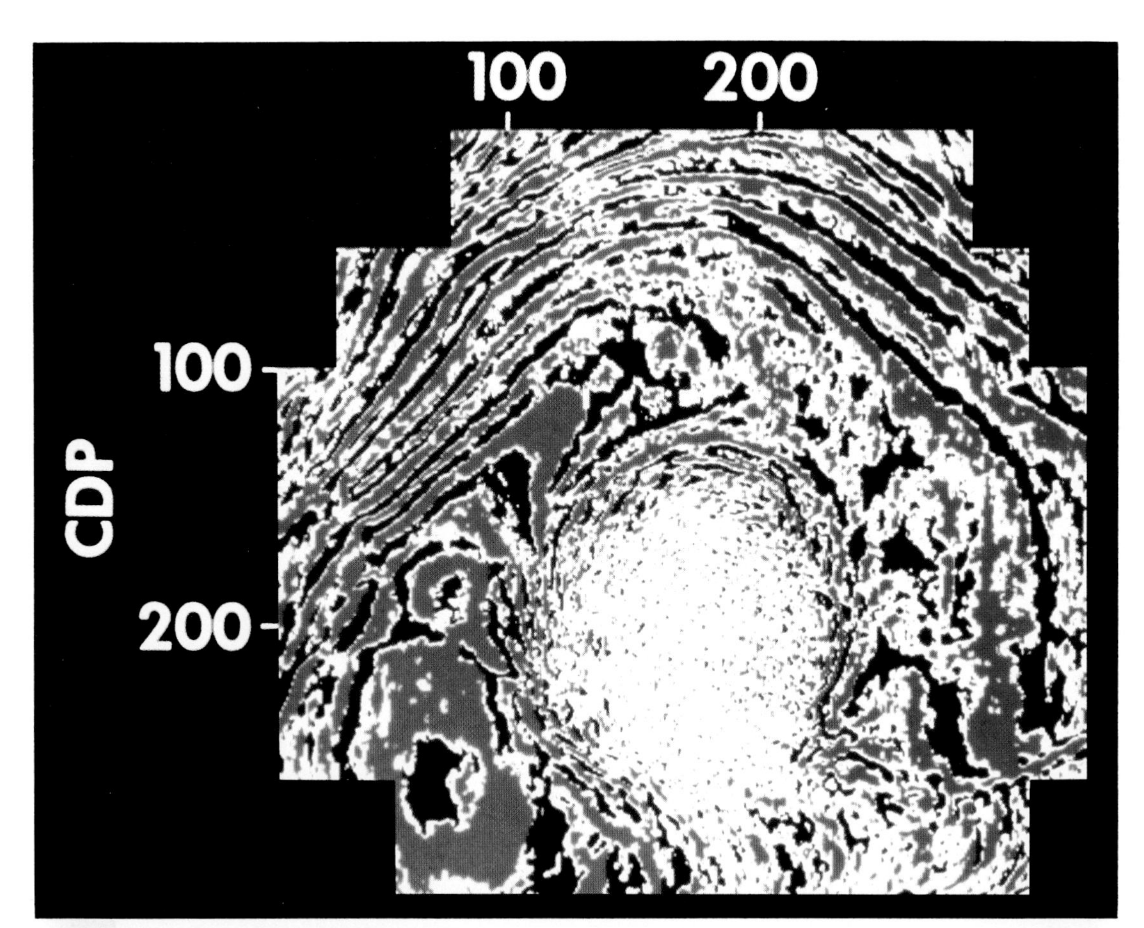

Fig. 3-14. Seiscrop section at 3760 ms from Eugene Island area of Gulf of Mexico. (Courtesy Hunt Oil Company.)

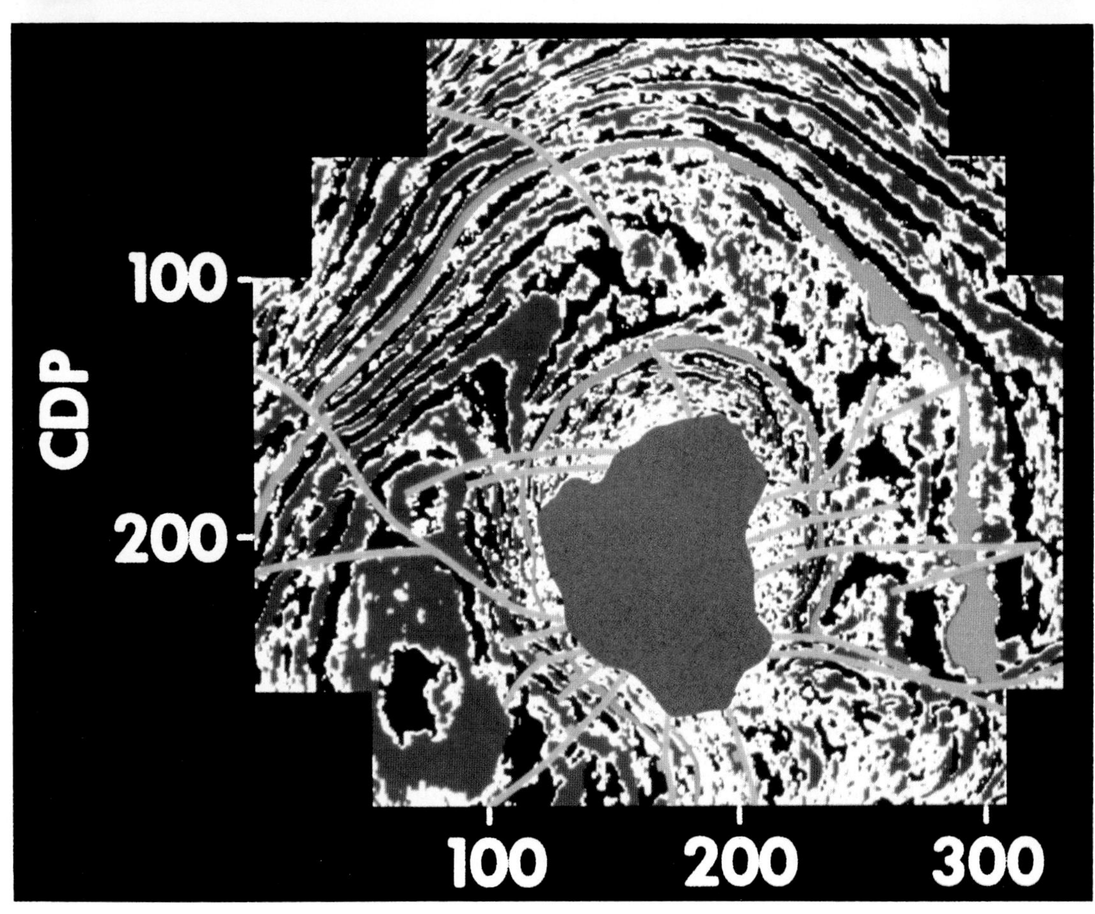

Fig. 3-15. Same Seiscrop section as Figure 3-14 with interpretation of faults and the Green horizon. (Courtesy Hunt Oil Company.)

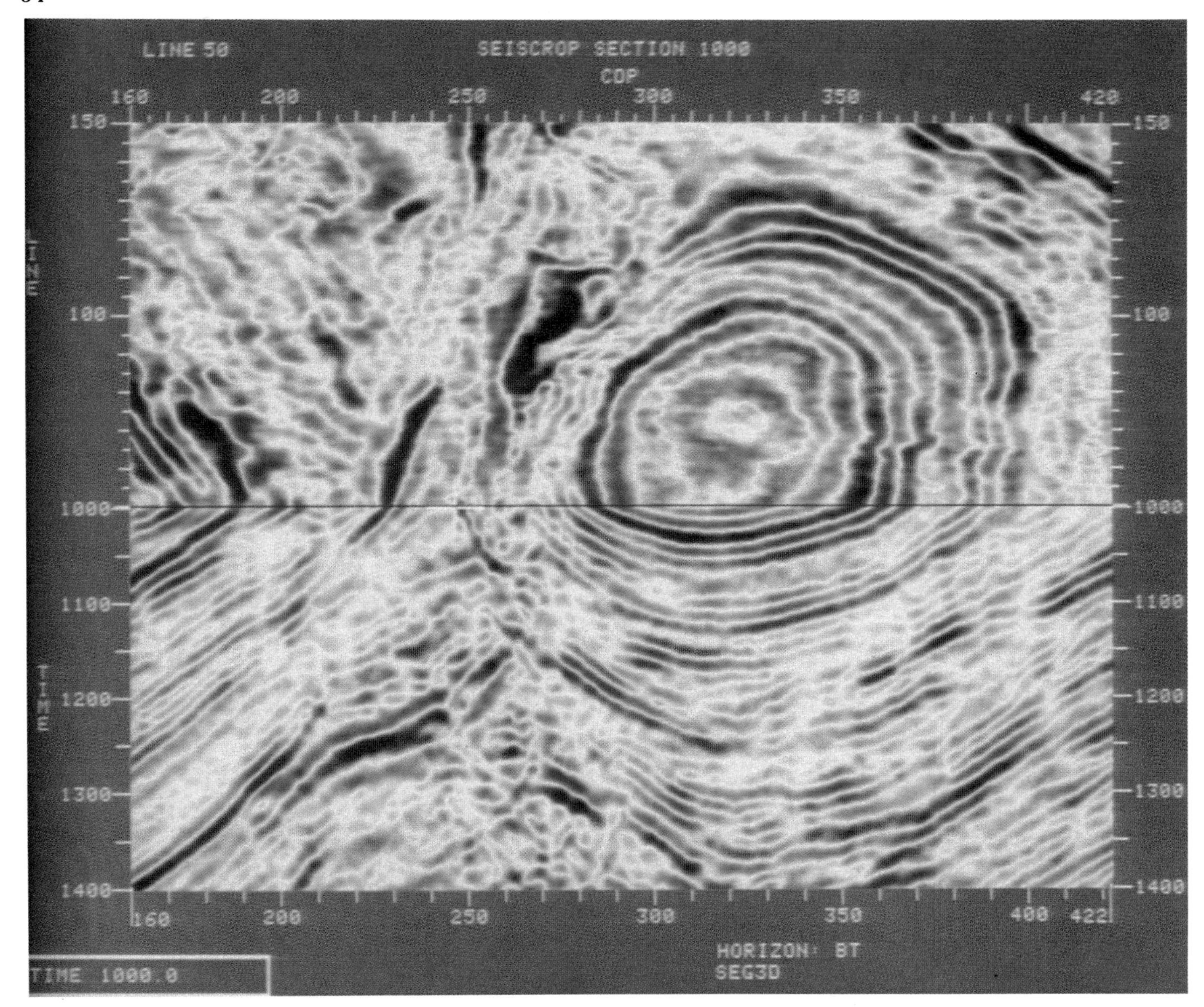

Fig. 3-16. Composite display of Seiscrop and vertical sections from onshore Europe. Vertical section segment lies beneath Seiscrop section.

then the increased dynamic range of full variable intensity color is required. The many ways of using color to interpretive advantage are discussed in Chapter 2. Gradational blue and red is a most useful application; this is illustrated in the middle row of sections in Figure 3-22 and explained in detail by the diagram of Figure 3-24. On such a display the interpreter can see the local amplitude maxima of a peak (or a trough) and draw a contour along the locus of those maxima, thus picking the crest of the seismic waveform.

A further option available to the structural interpreter is horizontal sections displayed in phase, using instantaneous phase derived from the complex trace (Taner, Koehler and Sheriff, 1979). This approach is illustrated by the lower row of sections in Figure 3-22 and explained in detail by the diagram of Figure 3-24.

Phase indicates position on the seismic waveform without regard to amplitude, making a phase section like one with fast AGC (Automatic Gain Control), destroying amplitude variations and enhancing structural continuity. A phase section is displayed with color encoded to phase over a given range, for example 30°. Color boundaries occur at significant phase values such as 0° (a peak), 180° (a trough), +90° and −90° (zero crossings). By following a chosen color

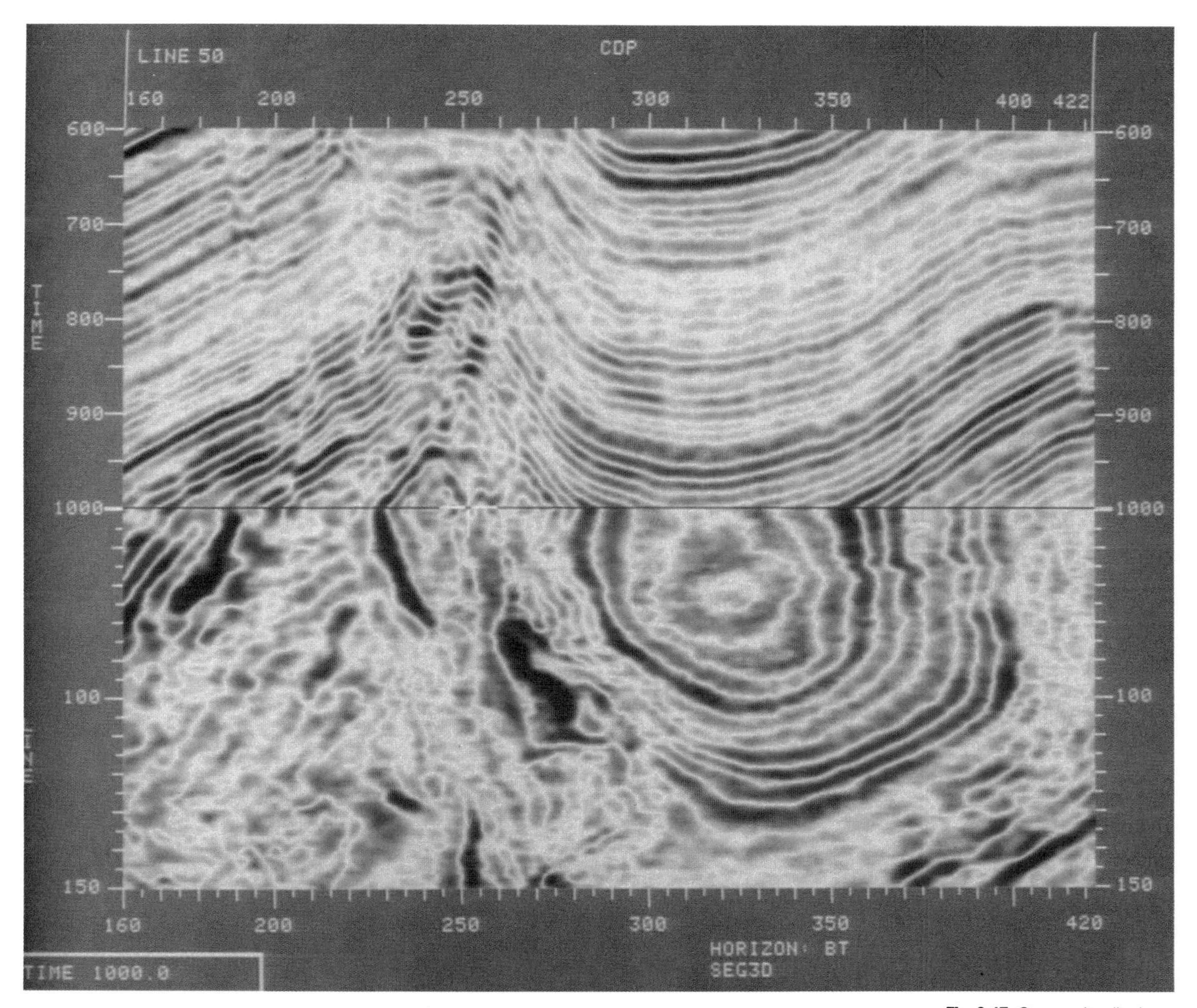

Fig. 3-17. Composite display of Seiscrop and vertical sections from onshore Europe. Vertical section segment lies above Seiscrop section.

boundary on a horizontal section displayed in this way, the interpreter is drawing a contour for his horizon map picked at a specific phase point. Thus the interpreter can also, if necessary, compensate for any estimated amount of phase distortion in the seismic wavelet.

Figure 3-25 is a horizontal phase section from a different area; the structural continuity is clear. Figure 3-26 shows the same section in edited phase, a simple modification of the display colors. A few degrees of phase centered on 0° have been colored black; a few degrees of phase centered on 180° have been colored red; and all other phases have been colored white. This gives the appearance of an automatically picked section with all the peaks and troughs at that level indicated. The interpreter simply selects the one he wants.

Horizontal Sections From Widely-Spaced Data

Because of the interpretive benefits of horizontal sections, the question of their possible construction from widely-spaced data naturally arises — data such as a conventional 2-D grid. The interpreter would like to slice all the data values at a single time from the 2-D lines, lay each down at its appropriate horizontal position and fill the grid cell spaces by some data interpolation procedure. In general, 2-D data will not satisfy the 3-D sampling requirements between lines as

Continued on page 60

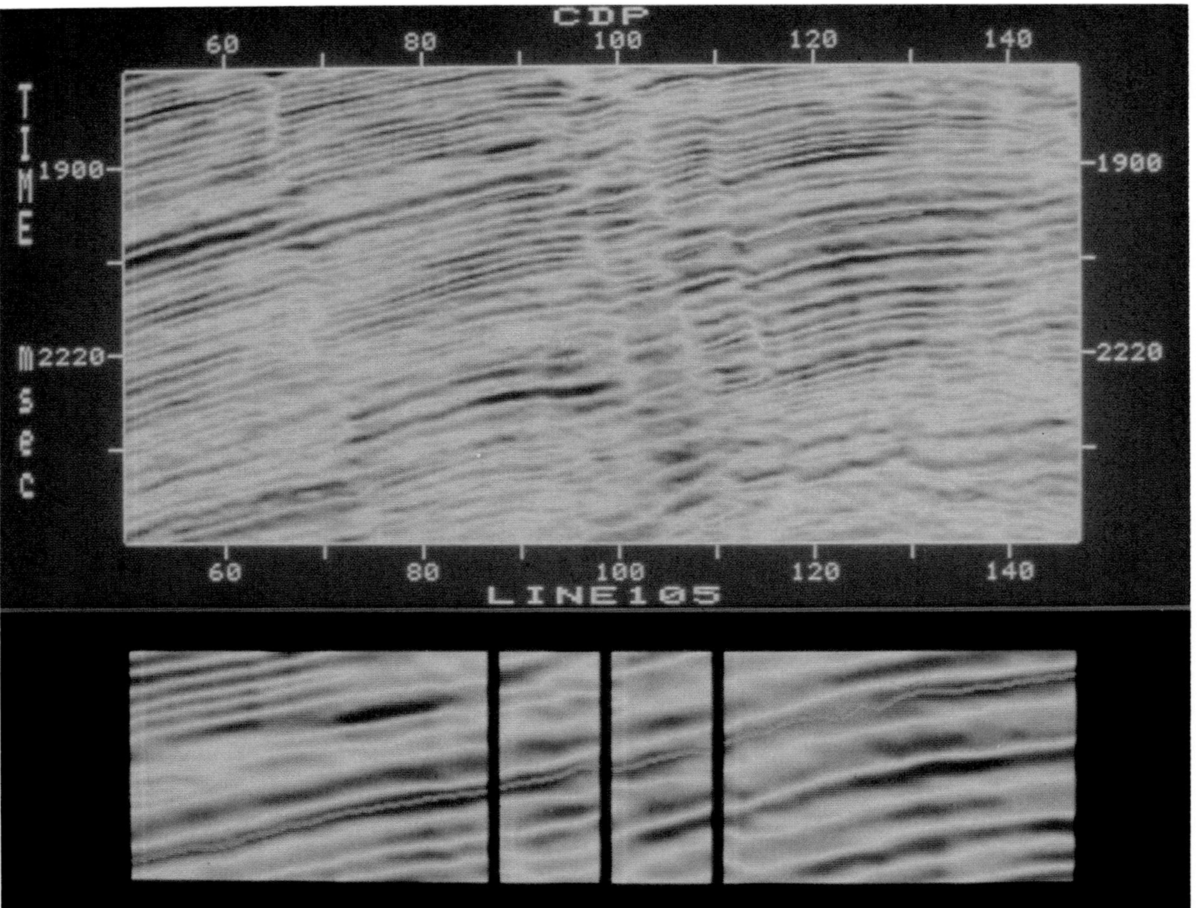

Fig. 3-18. Vertical section and magnified portions thereof designed to study fault correlations offshore Trinidad. (Courtesy Texaco Trinidad Inc.)

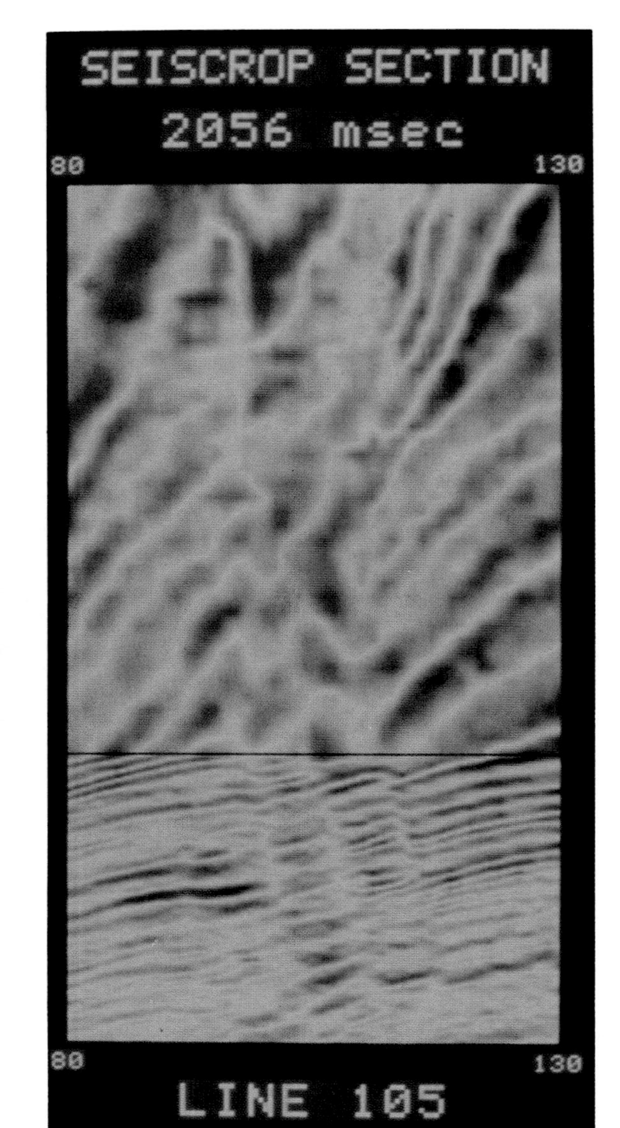

Fig. 3-19. Composite display of Seiscrop and vertical sections from offshore Trinidad showing horizontal extent of faults studied in Figure 3-18. (Courtesy Texaco Trinidad Inc.)

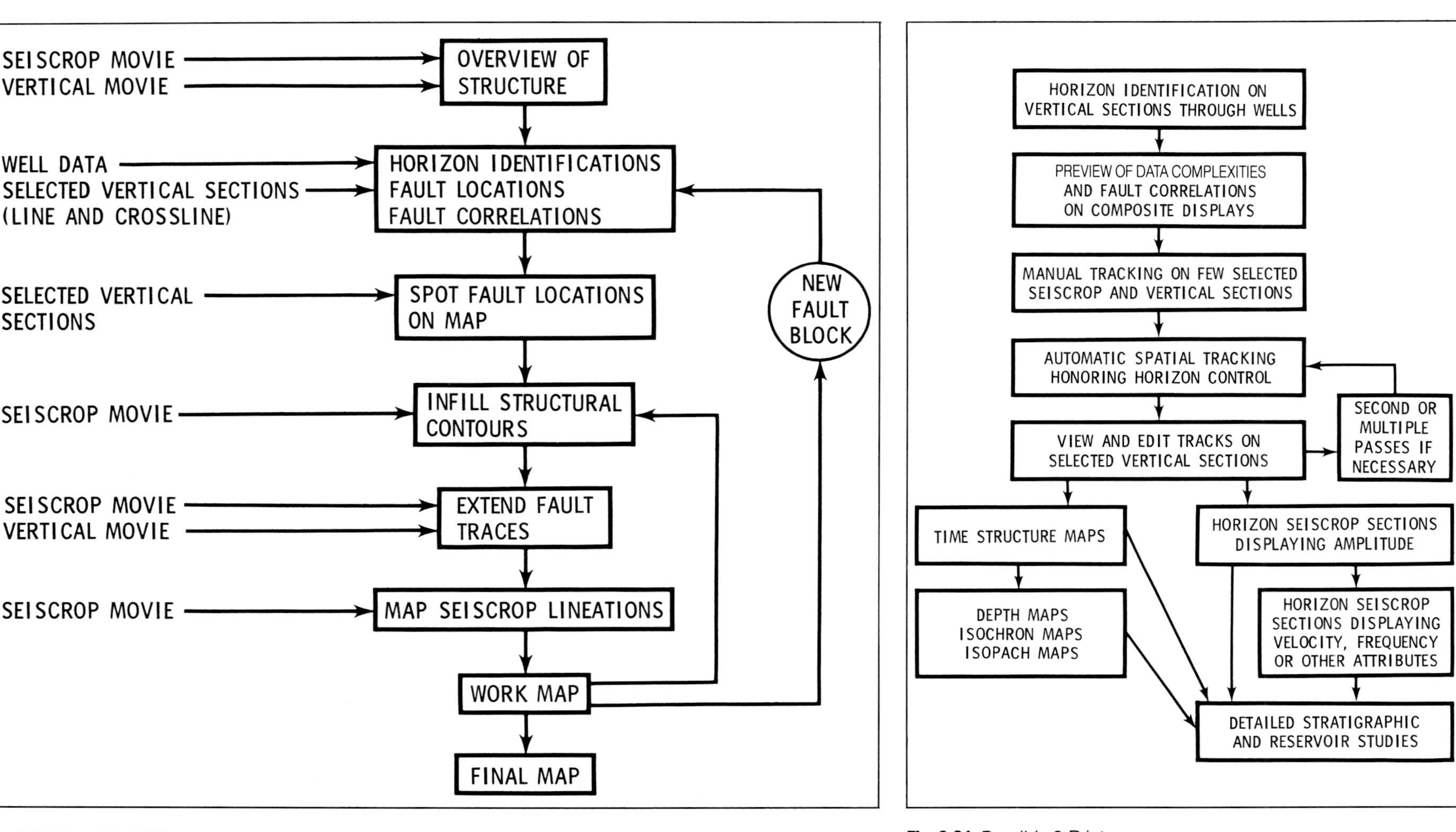

Fig. 3-20. Possible 3-D interpretation procedure using Seiscrop Interpretation Table.

Fig. 3-21. Possible 3-D interpretation procedure using interactive workstation.

Fig. 3-22. Seiscrop sections, 8 ms apart, from Gulf of Thailand displayed in dual polarity variable area (upper row), with seismic amplitude coded to color (middle row), and with instantaneous phase coded to color (lower row). (Courtesy Texas Pacific Oil Company Inc.)

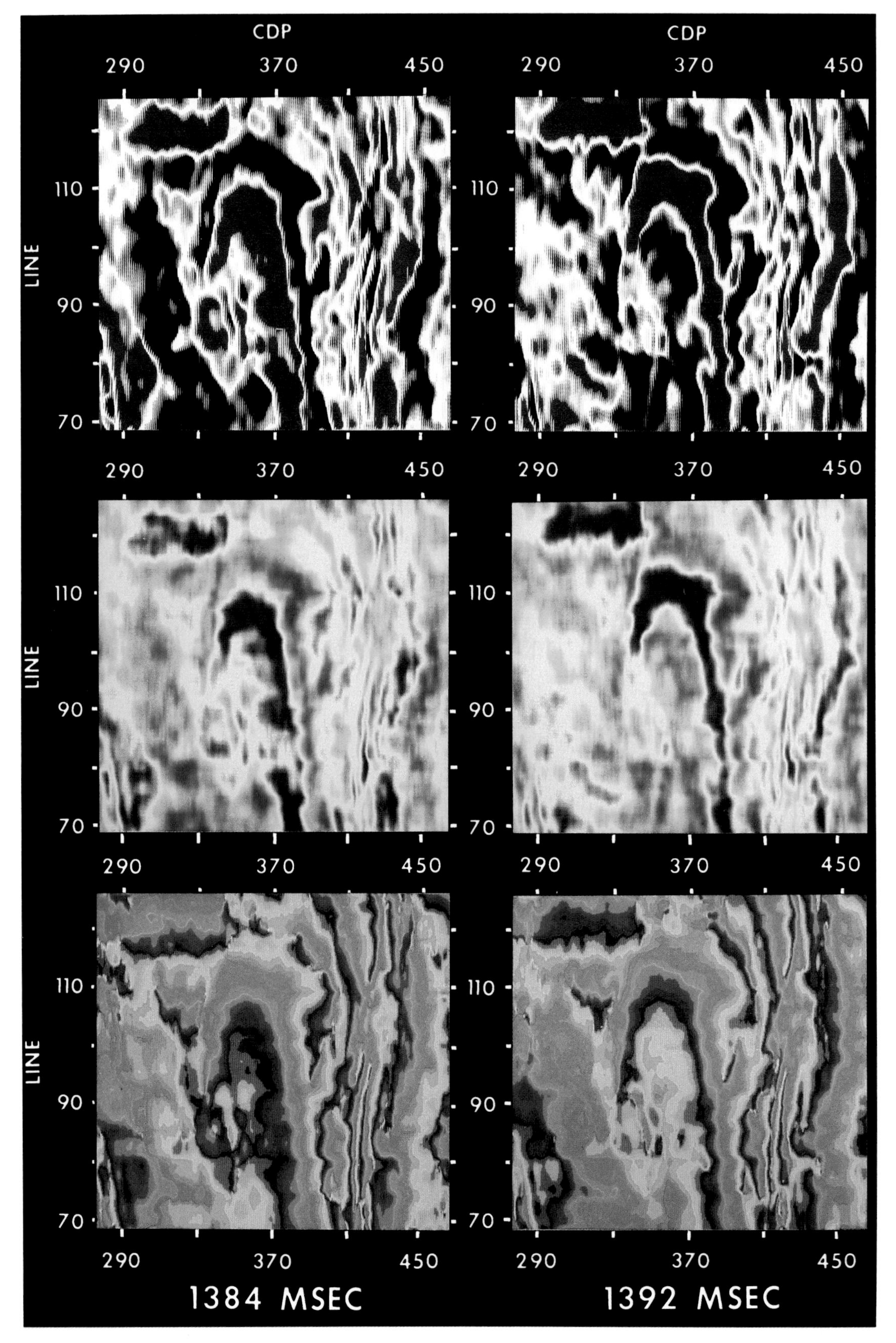

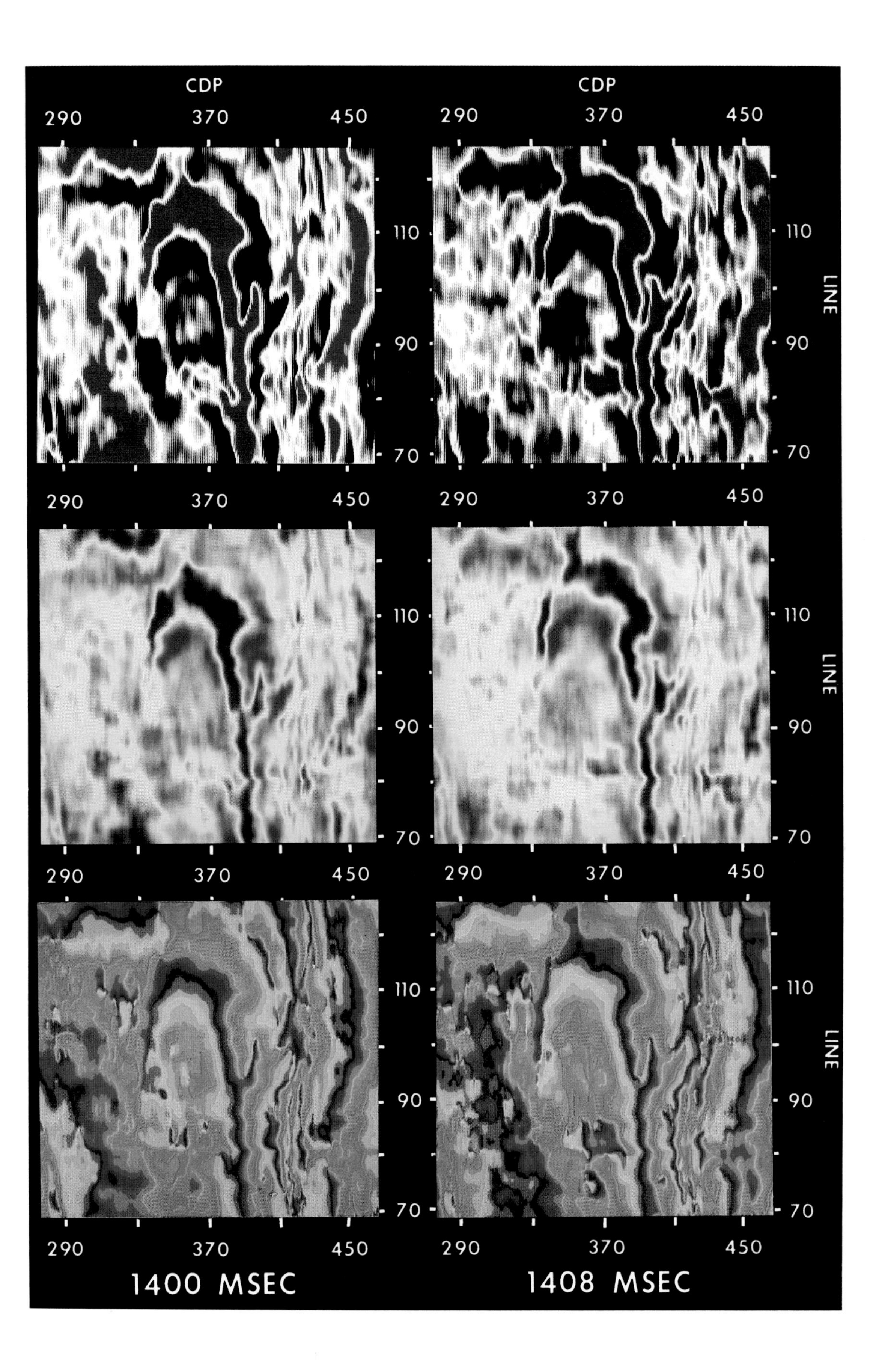
CDP
CDP
290
370
450
290
370
450
110
90
70
110
90
70
LINE
LINE
LINE
1400 MSEC
1408 MSEC

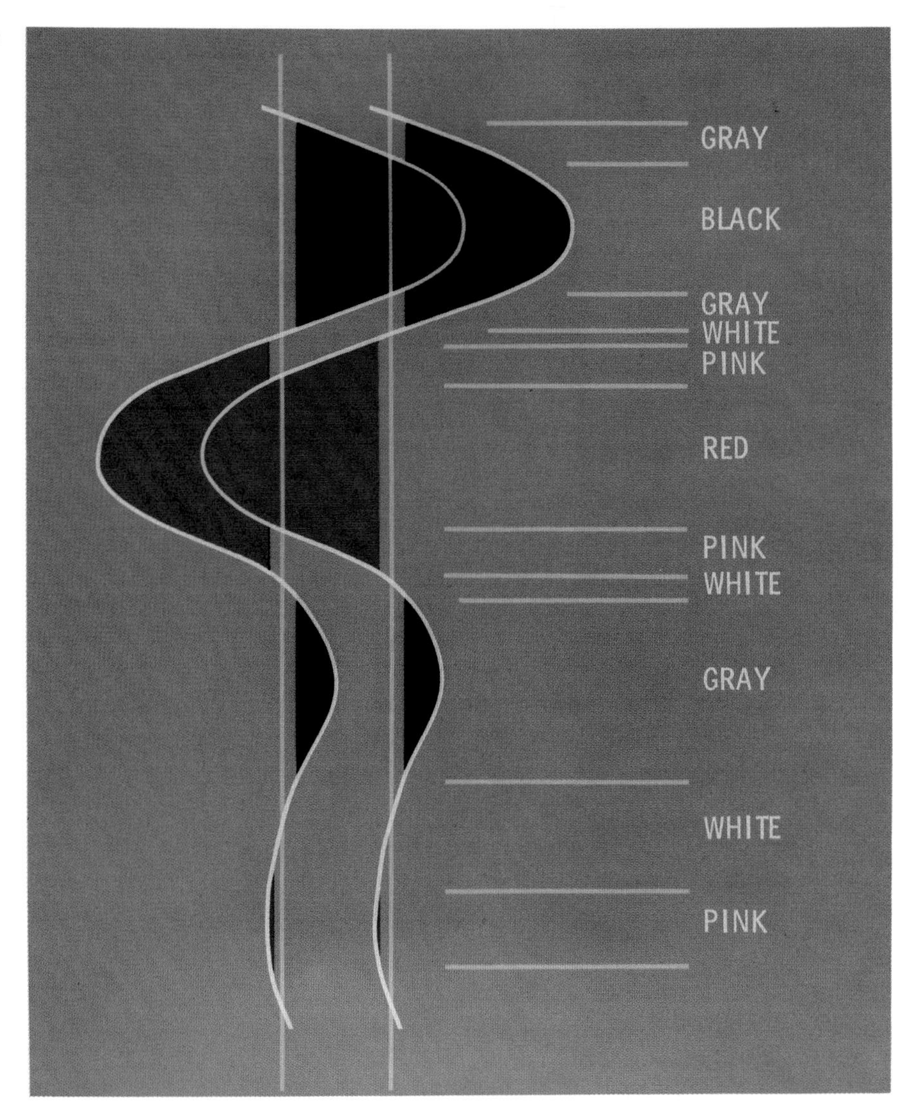

Fig. 3-23. Waveform definition using dual polarity variable area sections. The troughs are shown as excursions to the left; in practice sections are displayed with the troughs rectified and hence swinging to the right.

discussed in Chapter 1. Aliasing will therefore occur across the 2-D grid cells during the data interpolation. This may be reduced by filtering out the higher frequencies, and the amount of aliasing remaining will depend on the amount of dip across a grid cell in terms of the shorter wavelengths left in the seismic signal. This approach has been followed for several areas but it is normally useful only for studying broad structural trends.

Recently an adaptation of 3-D surveying has been introduced, in which the parallel recording lines are more widely spaced than the sampling theory demands. A new type of interpolation, generally known as intelligent interpolation, is then used during processing to fill the gaps with data at a suitably close spacing for 3-D migration and horizontal section generation. This type of horizontal section is useful for structural interpretation and should have no aliasing problems. However, the interpreter must recognize the spacing at which the original lines were collected and not expect to see structural detail with dimensions smaller than this. The concept of this new approach is that, after a part of the area has been deemed worthy of closer study, lines of data can be collected between the original ones at a closer spacing and the data reprocessed. Finer structural and stratigraphic information should then be interpretable from the resulting data.

Subtle Structural Features

Figures 3-27 and 3-28 are horizontal sections from a data volume in which a subtle, small-throw fault became a significant part of the interpretation at the target level. On both figures the subtle fault is seen as a minor discontinuity in one peak (black) and one trough (red) between Lines 720 and 760 and Crosslines 40 and 55. Crossline 45 (Figure 3-29) shows this fault,

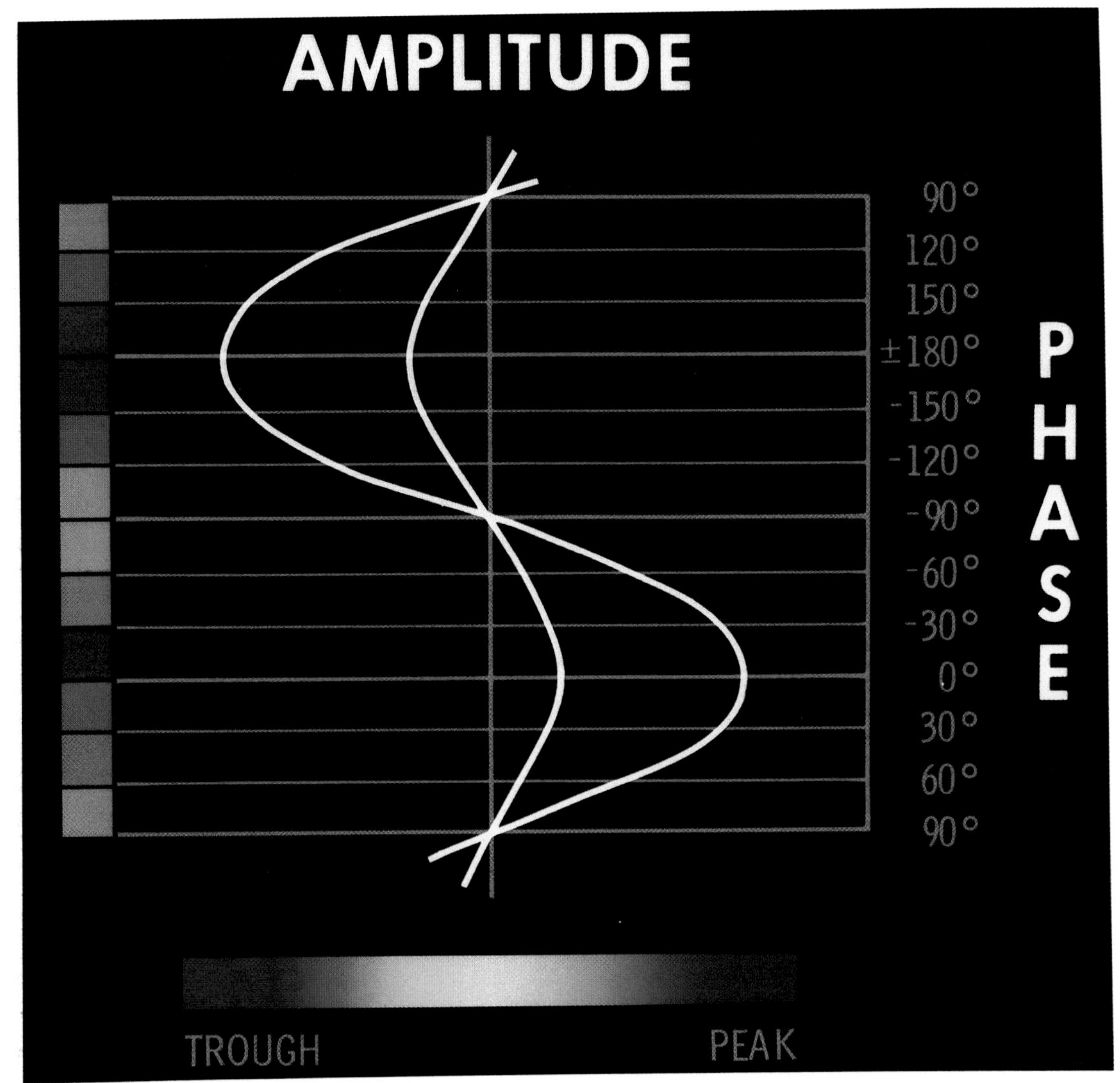

Fig. 3-24. Waveform definition using amplitude and phase color sections.

in the middle of the section between 2.3 and 2.4 seconds, to be a really subtle feature.

The interpreter working on this data first noticed the discontinuity on the horizontal sections and considered it a real geologic feature because it preserved its character over many contiguous slices. Hence, the interpretation of this nearly north-south fault was incorporated into the final structure map, as shown in Figure 3-30.

In an area of gentle dips, such as this one, the horizontal section events are broad and a discontinuity in them may be more easily visible than a discontinuity in the events on the vertical sections. This demonstrates that horizontal sections have a role in the identification of subtle faults, just as they do in the recognition of subtle stratigraphic features, as discussed in Chapter 4.

References

Blake, B. A., J. B. Jennings, M. P. Curtis, and R. M. Phillipson, 1982, Three-dimensional seismic data reveals the finer structural details of a piercement salt dome: Offshore Technology Conference Paper 4258, p. 403–406.

Johnson, J. P., and M. R. Bone, 1980, Understanding field development history utilizing 3-D seismic: Offshore Technology Conference Paper 3849, p. 473–475.

Taner, M. T., F. Koehler, and R. E. Sheriff, 1979, Complex seismic trace analysis: Geophysics, v. 44, p. 1041–1063.

Fig. 3-25. Seiscrop section at 1896 ms in instantaneous phase from offshore Trinidad. (Courtesy Texaco Trinidad Inc.)

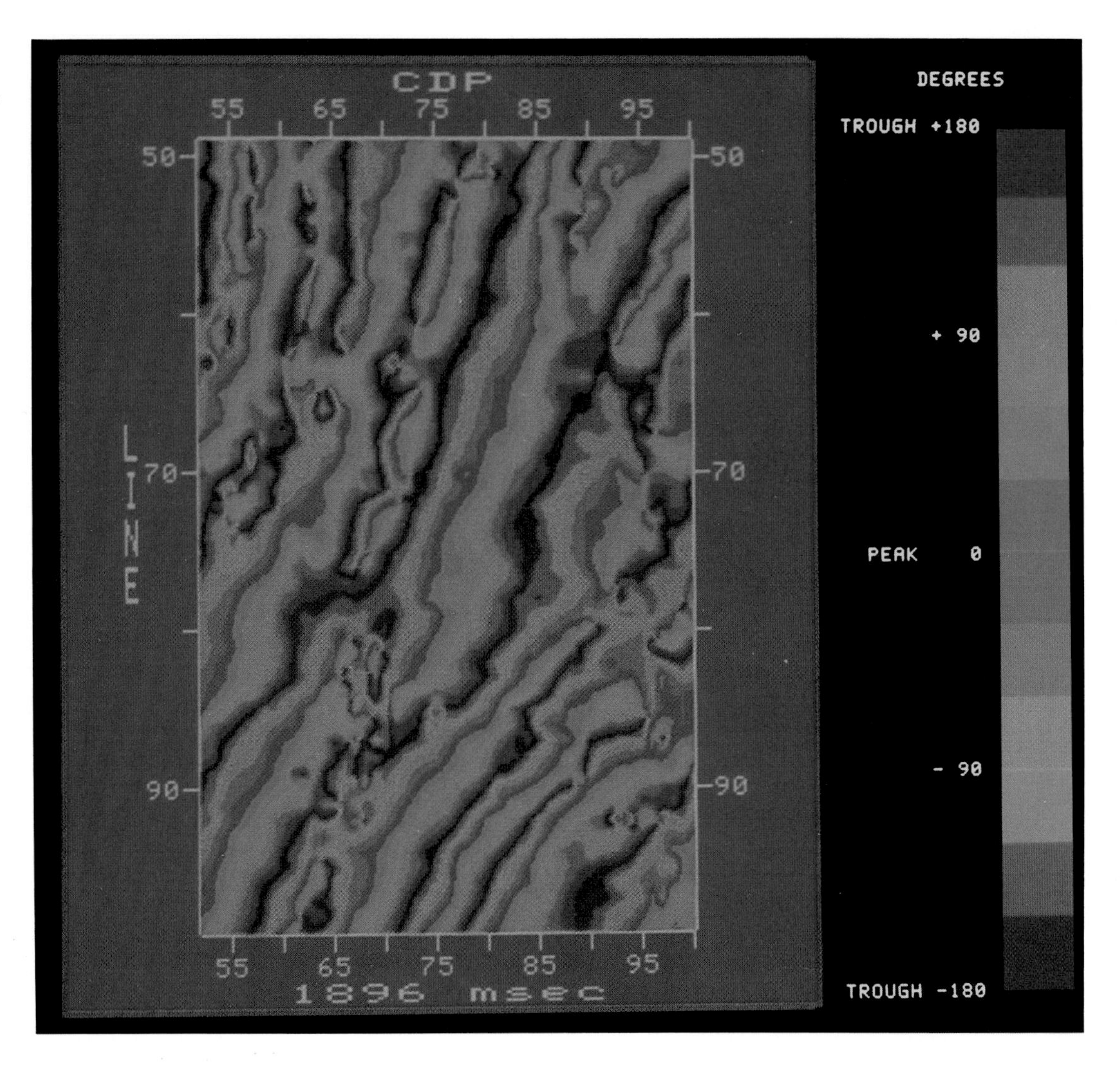

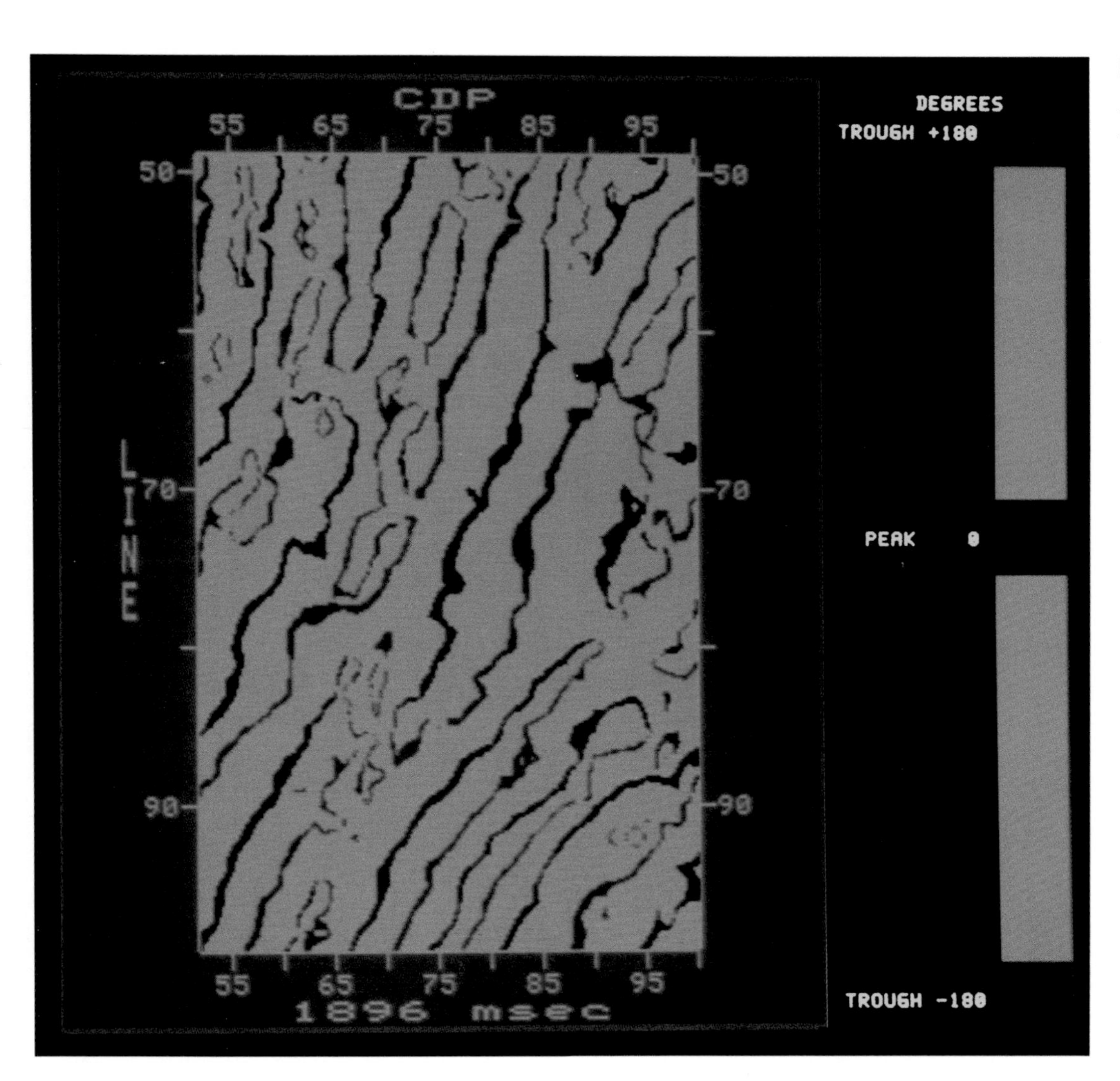

Fig. 3-26. Seiscrop section at 1896 ms in edited phase from offshore Trinidad. (Courtesy Texaco Trinidad Inc.)

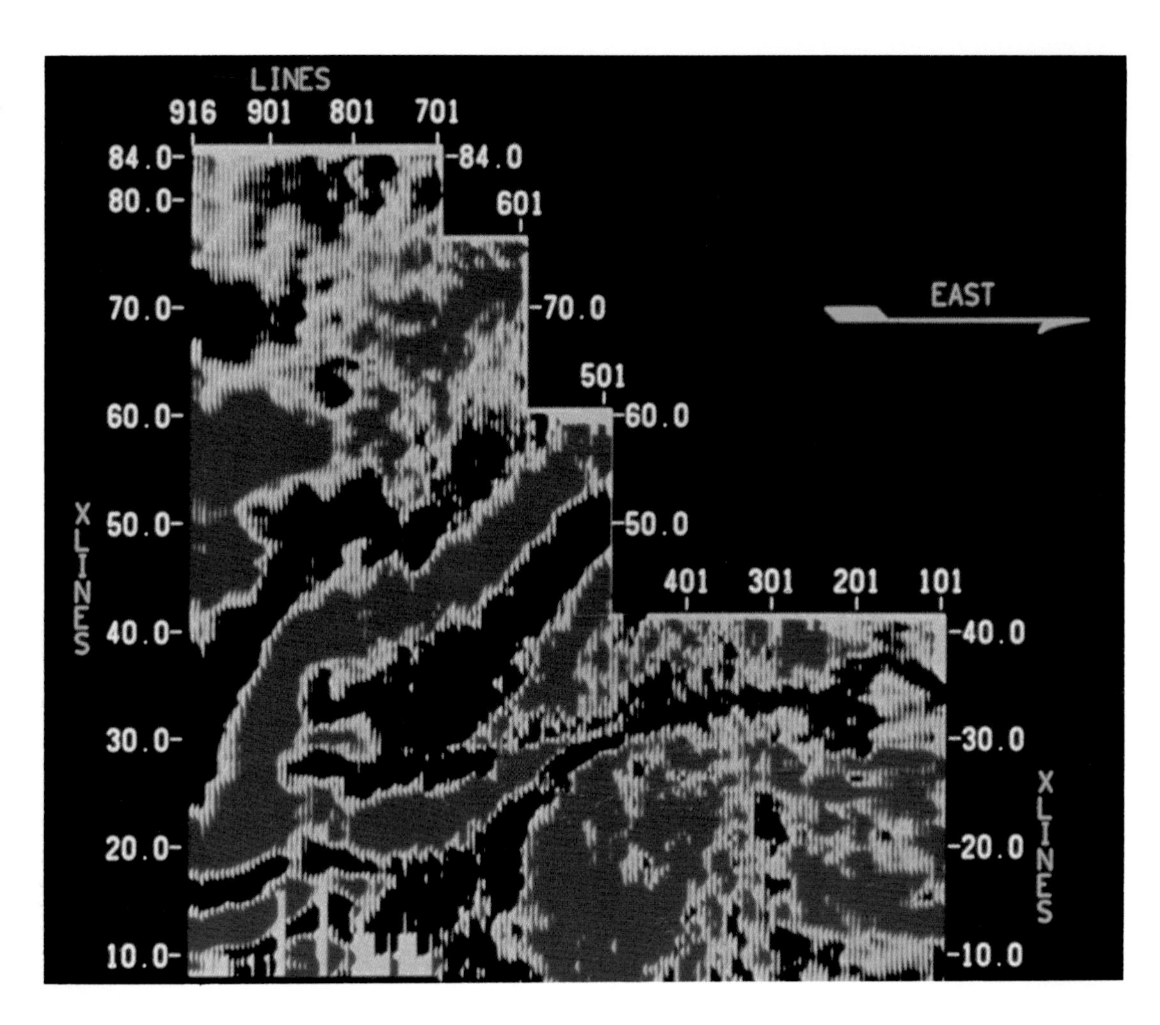

Fig. 3-27. Seiscrop section at 2332 ms from south Louisiana marsh terrain. (Courtesy Texaco Inc.)

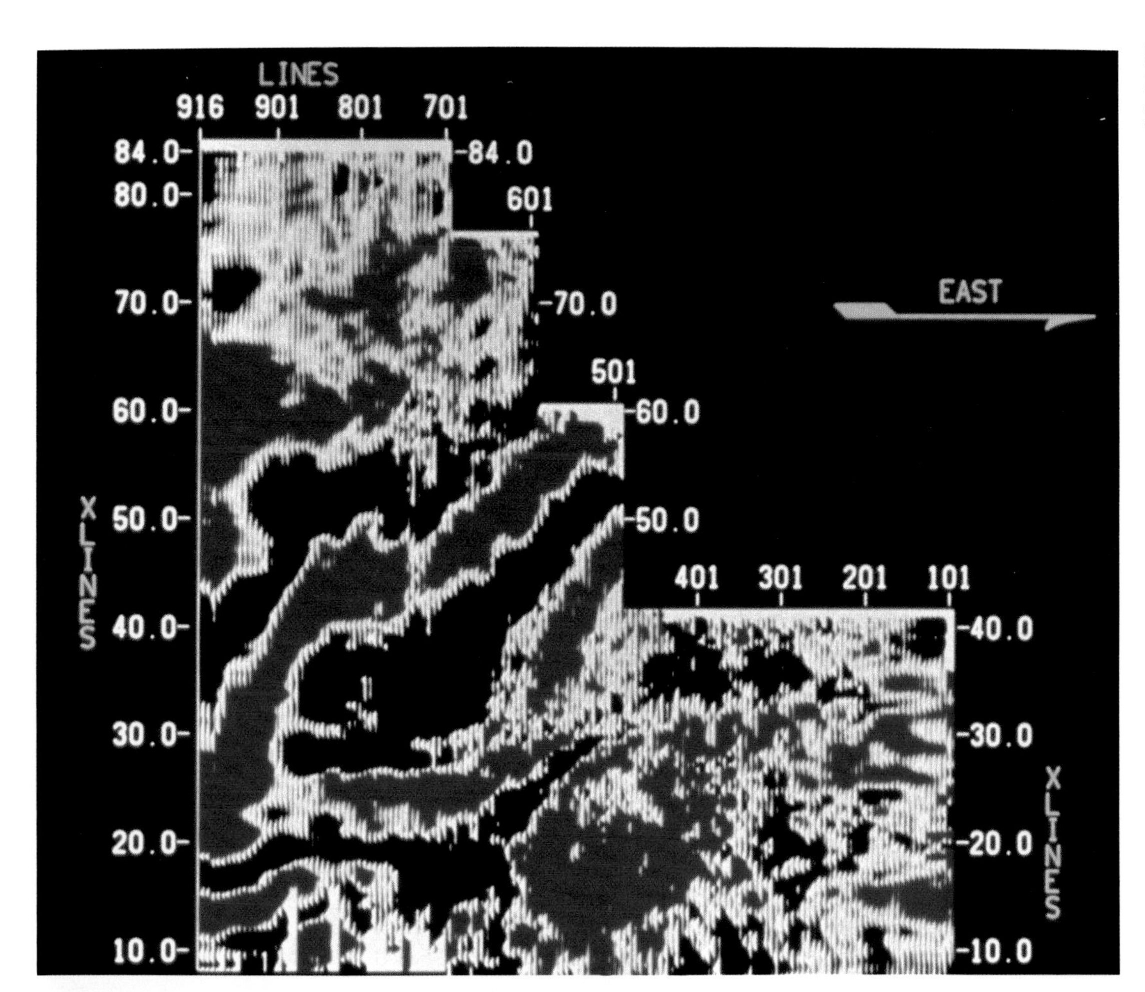

Fig. 3-28. Seiscrop section at 2340 ms from south Louisiana marsh terrain. (Courtesy Texaco Inc.)

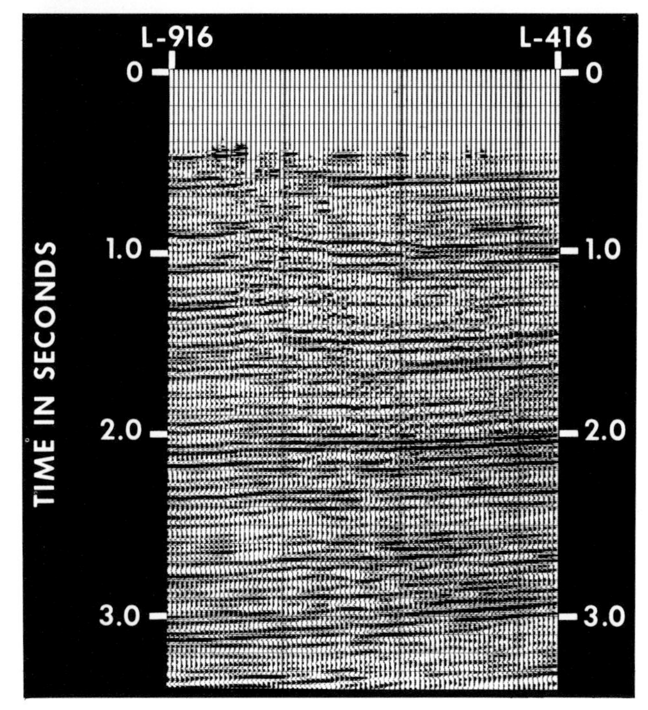

Fig. 3-29. (Left) Crossline 45 showing subtle fault identified on Seiscrop sections of Figures 3-27 and 3-28. (Courtesy Texaco Inc.)

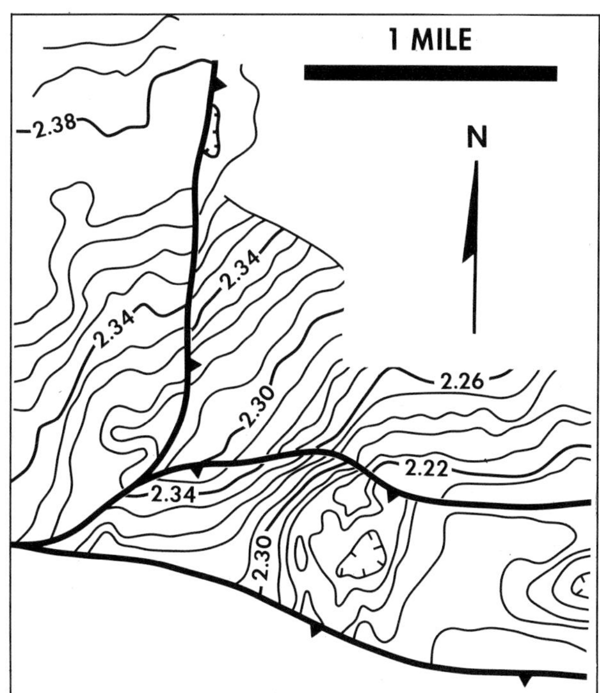

Fig. 3-30. (Right) Structural contour map showing subtle fault identified on Seiscrop sections of Figures 3-27 and 3-28. (Courtesy Texaco Inc.)

STRATIGRAPHY FROM HORIZONTAL SECTIONS

Recognition of Characteristic Shape

Where a vertical seismic section intersects a stratigraphic feature the interpreter can normally find a small amplitude or character anomaly. The expression of a sand-filled channel or bar, for example, is therefore normally so subtle that it takes a considerable amount of interpretive skill to detect it. In contrast, a horizontal section reveals the spatial extent of an anomaly. The interpreter can thus observe characteristic shape and relate what he sees to geologic experience. A shape or pattern which is unrelated to structure may prove to be interpretable as a depositional, erosional, lithologic or other feature of significance. Klein (1985) and Broussard (1975), among others, have provided depositional models on which the interpreter can base his recognition of depositional features.

Figure 4-1 shows five adjacent vertical seismic sections from a small 3-D survey in the Williston basin of North Dakota. Note that the reflections indicate largely flat-lying beds. At 1.8 seconds there is a very slight draping of reflections which is only just discernible. Figure 4-2 shows two single-polarity horizontal sections superimposed on each other. The data from both levels reveal the same almost circular shape. This is the outline of a carbonate buildup measuring approximately one kilometer in diameter.

Figures 4-3 and 4-4 are horizontal sections from a 3-D survey recorded in the Gippsland basin offshore southeastern Australia (Sanders and Steel, 1982). Many small circular features are strikingly evident. These appear as small depressions on the vertical sections which attract little attention. It is the characteristic circular shape when viewed horizontally that attracts the interpreter's eye. The circular features measure 200 to 500 m in diameter and are interpreted as sinkholes in a Miocene karst topography. The beds in which these features exist are dipping from upper left to lower right (east) in Figures 4-3 and 4-4. The width of the reflection is a function of seismic frequency and structural dip (see Chapter 3). The visibility of the sinkholes in the presence of this structure is because their diameters are each less than the reflection width.

Figure 4-5 shows a bifurcating channel close to a Gulf of Mexico salt dome. The salt dome's semi-circular expression results from the intersection of the horizontal section at 416 ms with the dipping structural reflections adjacent to the dome. Away from the salt dome the beds are close to flat-lying, so the horizontal section is sliced along the bedding plane. As a result, the channel is almost completely visible. In fact, the bedding is not exactly flat and some parts of the channel are more clearly seen on the adjacent section at 412 ms. Simple addition of these two horizontal sections improved the continuity of the channel (Figure 4-6). Adding together of Seiscrop sections is a useful approach to the enhancement of stratigraphic features if, *but only if*, the structural variation across the feature is less than half a period of the appropriate seismic signal.

Figure 4-7 shows another channel deeper in the same data volume. Enhancement again resulted from adding together the horizontal sections from 812 and 816 ms. The channel branches at Line 70, CDP 470, but the eastern branch is not visible. Figure 4-8 shows just the portion of the survey area covering the channel system and includes the horizontal section at 820 ms. Here the eastern branch is clearly visible showing that it is structurally slightly deeper

Continued on page 76

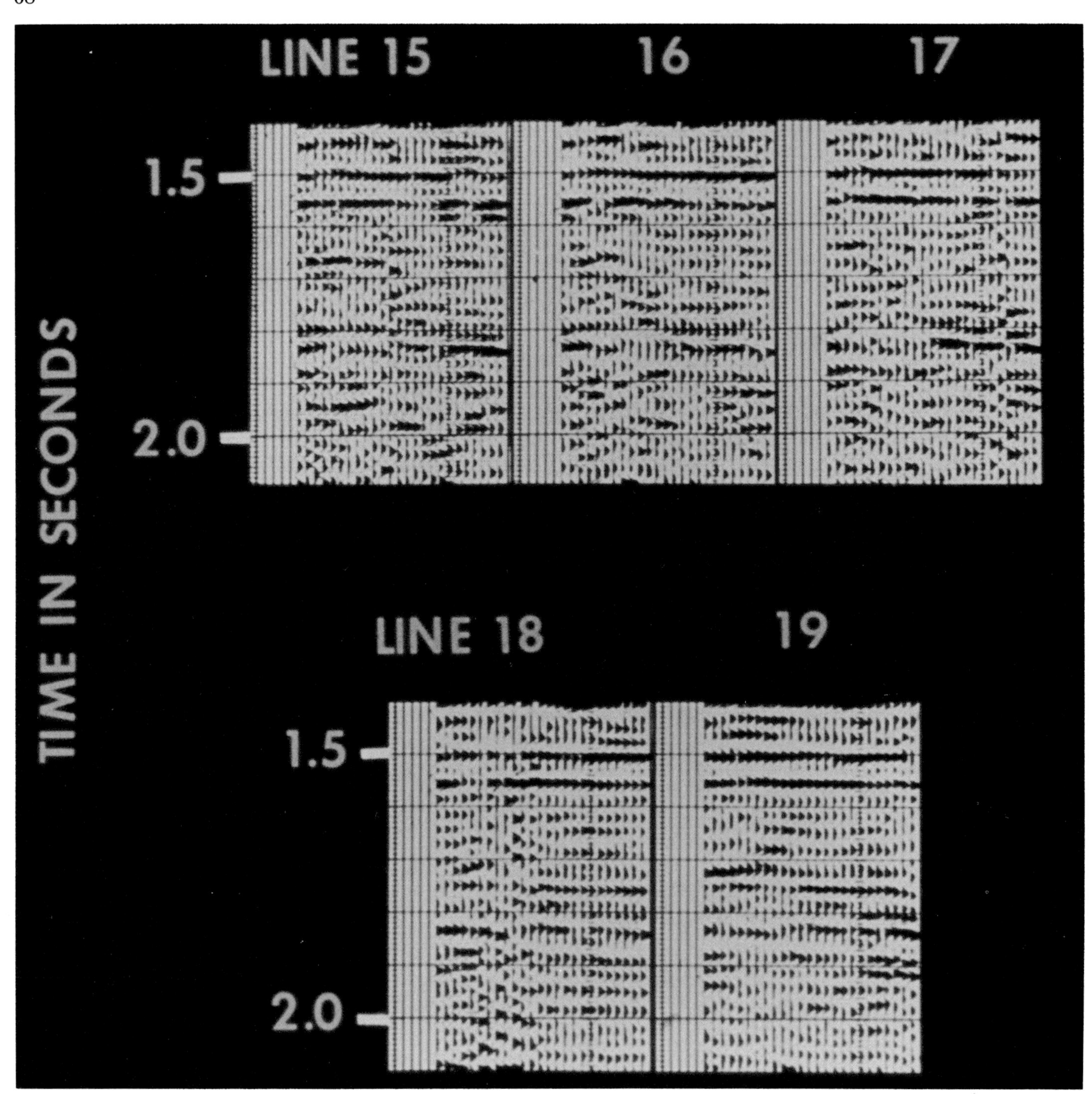

Fig. 4-1. Five adjacent vertical sections from 3-D survey in the Williston basin of North Dakota.

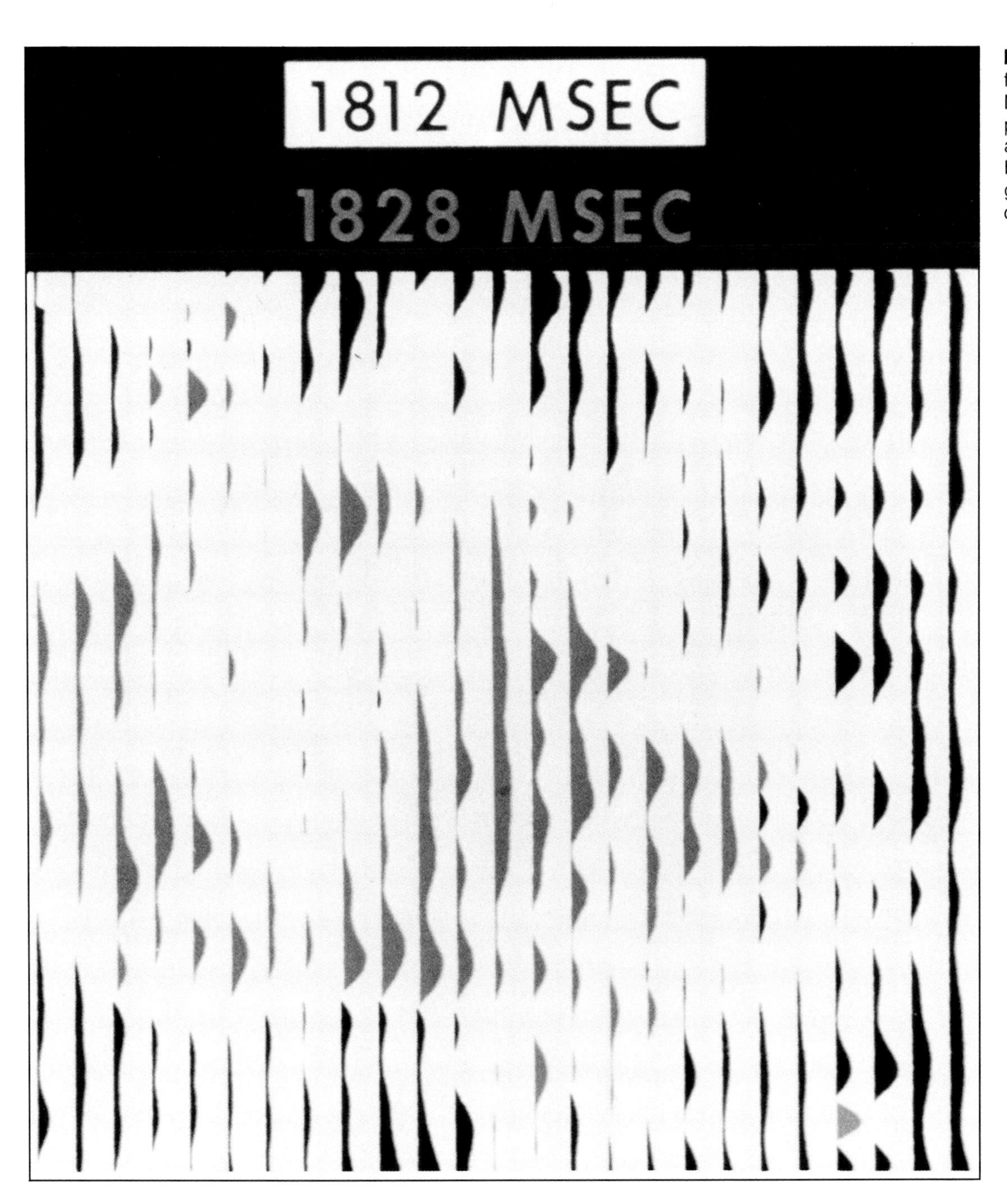

Fig. 4-2. Seiscrop sections from 1812 and 1828 ms from North Dakota each showing positive amplitudes only. The approximately circular outline between the black and the gray indicates the shape of a carbonate buildup.

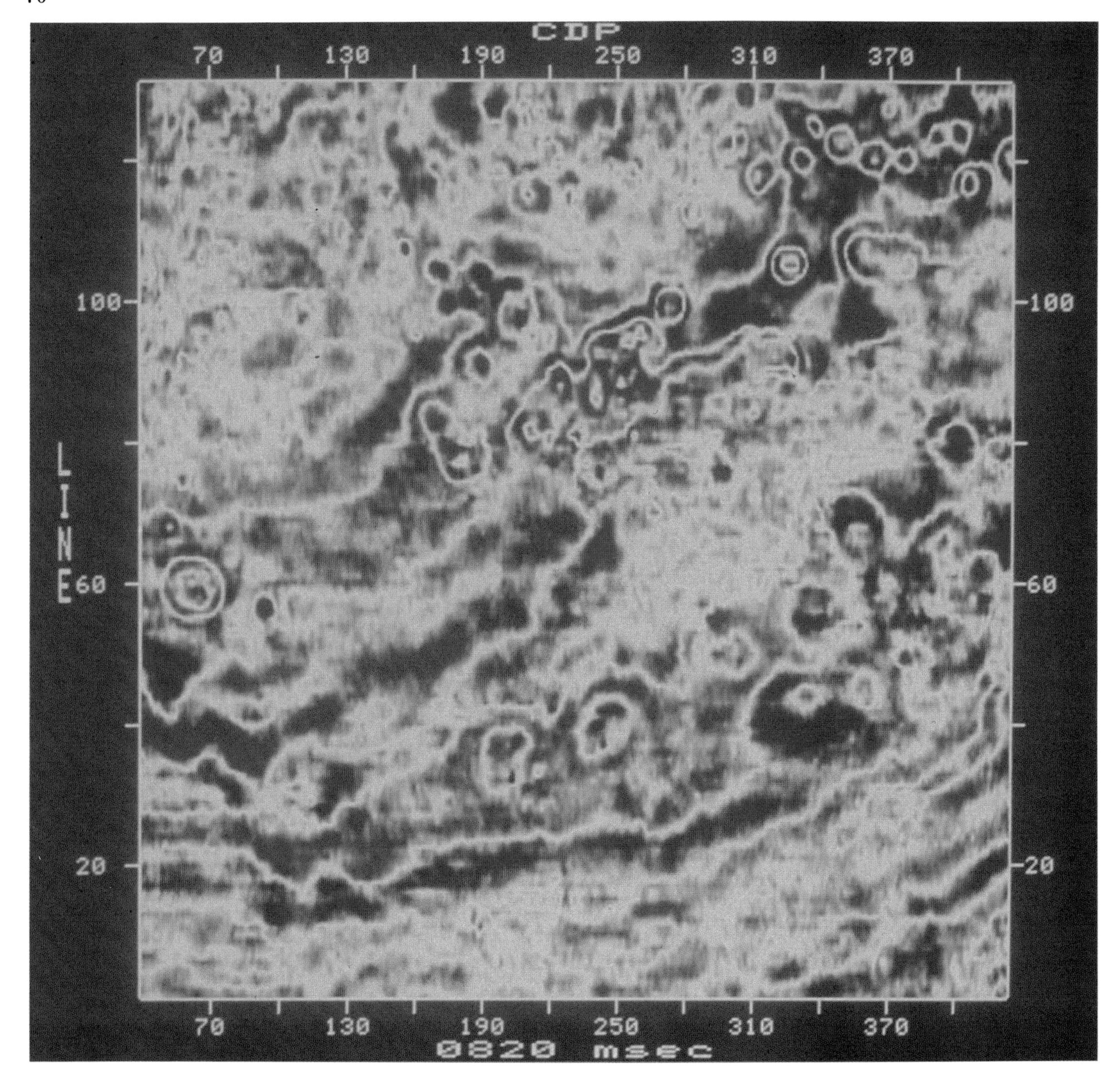

Fig. 4-3. Seiscrop section at 820 ms from 3-D survey over Mackerel field in offshore Gippsland Basin, southeastern Australia. Circular objects are interpreted as sinkholes in karst topography. (Courtesy Esso Australia Ltd.)

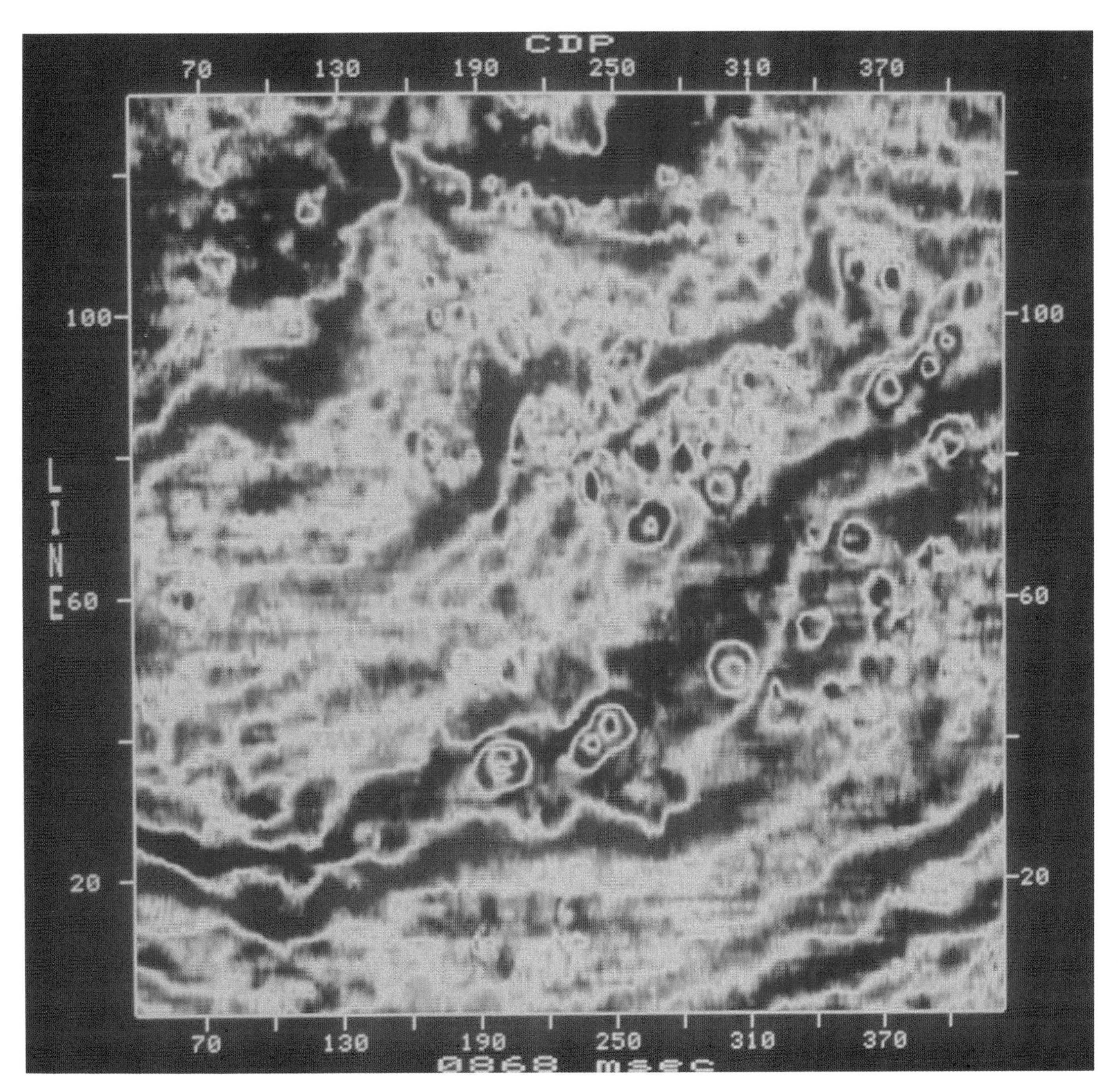

Fig. 4-4. Seiscrop section at 868 ms from 3-D survey over Mackerel field in offshore Gippsland basin, southeastern Australia. Circular objects are interpreted as sinkholes in karst topography. (Courtesy Esso Australia Ltd.)

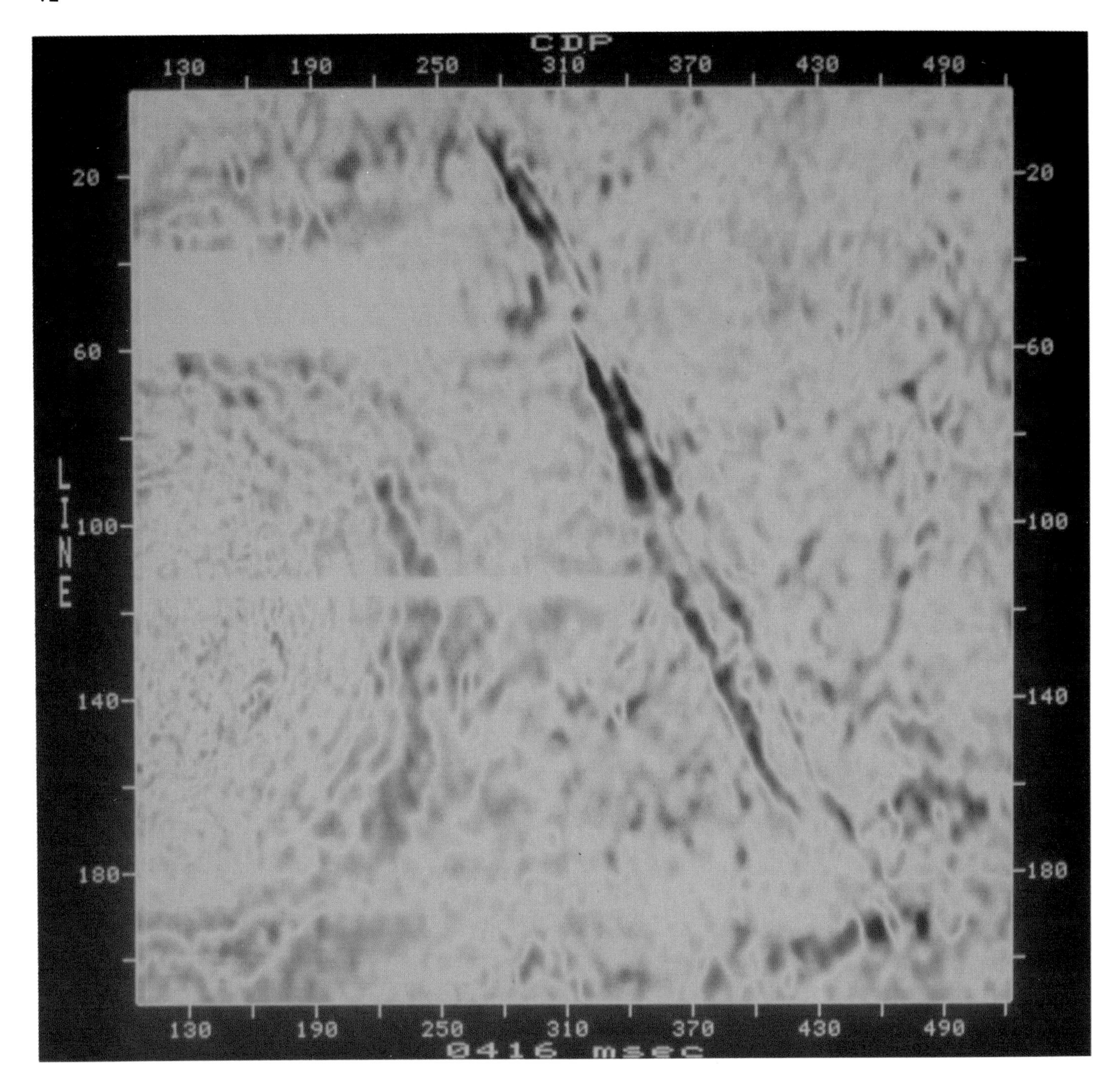

Fig. 4-5. Seiscrop section at 416 ms from 3-D survey in the Gulf of Mexico. The bifurcating channel is seen close to the edge of a salt dome. (Courtesy Chevron U.S.A. Inc.)

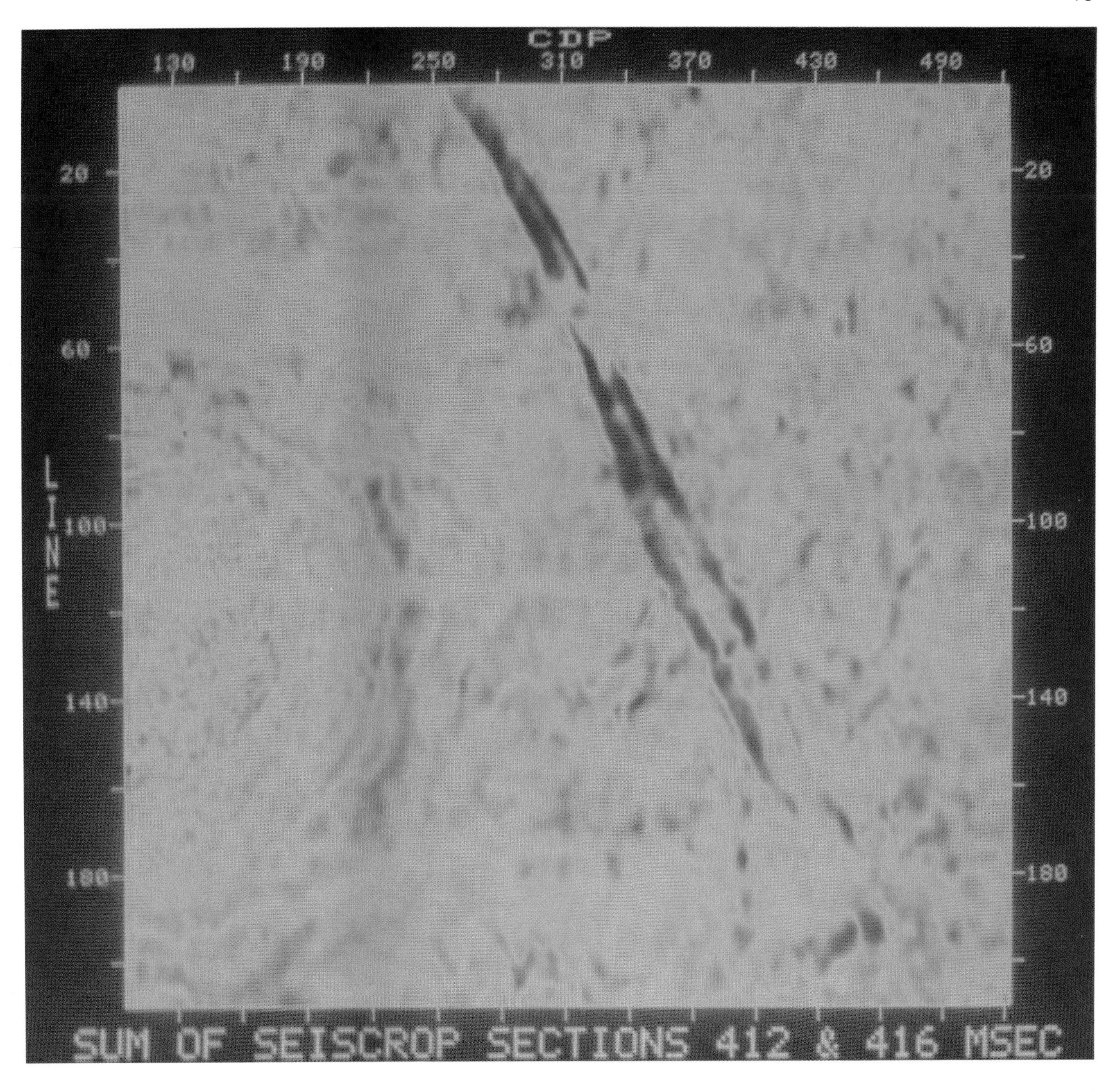

Fig. 4-6. Sum of Seiscrop sections at 412 and 416 ms from same survey as Figure 4-5 showing enhancement of the channel. (Courtesy Chevron U.S.A. Inc.)

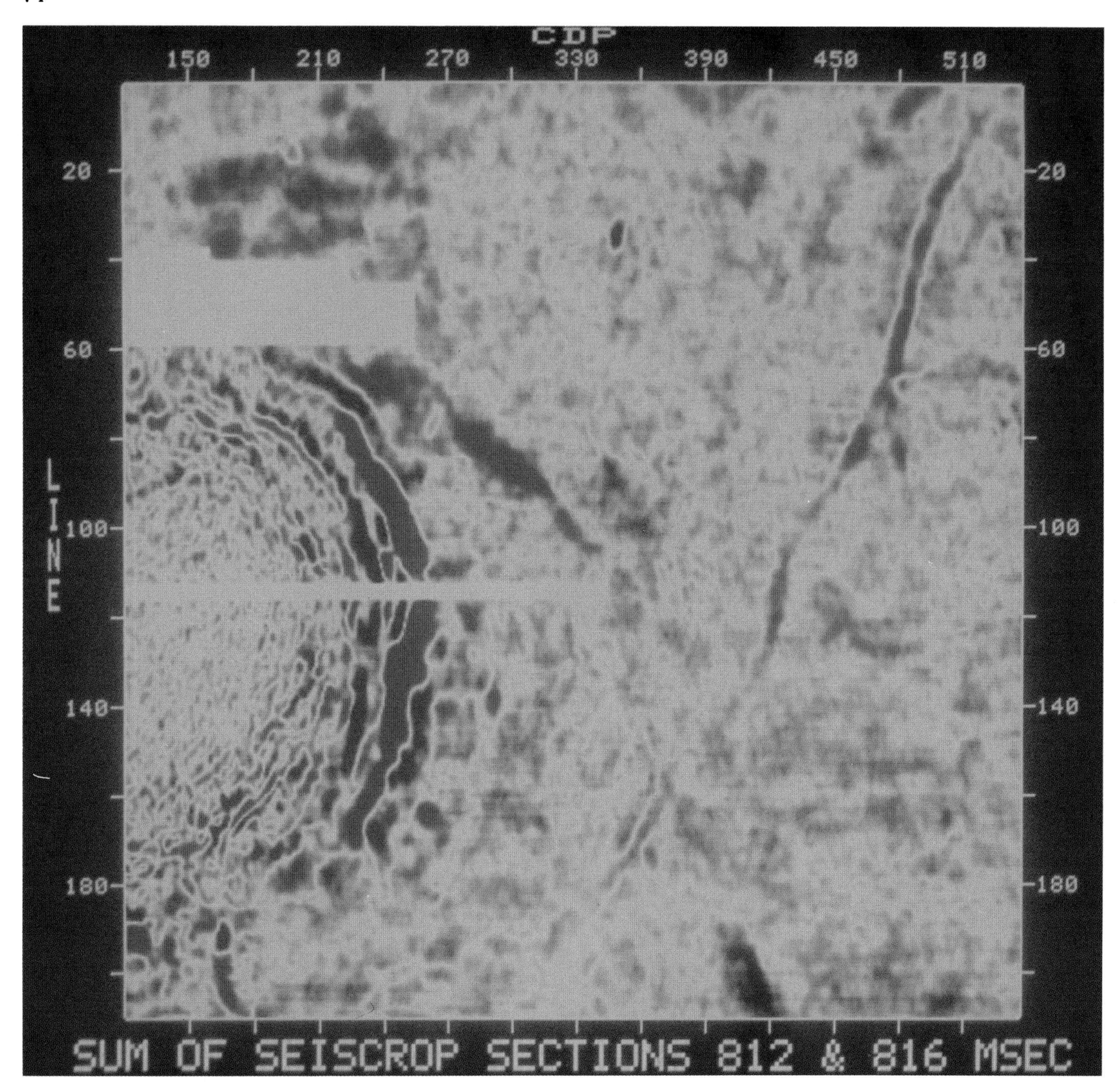

Fig. 4-7. Sum of Seiscrop sections at 812 and 816 ms from same survey as Figure 4-5 showing a branching channel. (Courtesy Chevron U.S.A. Inc.)

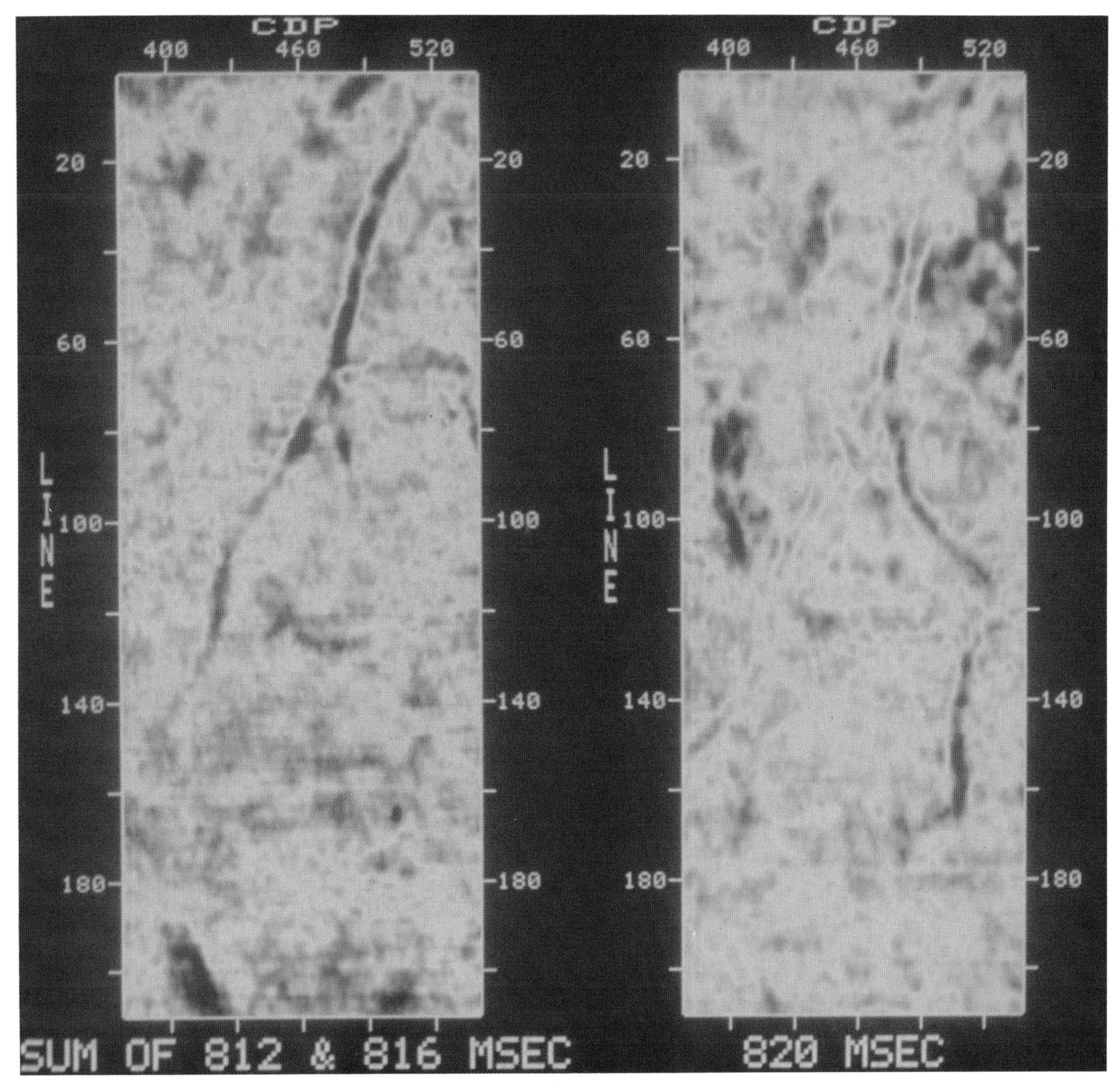

Fig. 4-8. Composite display of Seiscrop sections at 812 and 816 ms showing western branch of channel and at 820 ms showing eastern branch. (Courtesy Chevron U.S.A. Inc.)

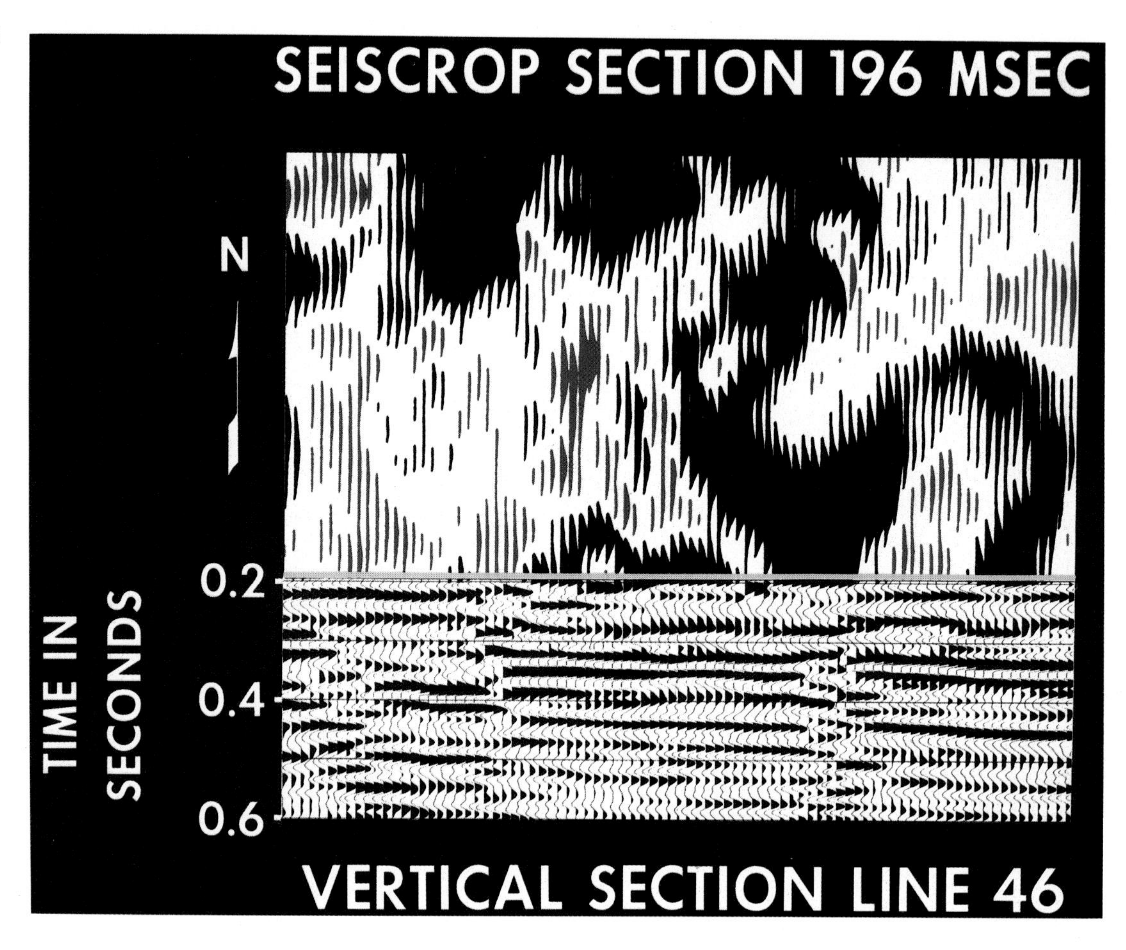

Fig. 4-9. Composite display of vertical and Seiscrop sections from Gulf of Thailand showing spatial continuity of vertical section event segments. (Courtesy Texas Pacific Oil Company Inc.)

than the western branch. This indicates that the depositional surface containing this channel dips away from the salt dome, which dip was presumably induced by the movement of the salt. Thus, in order to view the entire channel system, several horizontal sections covering the structural range of this depositional surface are required.

Figures 4-9 through 4-16 show examples of depositional features observed on horizontal sections through flat-lying beds in the Gulf of Thailand. The vertical section in Figure 4-9 shows that the beds are flat-lying and that around 200 ms there are some abrupt character changes. The attached horizontal section shows that these reflection segments have spatial continuity. Figure 4-10, covering the whole prospect area, makes it clear that the continuity is part of a meandering channel system. Anyone who has flown over the Mississippi River will immediately relate Figure 4-10 to observations made from the airplane window.

In the Gulf of Thailand there is a regional unconformity in the mid-Miocene and above that unconformity the beds in this prospect area are largely flat-lying. Therefore, many horizontal sections above 900 ms directly reveal depositional features because the sections are parallel to bedding planes. Figure 4-11 is a schematic composite of the features observed. The interpretation of these in sequence indicated a delta prograding across the survey area from southwest to northeast during the mid-Miocene to Pleistocene.

Reconstituting a Depositional Surface

In general, stratigraphic features, after being deposited on a flat-lying surface, will be bent and broken by later tectonic movements. Stratigraphy and structure then become confused and the interpretive task comes in separating them. The structure must be interpreted before stratigraphy can be appreciated. There are several ways to do this.

Figure 4-17 illustrates schematically how a channel can be recognized and delineated in the presence of structure. In this example the interpreter has horizontal sections at 4 ms intervals from 1240 to 1260 ms. The selected event at 1240 ms for the horizon under study is traced to

Continued on page 85

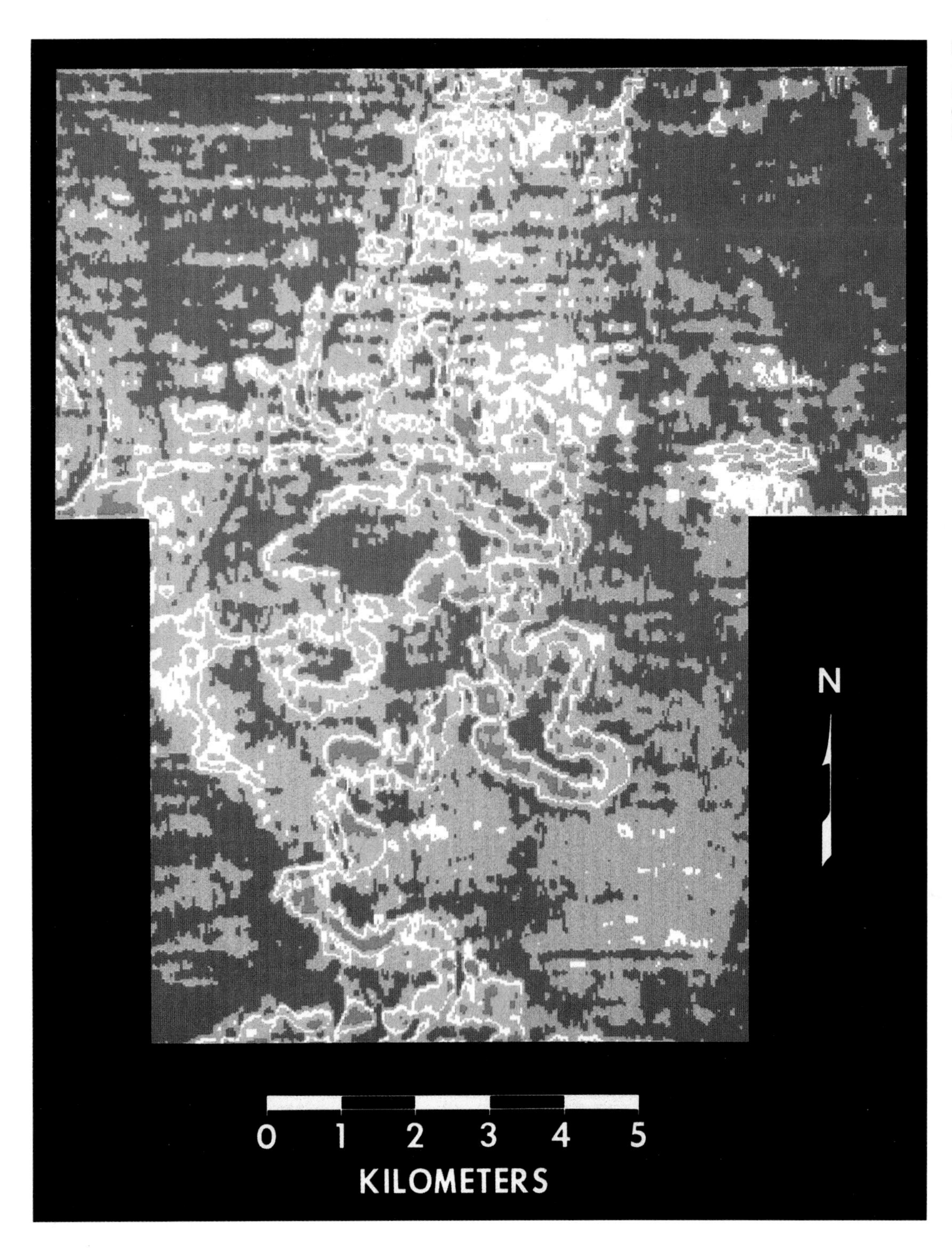

Fig. 4-10. Seiscrop section at 196 ms from Gulf of Thailand showing meandering stream channel. (Courtesy Texas Pacific Oil Company Inc.)

Fig. 4-11. Schematic diagram of delta prograding across the Gulf of Thailand 3-D survey area between mid-Miocene and Pleistocene.

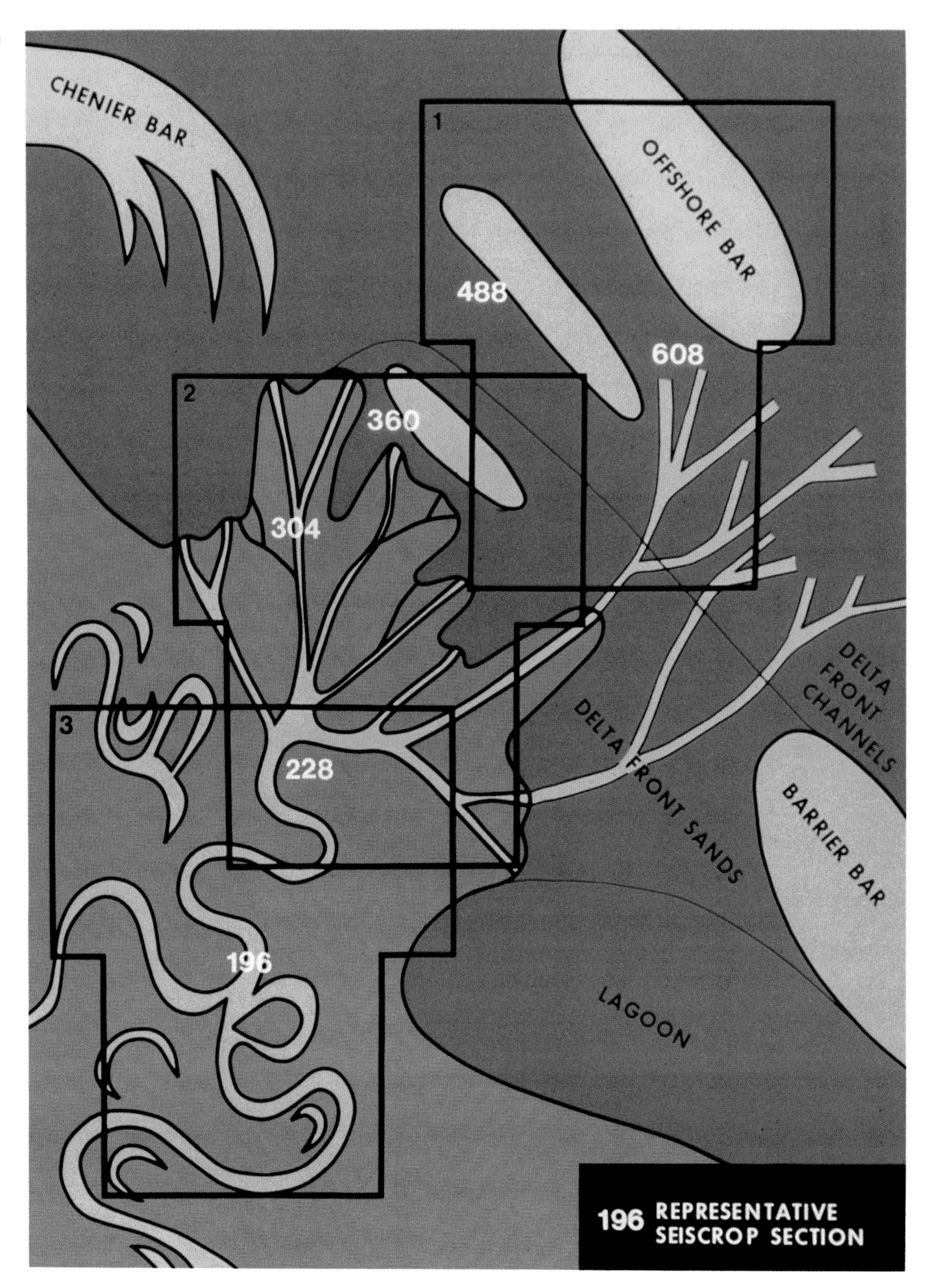

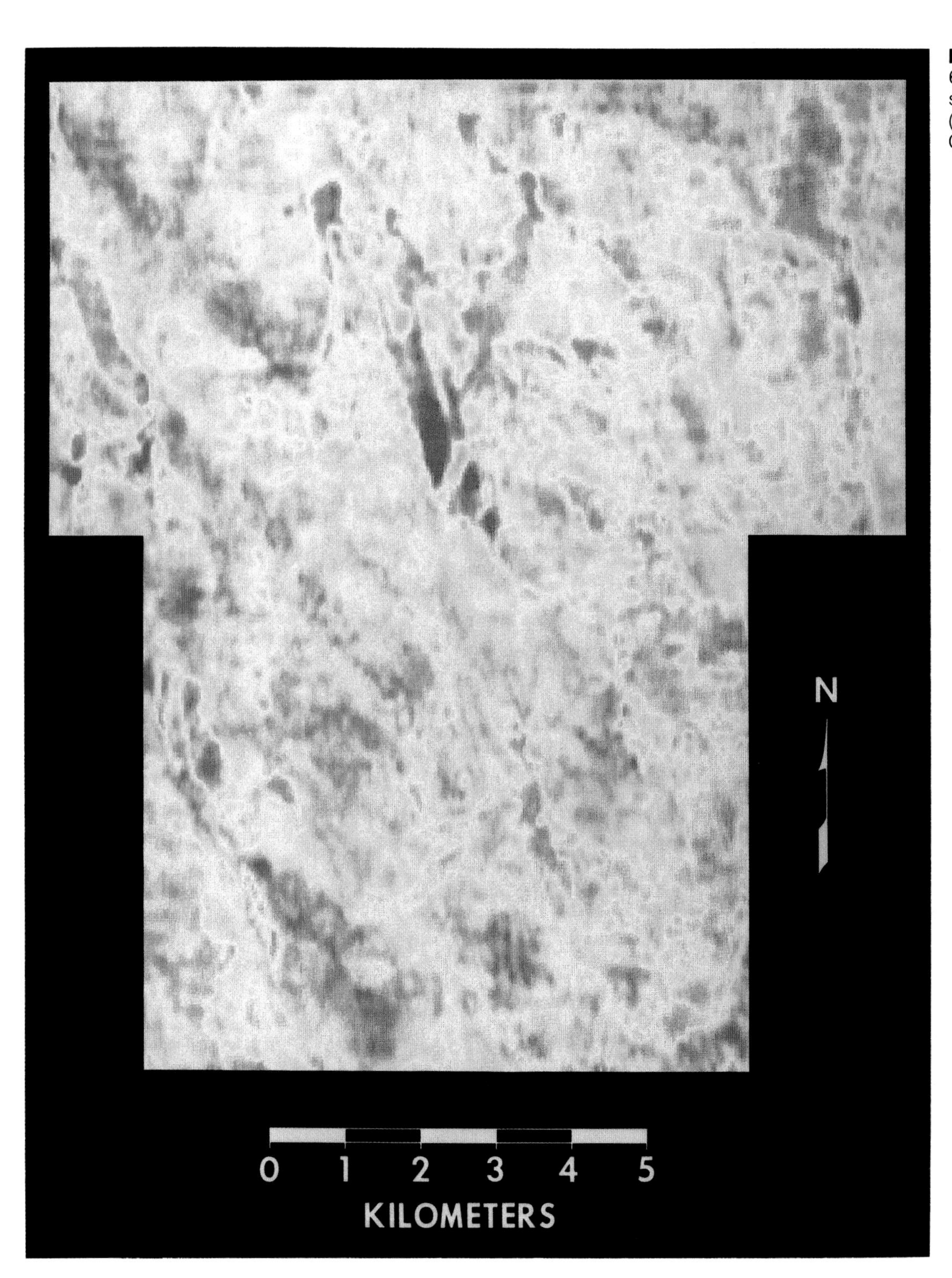

Fig. 4-12. Seiscrop section at 608 ms from Gulf of Thailand showing delta front channel. (Courtesy Texas Pacific Oil Company Inc.)

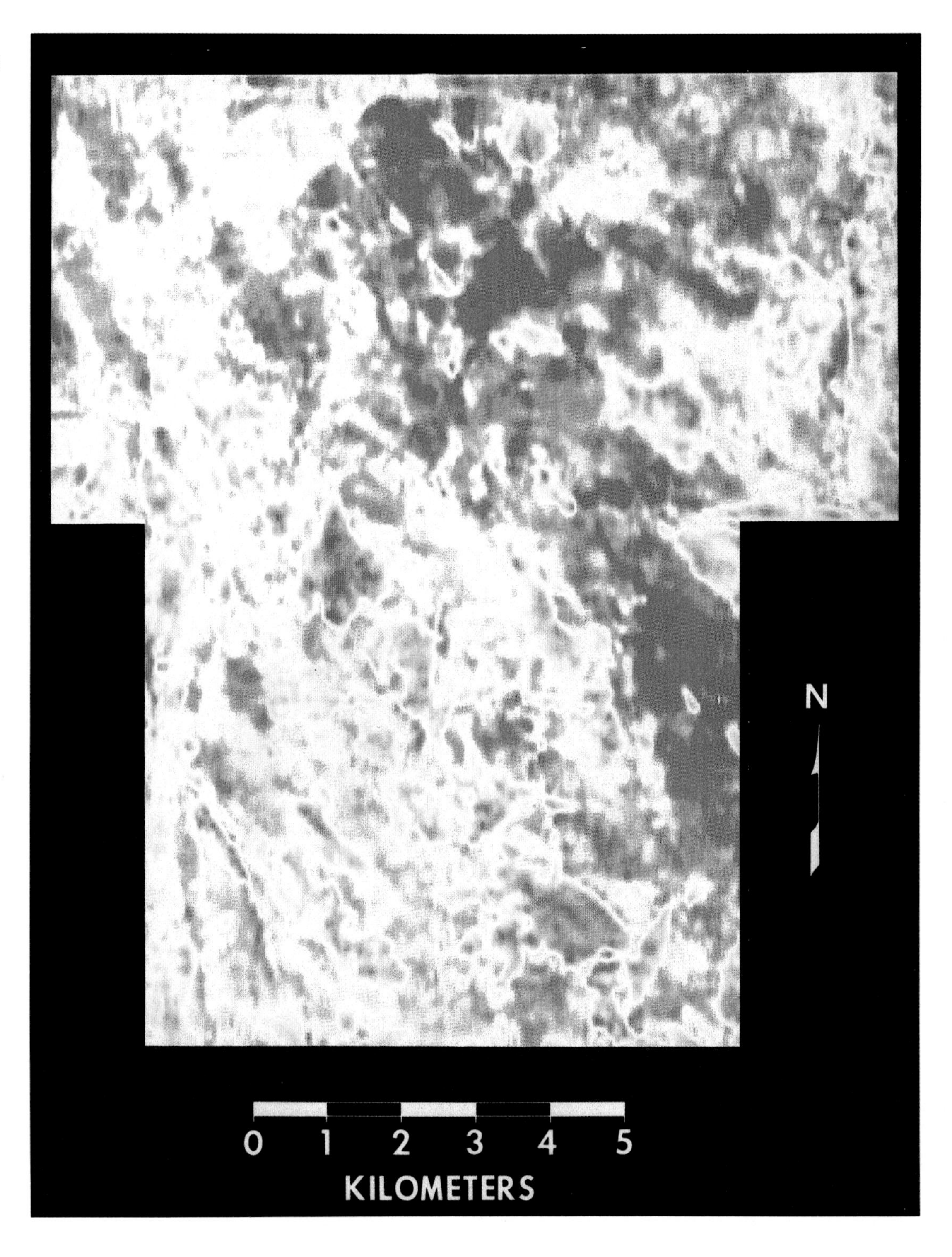

Fig. 4-13. Seiscrop section at 488 ms from Gulf of Thailand showing large offshore sand bar. (Courtesy Texas Pacific Oil Company Inc.)

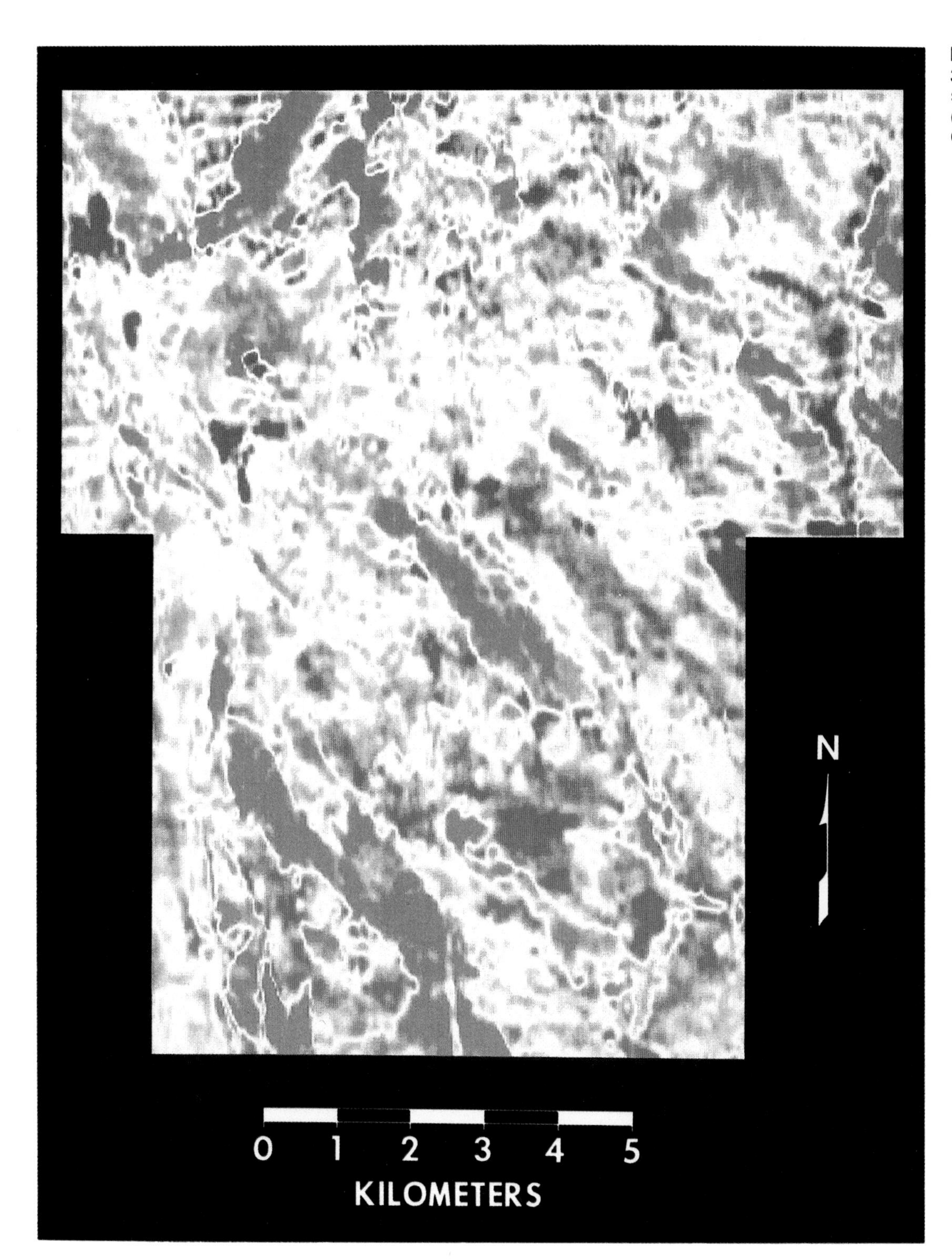

Fig. 4-14. Seiscrop section at 360 ms from Gulf of Thailand showing small sand bars. (Courtesy Texas Pacific Oil Company Inc.)

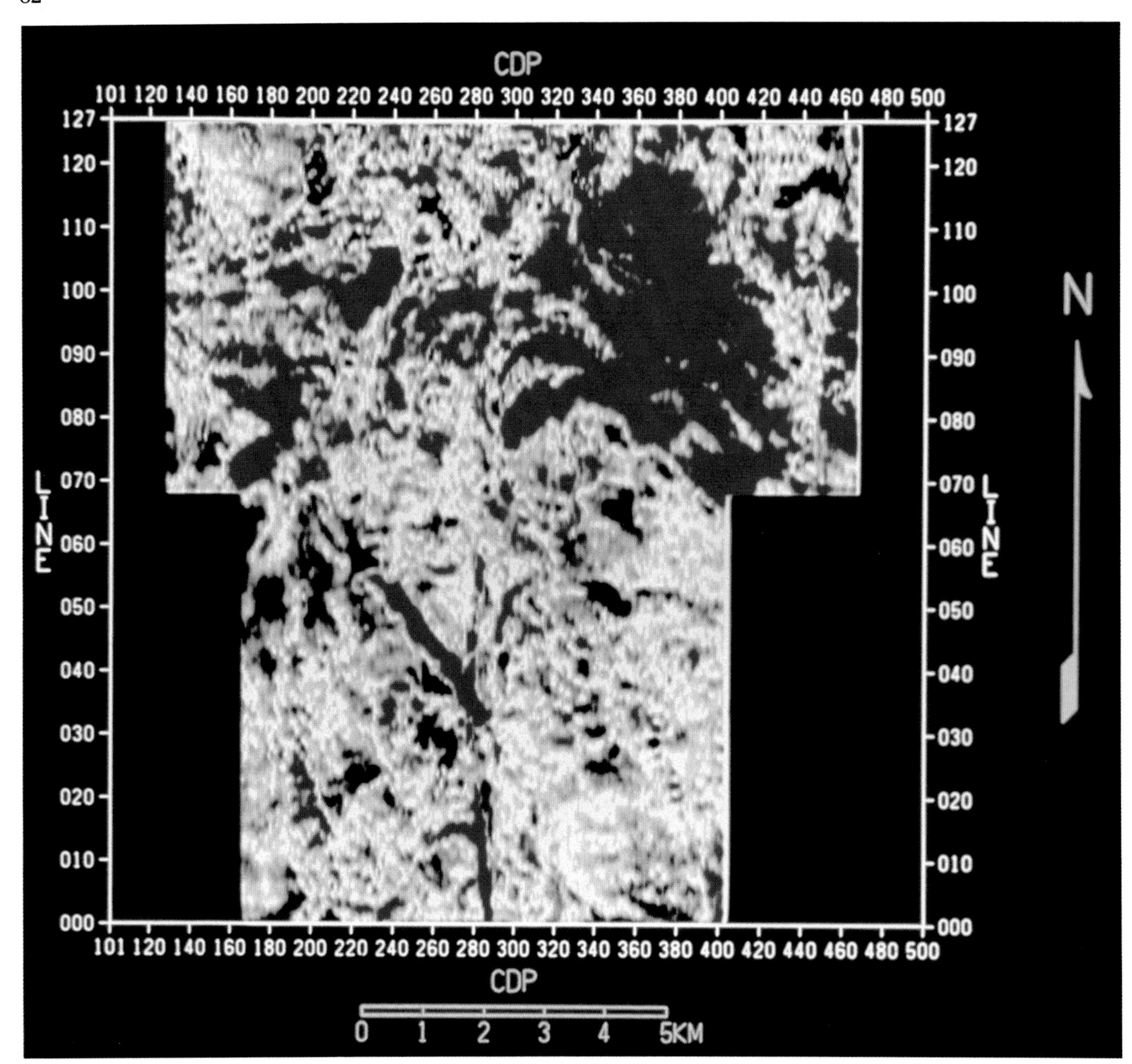

Fig. 4-15. Seiscrop section at 304 ms from Gulf of Thailand showing a Chenier bar and distributary channels (Courtesy Texas Pacific Oil Company Inc.)

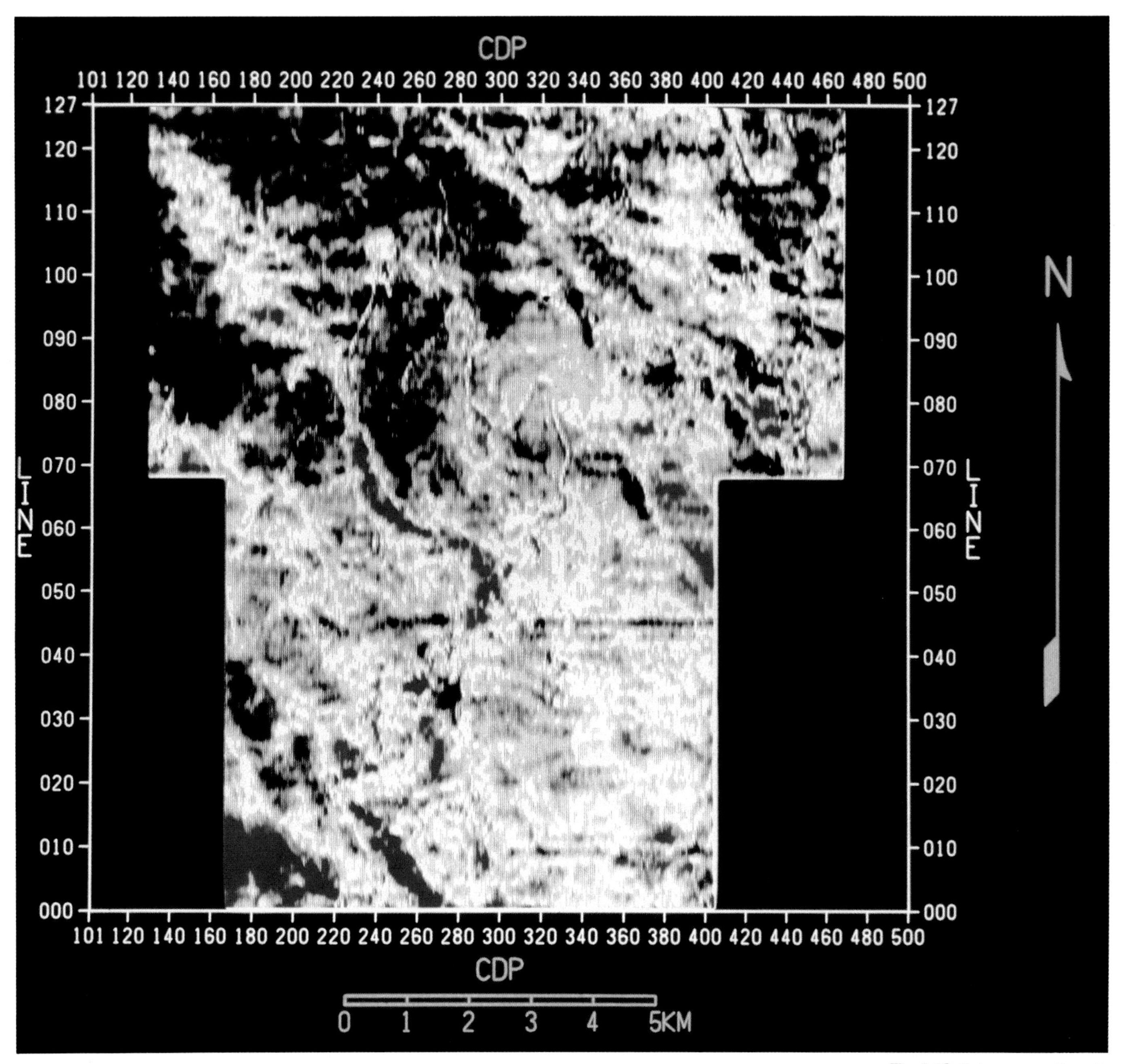

Fig. 4-16. Seiscrop section at 228 ms from Gulf of Thailand showing several channels, large and small. (Courtesy Texas Pacific Oil Company Inc.)

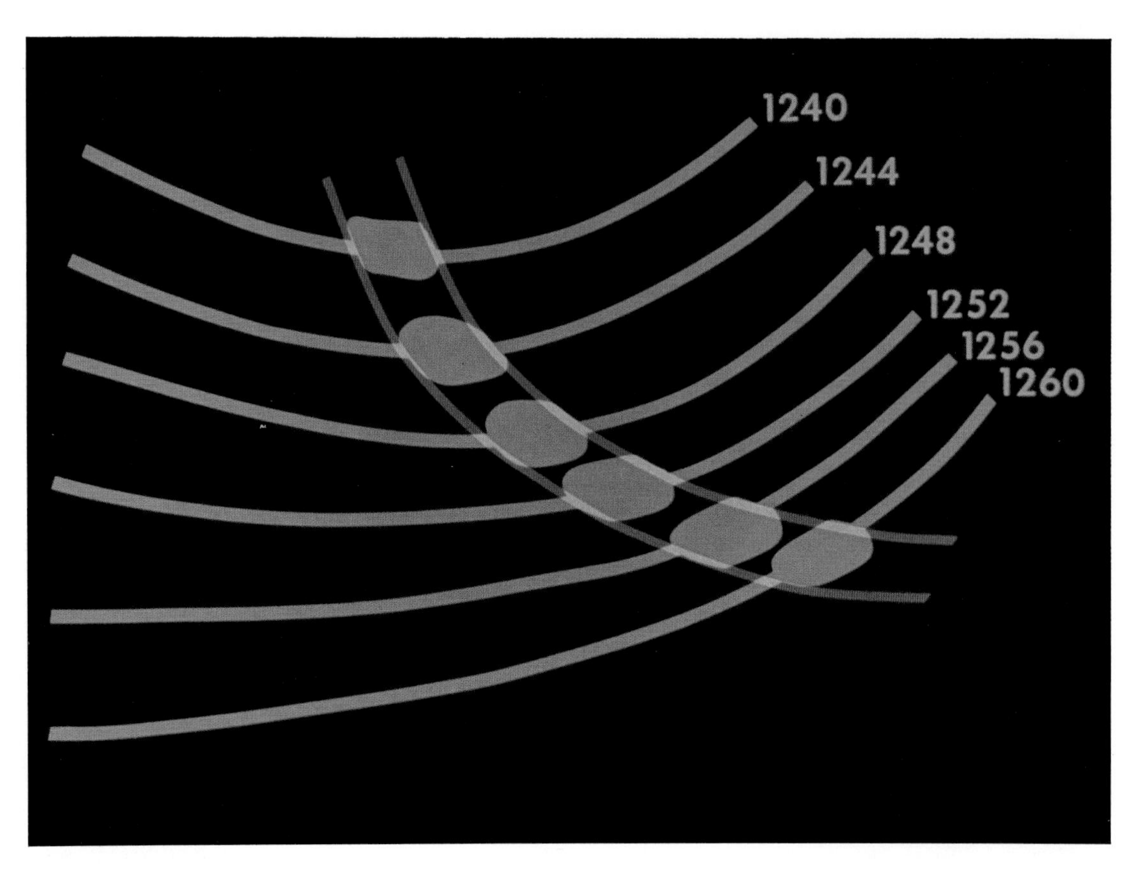

Fig. 4-17. How to follow an anomalous amplitude feature in the presence of structure on a sequence of Seiscrop sections.

provide the contour as shown for 1240 ms. A high amplitude anomaly is recognized and marked at the position of the green blob. This procedure is repeated at 1244, 1248, 1252, 1256, and 1260 ms. At each of these levels the interpreter found an amplitude anomaly; together these arranged themselves into the curvilinear feature marked by the orange lines in Figure 4-17. The interpreter deemed this to be a channel.

A similar approach to this is illustrated in Figures 4-18 through 4-20 for the mapping of a sand bar. The two horizontal sections in Figures 4-18 and 4-19 are each designated in feet as the whole data volume had previously been depth-converted. The high amplitude event under study lies on each of these sections between approximately Line 100, CDP 240 and Line 70, CDP 300. This bright event was outlined on these two sections and on several adjacent ones to provide the contour map of the sand bar shown in Figure 4-20. Two wells penetrate this bar as shown, one indicating gas, one not. Assuming the gas cap is structurally controlled within the bar, the extent of the gas is shown by the pink color.

This approach, however, does have shortcomings. The interpreter must recognize a feature of interest from its intersection on either a horizontal or vertical section, and is thus deprived of the benefits of characteristic shape advocated so strongly in the previous section of this chapter.

Figure 4-21 shows a vertical section interpreted on three horizons. The Shallow Horizon, marked in blue, was selected on the basis of both structural and stratigraphic objectives. Figure 4-22 shows the structural contour map of the Shallow Horizon resulting from a full-scale structural interpretation of all the 3-D data. The desire then was to slice through the data volume along this structurally interpreted horizon in order to gather up all the seismic amplitudes associated with it. This was actually accomplished by flattening the data volume on the Shallow Horizon, as structurally interpreted in Figure 4-22, and then slicing horizontally through the flattened volume at the level of the interpreted horizon.

The resultant section is known as a horizon slice or horizon Seiscrop section, where the critical word is **horizon**. This type of section, following one horizon, must be parallel to bedding planes or it loses its value for stratigraphic interpretation. The importance of this approach was first stressed by Brown, Dahm, and Graebner (1981).

Figures 4-23 and 4-24 are horizon slices through adjacent conformable horizons both following the structural configuration of Figure 4-22. Both were sliced through peaks and hence all amplitudes are positive and show as varying intensities of blue; the darker blues indicate the higher amplitudes. The approximately north-south light-colored streaks are the faults; the width of a streak gives an indication of fault heave.

Figure 4-23 shows a broad high amplitude trending northwest-southeast toward the left of the section. This is interpreted as a sand bar. It is evident that this inferred bar has been dissected by several faults. The process of constructing the horizon slice has put the bar back together. Hence the construction of a horizon slice amounts to the reconstitution of a depositional surface.

Figure 4-24 shows more spatial consistency of the darker blues, indicating that this horizon follows a sheet sand. There is a curvilinear feature, somewhat the shape of a shepherd's crook, which runs northwest-southeast just to the west of well 5X. This is interpreted as an erosion channel in the sheet sand. The fact that this inferred channel is continuous across the fault just west of well 5X lends support that this horizon slice has correctly reconstituted the depositional surface into which the channel was cut.

Figure 4-25 indicates by two black arrows the two seismic horizons followed in the construction of the horizon slices of Figure 4-26. The high amplitude feature shaped somewhat like a hockey stick appears very similar on the two sections. It is invisible on other adjacent horizon slices (not shown). Hence the seismic signature of this inferred channel is trough-over-peak, which implies high velocity material, given the polarity convention implicit in these data. After inverting the whole data volume to seismic logs, a horizon slice through this velocity volume positioned between the horizon slices of Figure 4-26 generated the horizon slice of Figure 4-27. The darker colors indicate the high velocity channel fill. This type of section is known as a horizon seiscrop velocity section.

Automatic horizon tracking, now commonplace in interactive interpretation systems, has greatly facilitated the generation of selected horizon slices. When a horizon is tracked, the extreme amplitude as well as its time is stored in the digital database. Mapping of the times produces a structure map; mapping of the amplitudes produces a horizon slice or horizon

Continued on page 90

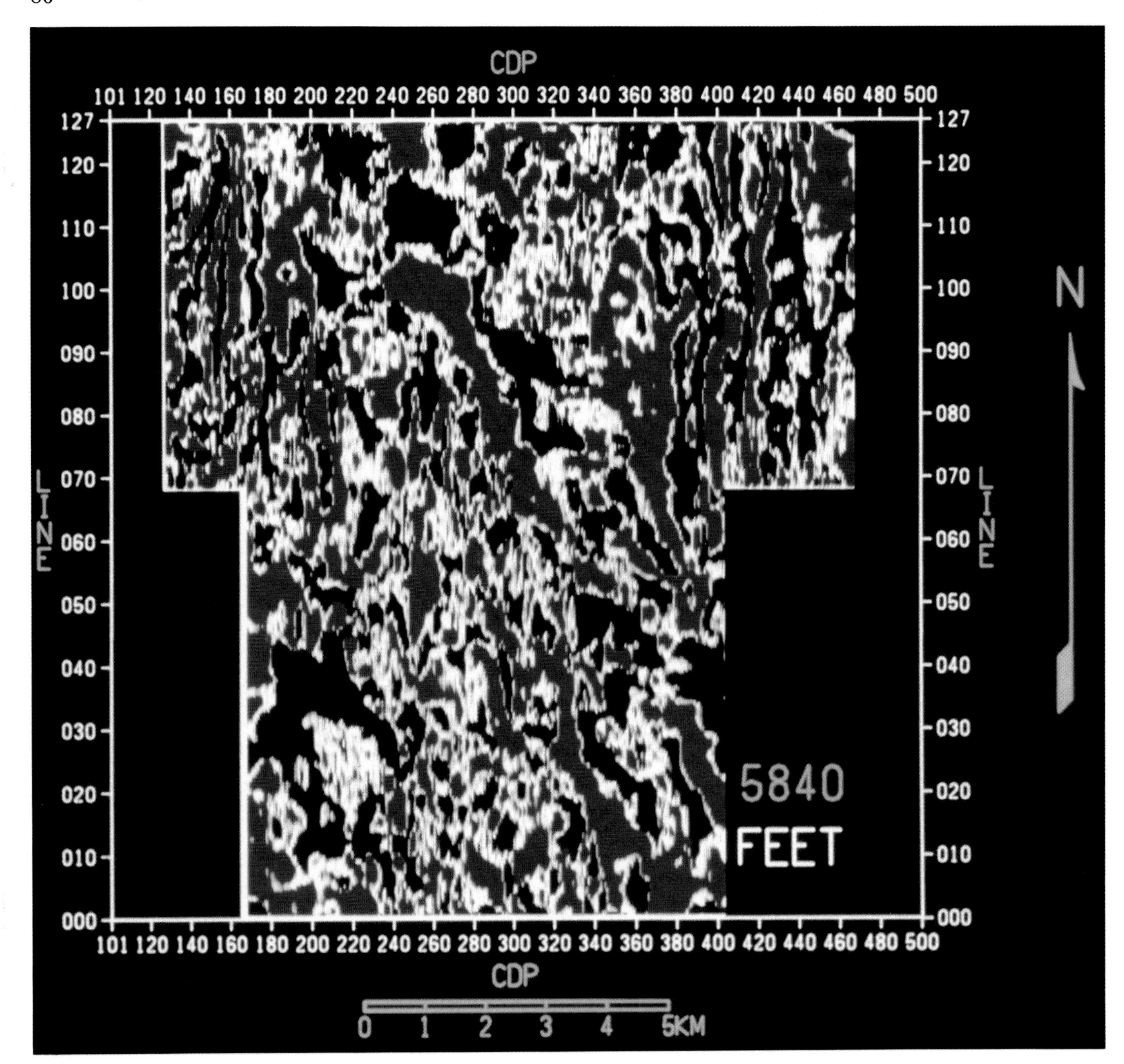

Fig. 4-18. Seiscrop section at 5,840 feet (1,780 m) from Gulf of Thailand. (Courtesy Texas Pacific Oil Company Inc.)

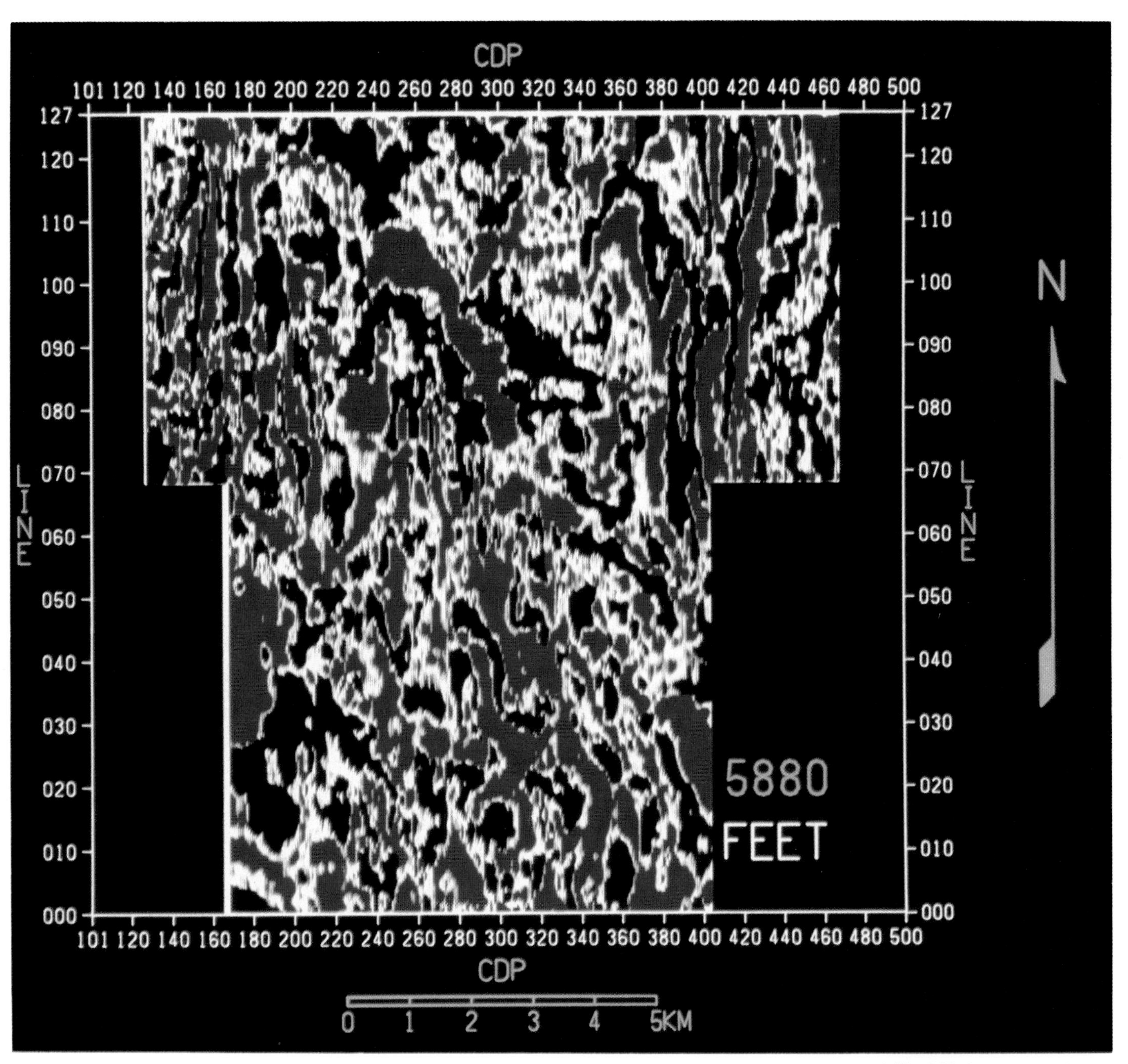

Fig. 4-19. Seiscrop section at 5,880 feet (1,790 m) from Gulf of Thailand. (Courtesy Texas Pacific Oil Company Inc.)

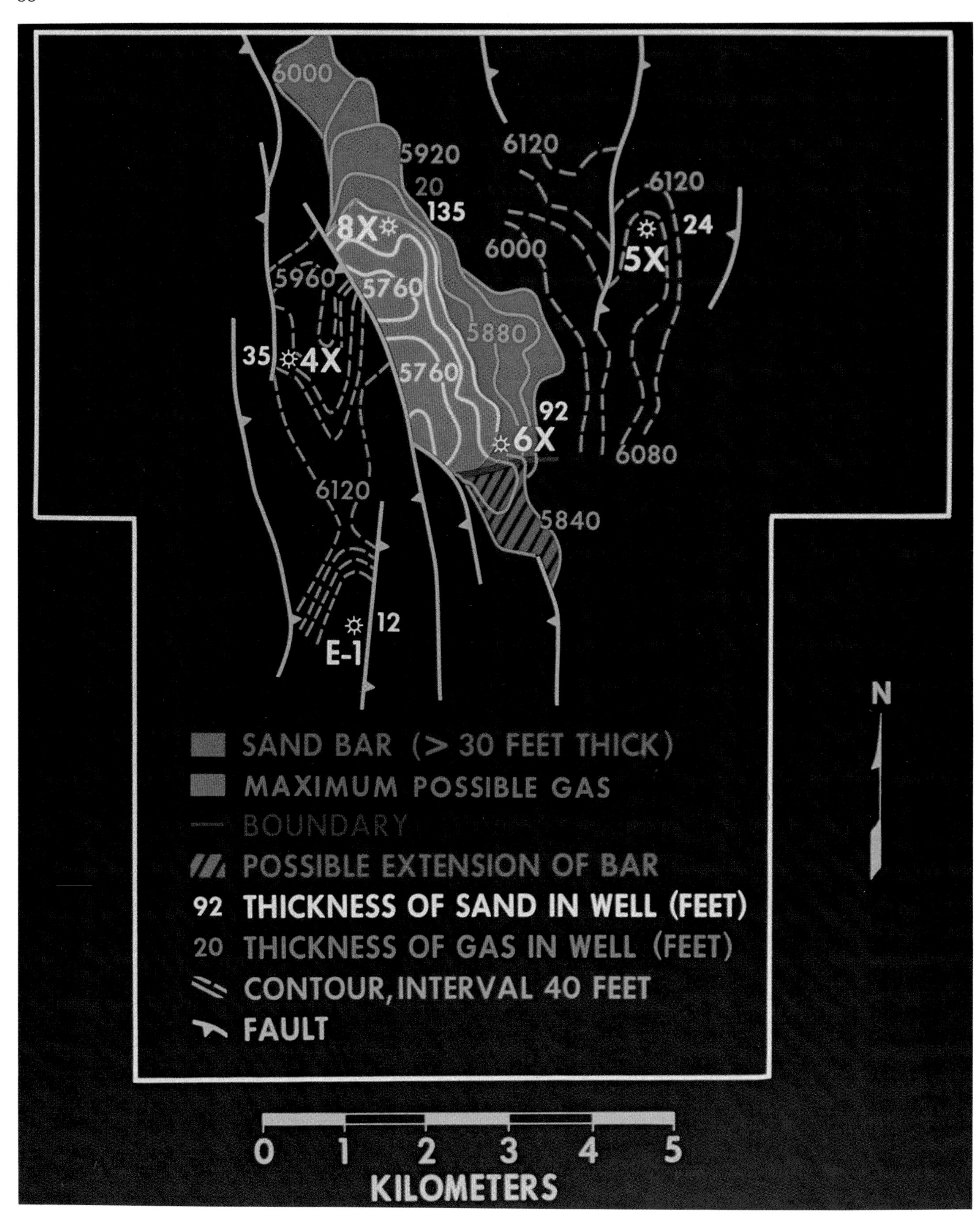

Fig. 4-20. Sand bar with gas mapped from Gulf of Thailand Seiscrop sections.

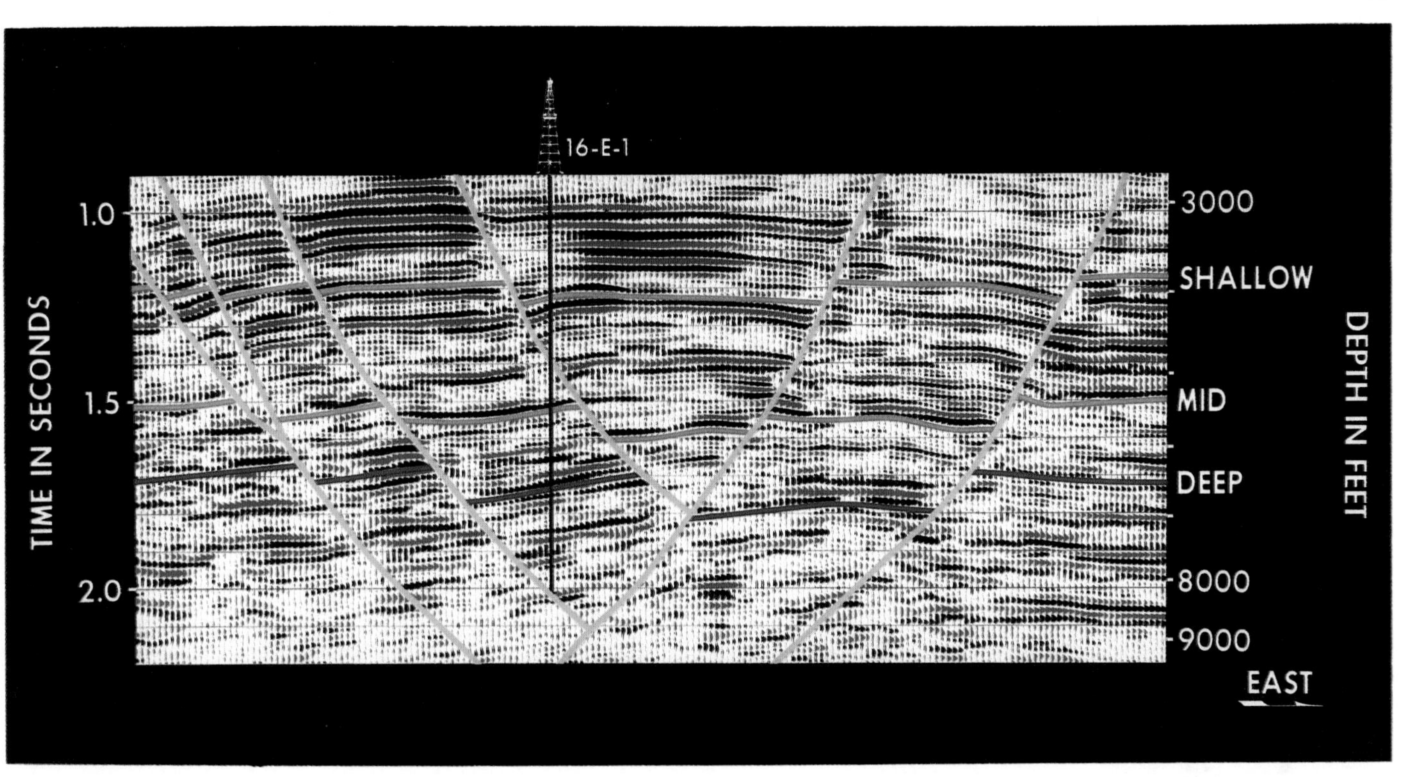

Fig. 4-21. Line 55 interpreted showing structure of Shallow Horizon. (Courtesy Texas Pacific Oil Company Inc.)

Seiscrop section. In addition, it is possible to extract the amplitudes not coincident with the tracked horizon but parallel to it and shifted by a chosen number of milliseconds.

Figure 4-28 shows two lines from a Gulf of Mexico 3-D prospect, where a horizon is tracked 1 1/2 periods above a red blob considered to be of stratigraphic interest. The structural continuity is better for the horizon being tracked than for the blob, so the structure was defined at this level and the horizon slice made parallel to it through the blob at a fixed time increment deeper. This approach is viable only when conformability can be assumed between the tracked horizon and the slicing level.

The resulting horizon slice is shown in Figure 4-29 and the interpreter can readily infer the existence of another channel. The black horizontal lines indicate the positions of the two vertical sections of Figure 4-28. The amplitude of the channel reflection is greater to the northeast; a discussion of this relative to implied gas content appears in Chapter 5.

Figure 4-30 shows a sequence of faults affecting one horizon interpreted on a vertical section from a 3-D survey in the Gulf of Thailand. Figure 4-31 shows the time structure map resulting from the complete structural interpretation of the same horizon. The faults trending north-northwest to south-southeast divide the area into seven fault blocks. The corresponding horizon slice is shown in Figure 4-32. A meandering stream channel is evident and gas production from the channel has been established in two of the fault blocks.

The continuity of the channel confirms that the depositional surface has been correctly reconstituted. Clearly the value of such a horizon slice for stratigraphic purposes is critically dependent on the accuracy of the structural interpretation that was involved in its derivation. Here the stratigraphic and structural interpretation actually impacted each other iteratively. The first horizon slice generated for this level did not show the channel continuity of Figure 4-32 in one of the fault blocks. This suggested miscorrelation into that block. After re-examining the correlation and retracking the data in that block, the horizon slice shown as Figure 4-32 was obtained. The improved channel continuity indicated the relative correctness of the updated structural interpretation.

Figure 4-33 and 4-34 show the time structure map and horizon slice for one interpreted horizon in a Gulf of Mexico shallow water prospect. Two channels are evident, one of them intersected by a fault. The deeper channel lies between 2100 and 2200 ms which converts to depths around 2500 m (8,200 ft).

Unconformities, Faults and Multiple Horizon Mapping

Figure 4-35 shows a horizon slice from yet another Gulf of Mexico prospect. The amplitudes are in shades of blue and the time structure is superimposed as contour lines with an interval of 100 ms. Several amplitude lineations are apparent. The ones running approximately east-west are faults as evidenced by the displacement of the contours.

The major lineation running northwest-southeast is apparently unrelated to the faulting. It is interpreted as the truncation of a sand dipping up from the east. It is probably a depositional edge but the erosional truncation of a sand at an unconformity would show in exactly the same manner. It is this lineation on the horizon surface which caught the interpreter's eye and thus begged for an explanation.

The value of horizon slices for studying the spatial distribution of horizon properties and characteristic patterns therein is well established. Less established is the easy availability of multiple horizon slices for a prospect under study. The solution is in fact emerging through automatic **spatial** tracking, provided by some interactive interpretation systems. Given some horizon control, the computer attempts to track the chosen horizon in three dimensions throughout the 3-D data volume. This has been very successful in some prospects but depends on data quality and requires that the interpreter checks the result. The horizon slice of Figure 4-35 was a product of automatic spatial tracking.

Figure 3-21 sets forth a scenario for combined structural and stratigraphic interactive interpretation of 3-D data. Automatic spatial tracking is central to this approach as it provides the possibility of studying many horizons in a reasonable length of time. Tracking dozens of horizons in the manner outlined in Figure 3-21 constitutes a method of data reduction that the interpreter may use routinely in the future. In this approach, the interpreter would: (1) initially identify all the horizons which could conceivably be of interest, (2) track them all, (3) apply some quality checks and then (4) scan the resulting horizon products, time and amplitude, for features of

Continued on page 98

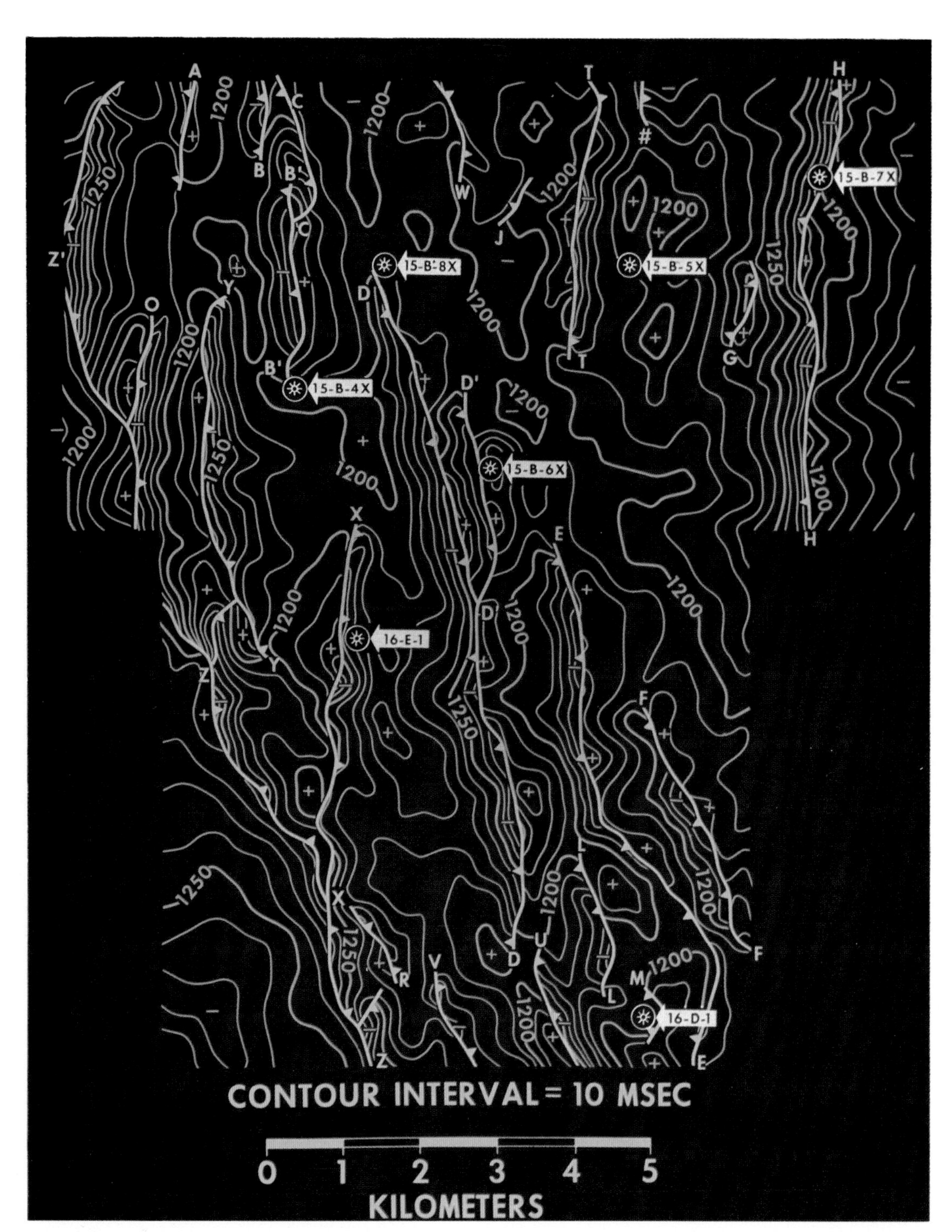

Fig. 4-22. Time structure map of Shallow Horizon. (Courtesy Texas Pacific Oil Company Inc.)

Fig. 4-23. Horizon Seiscrop section 180 feet (60 m) below Shallow Horizon showing northwest- southeast-trending high amplitude interpreted as a sand bar. (Courtesy Texas Pacific Oil Company Inc.)

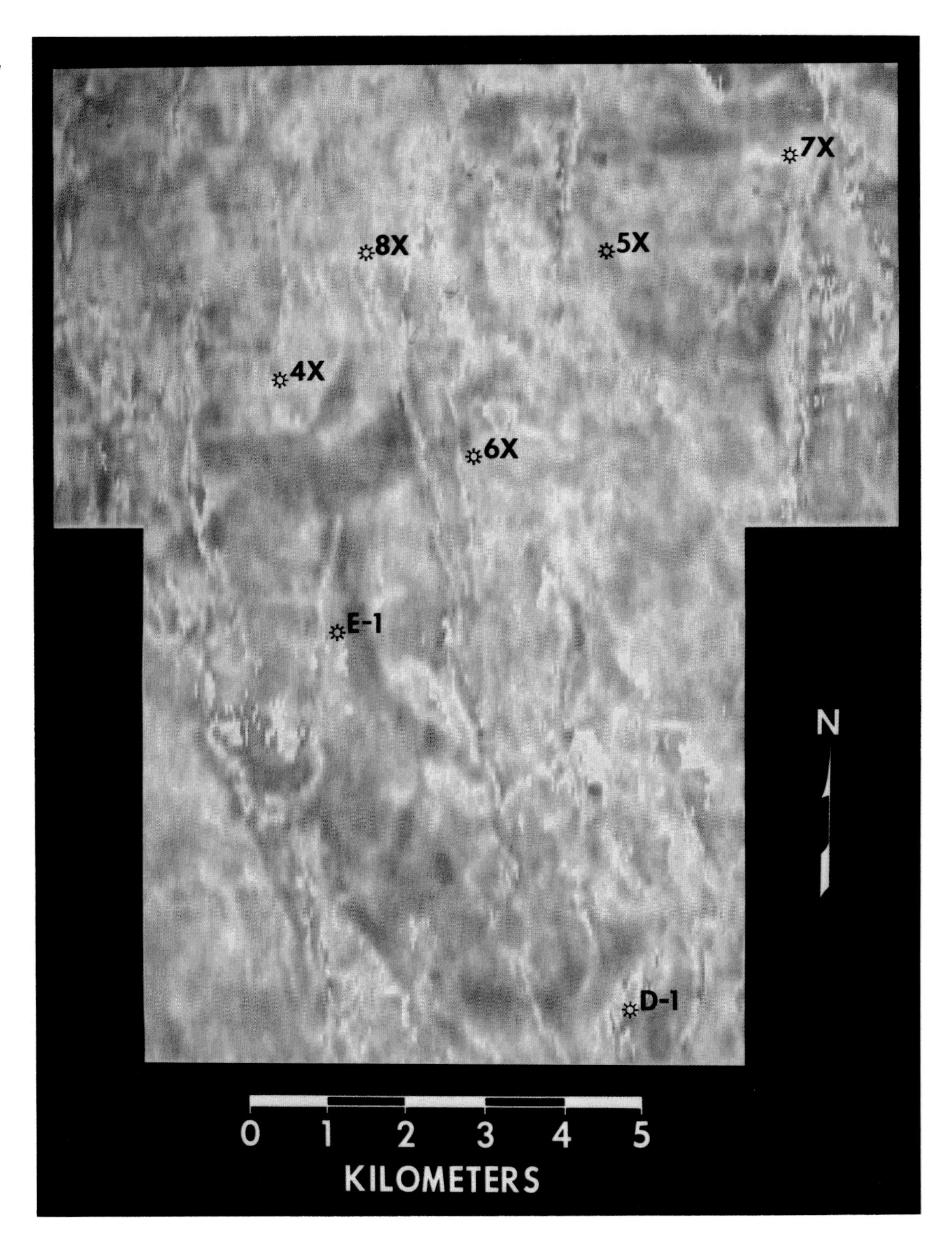

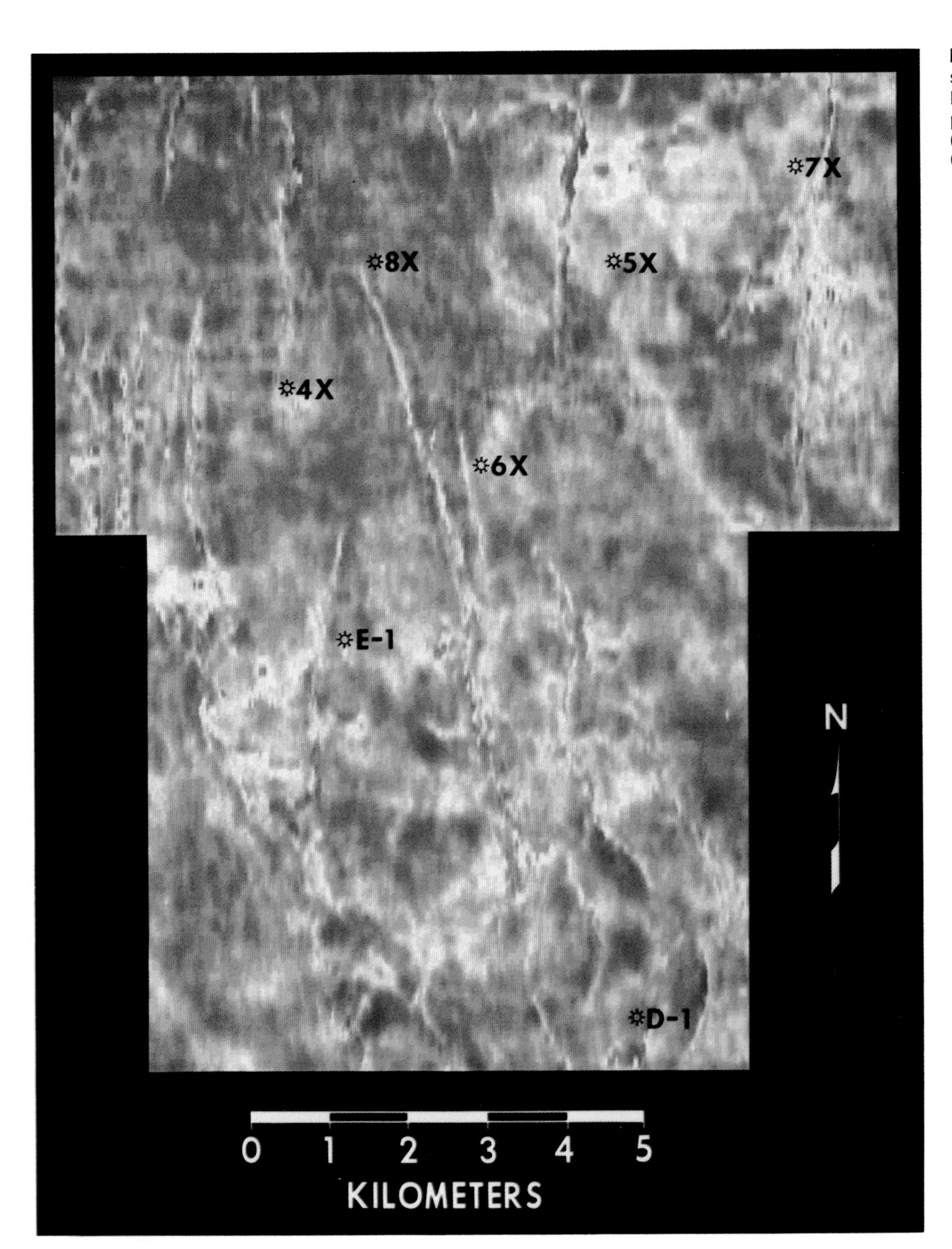

Fig. 4-24. Horizon Seiscrop section through Shallow Horizon showing a partly-eroded sheet sand. (Courtesy Texas Pacific Oil Company Inc.)

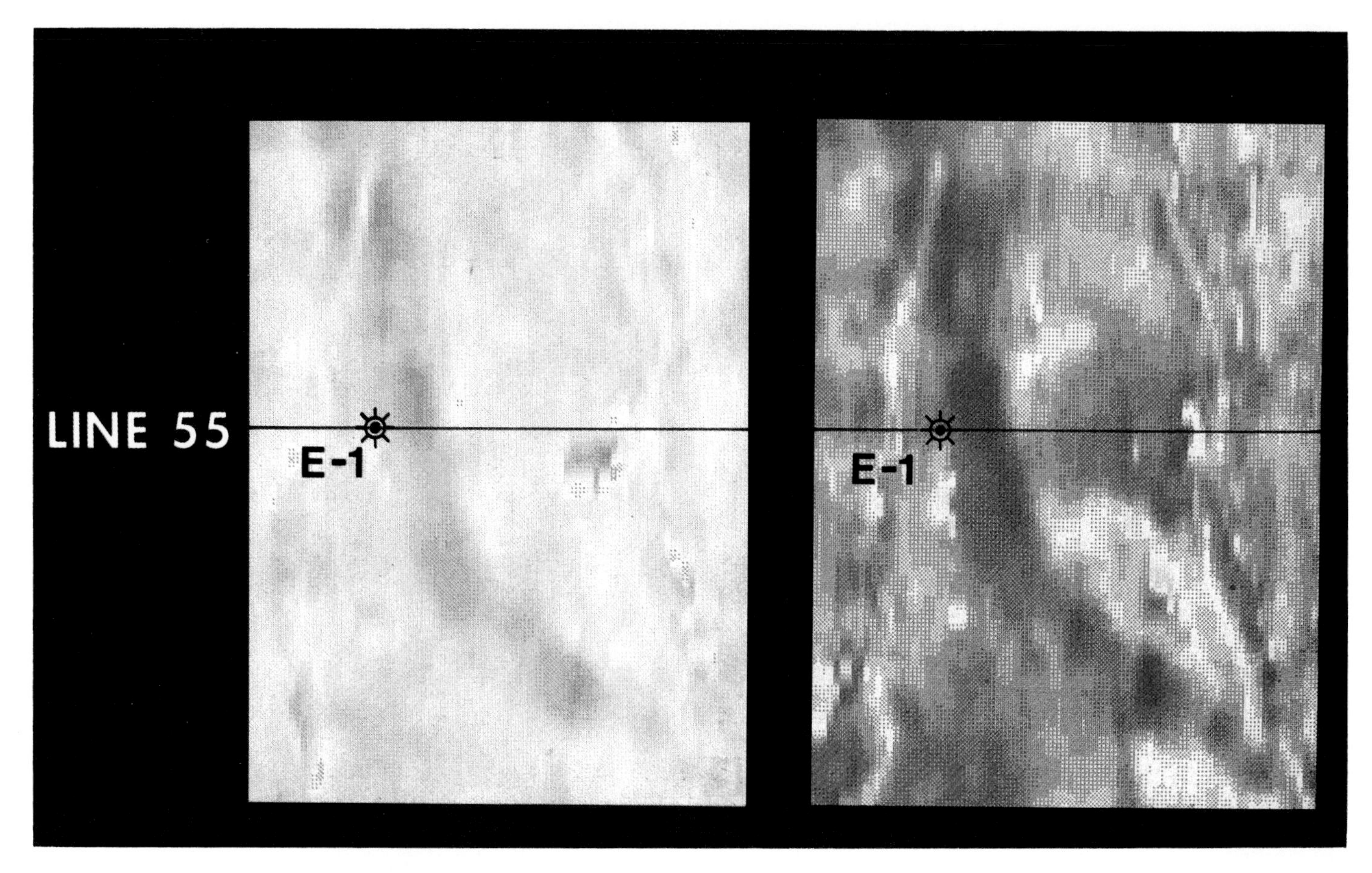

Fig. 4-26. Horizon Seiscrop sections through the two events marked with black arrows on Figure 4-25. The curvilinear features are interpreted as the reflections from the top and base of a channel. (Courtesy Texas Pacific Oil Company Inc.)

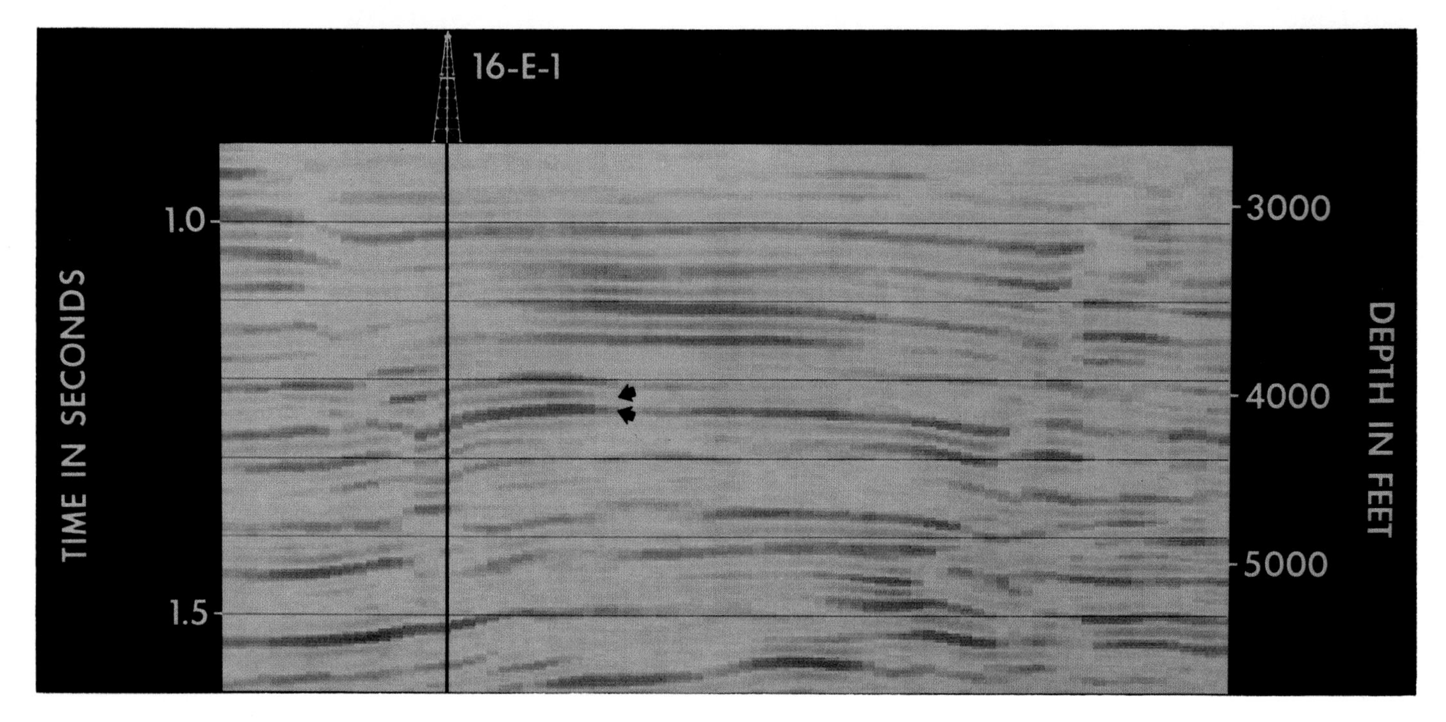

Fig. 4-25. A portion of Line 55 through the central graben of the 3-D prospect. (Courtesy Texas Pacific Oil Company Inc.)

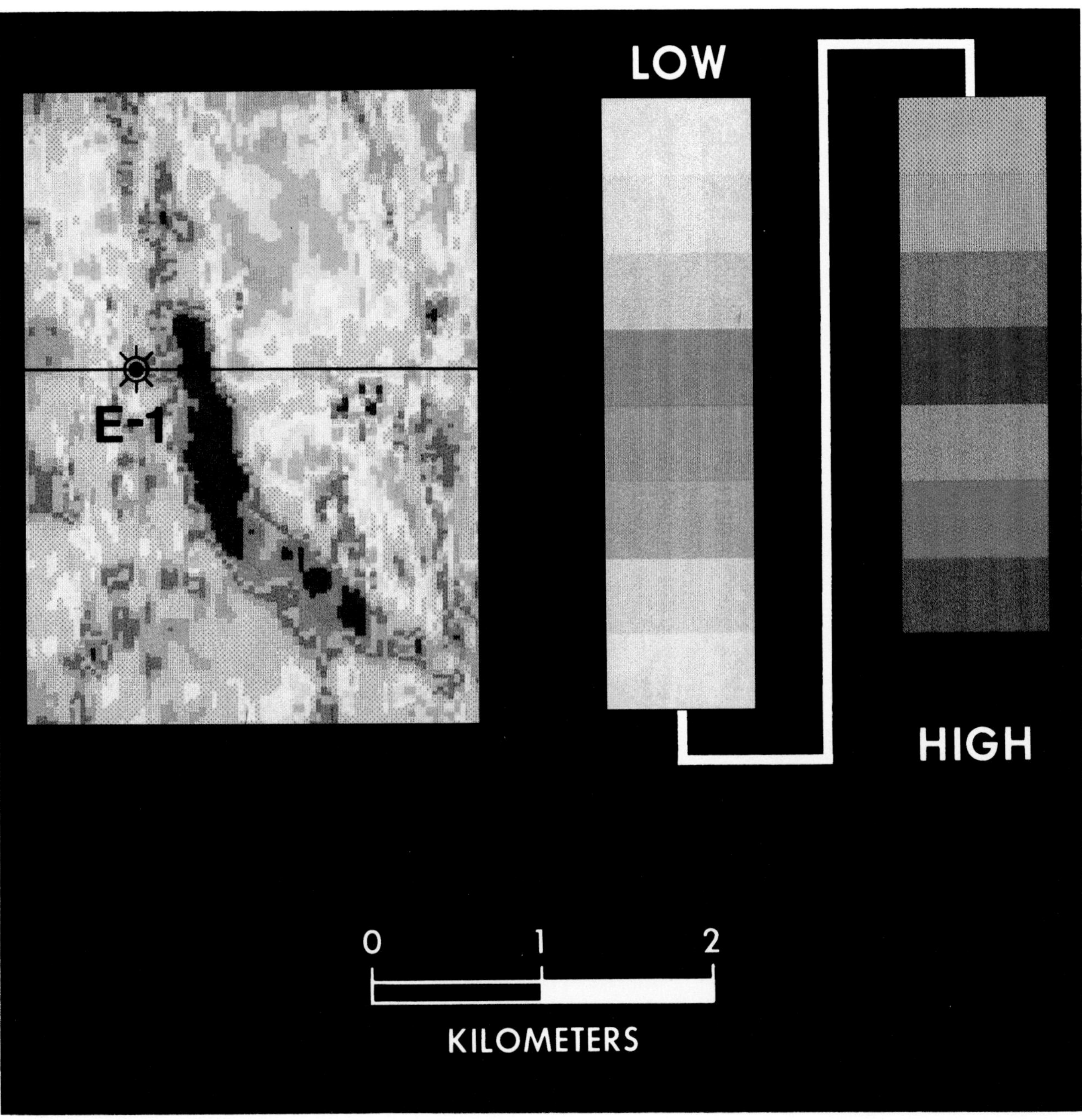

Fig. 4-27. Horizon Seiscrop velocity section positioned between the sections of Figure 4-26 and showing the extent of the high velocity channel fill. (Courtesy Texas Pacific Oil Company Inc.)

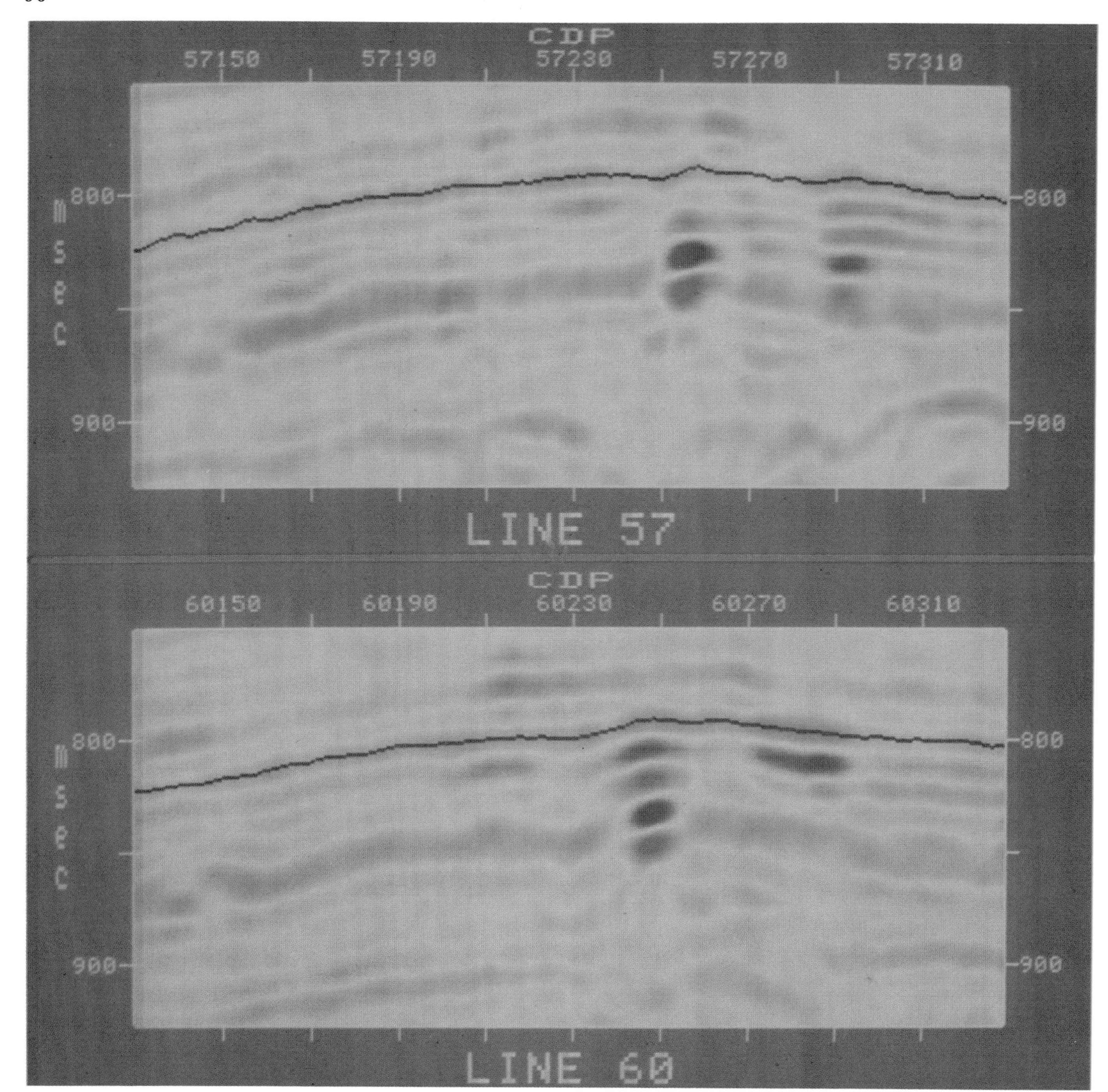

Fig. 4-28. Lines 57 and 60 from a 3-D survey in the Gulf of Mexico showing a tracked horizon above bright events indicating channel intersections. (Courtesy Chevron U.S.A. Inc.)

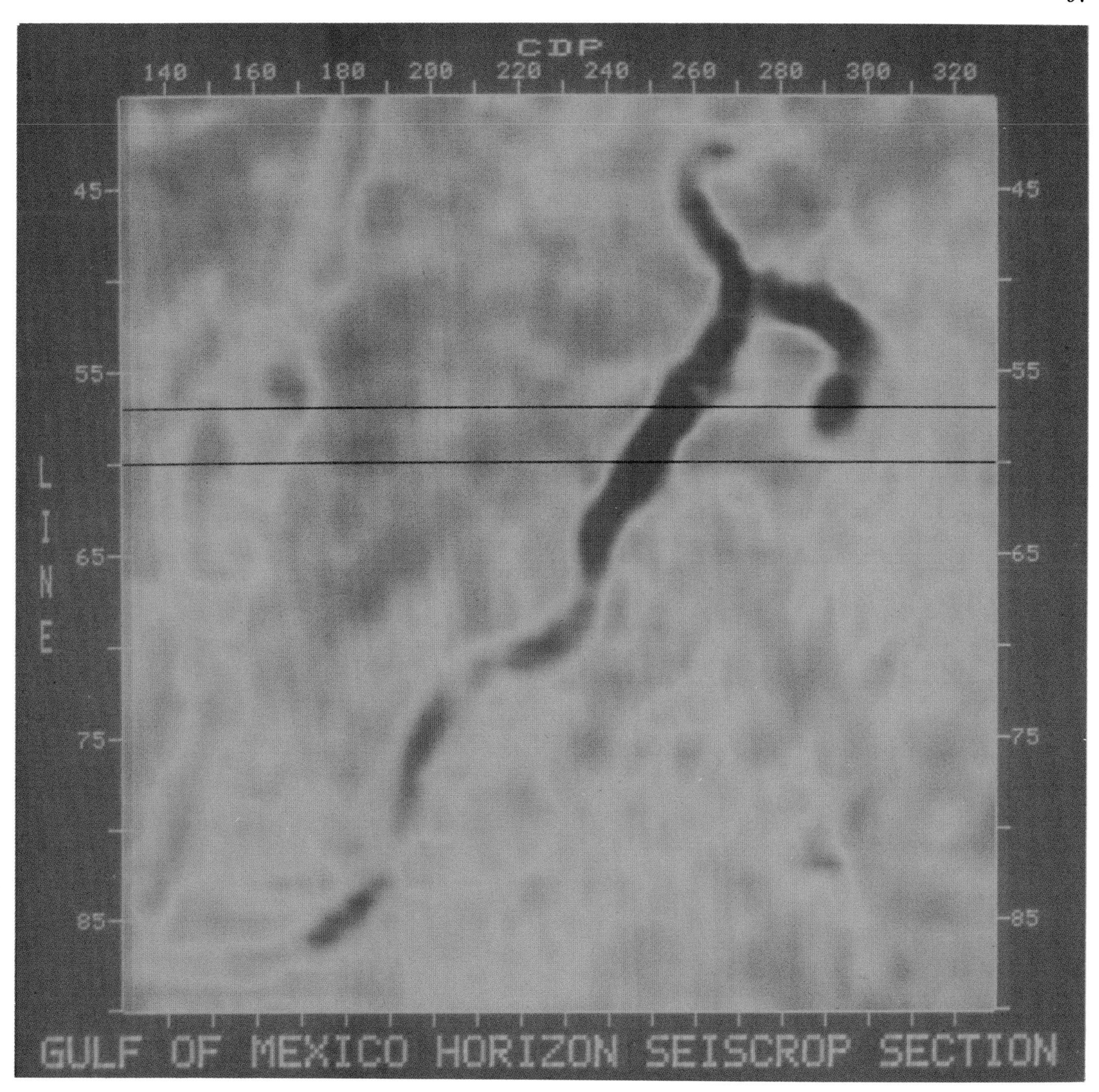

Fig. 4-29. Horizon Seiscrop section showing channel intersected in Figure 4-28. (Courtesy Chevron U.S.A. Inc.)

interest. Edge and anomaly detection can be used to identify faults as discontinuities in time, and stratigraphic features as bounded by discontinuities in amplitude.

References

Broussard, M. L., ed., 1975, Deltas: Houston Geological Society, 555 p.

Brown, A. R., C. G. Dahm, and R. J. Graebner, 1981, A stratigraphic case history using three-dimensional seismic data in the Gulf of Thailand: Geophysical Prospecting, v. 29, p. 327–349.

Klein, G. deV., 1985, Sandstone depositional models for exploration for fossil fuels, third edition: Boston, Massachusetts, International Human Resources Development Corporation, 209 p.

Sanders, J. I., and G. Steel, 1982, Improved structural resolution from 3D surveys in Australia: Australian Petroleum Exploration Association (APEA) Journal, v. 22, p. 17–41.

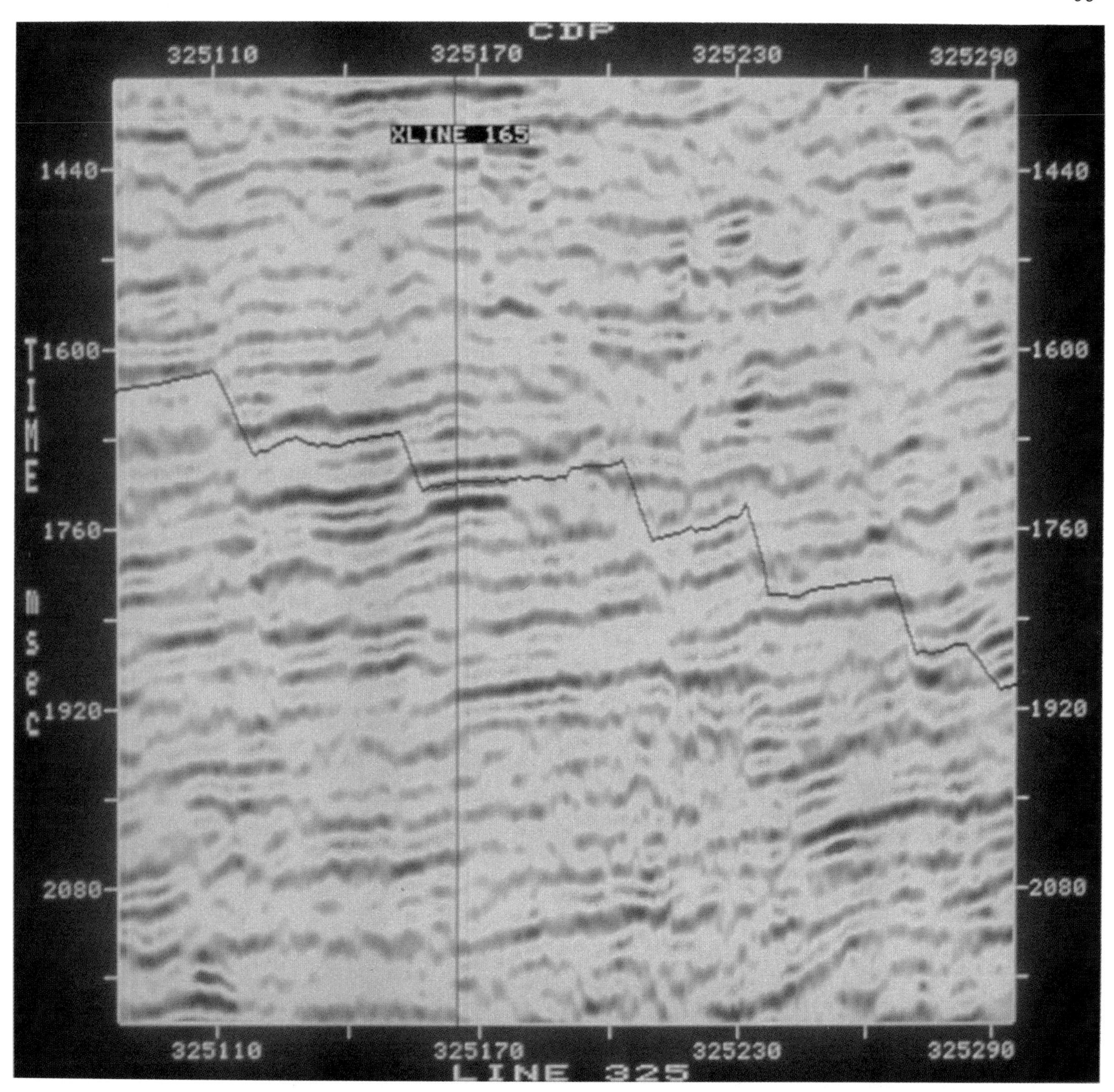

Fig. 4-30. Line 325 from 3-D survey in the Gulf of Thailand showing interpreted horizon through many fault blocks. (Courtesy Unocal Thailand Inc.)

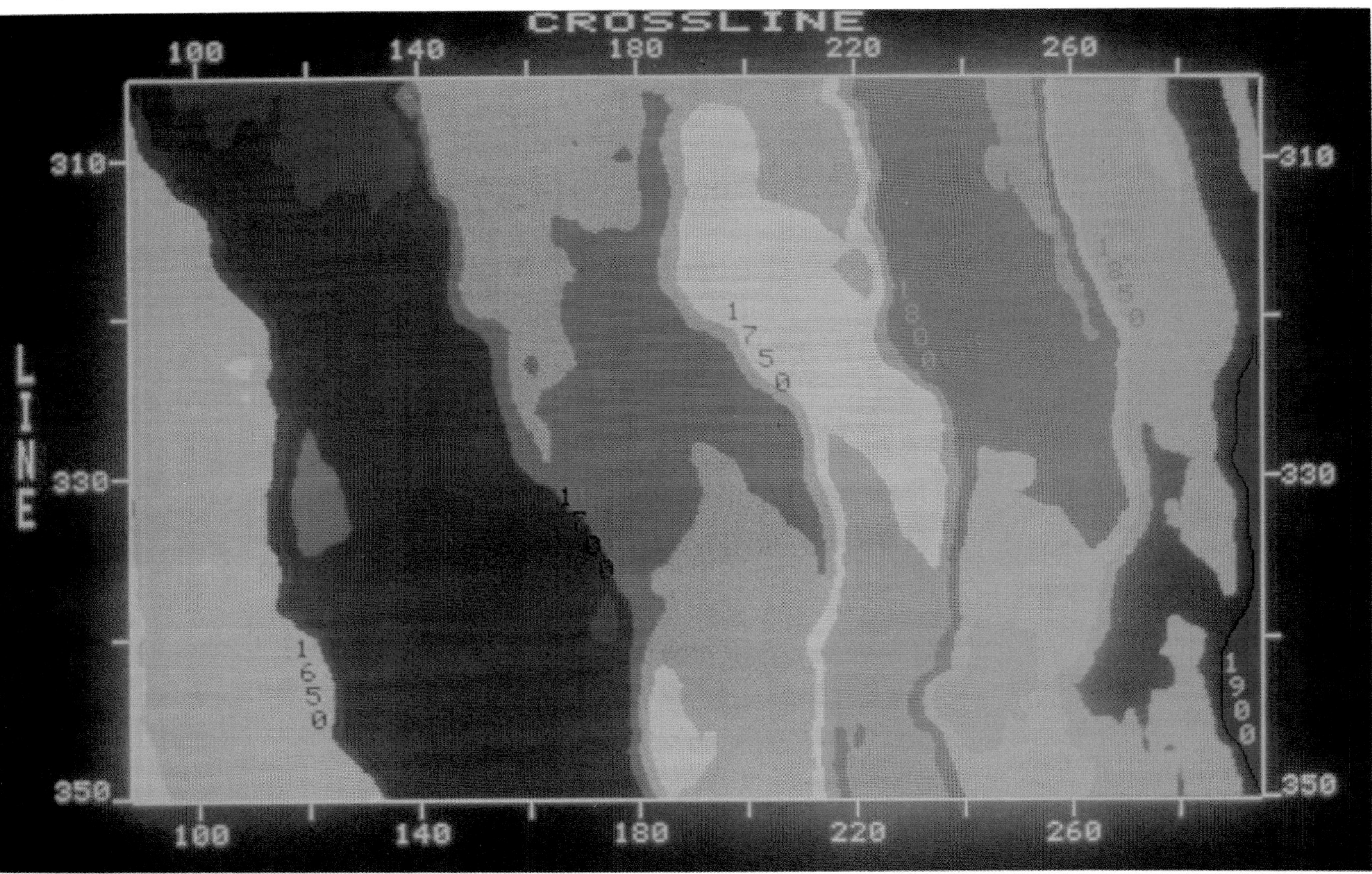

Fig. 4-31. Time structure map of horizon tracked in Figure 4-30 (Courtesy Unocal Thailand Inc.)

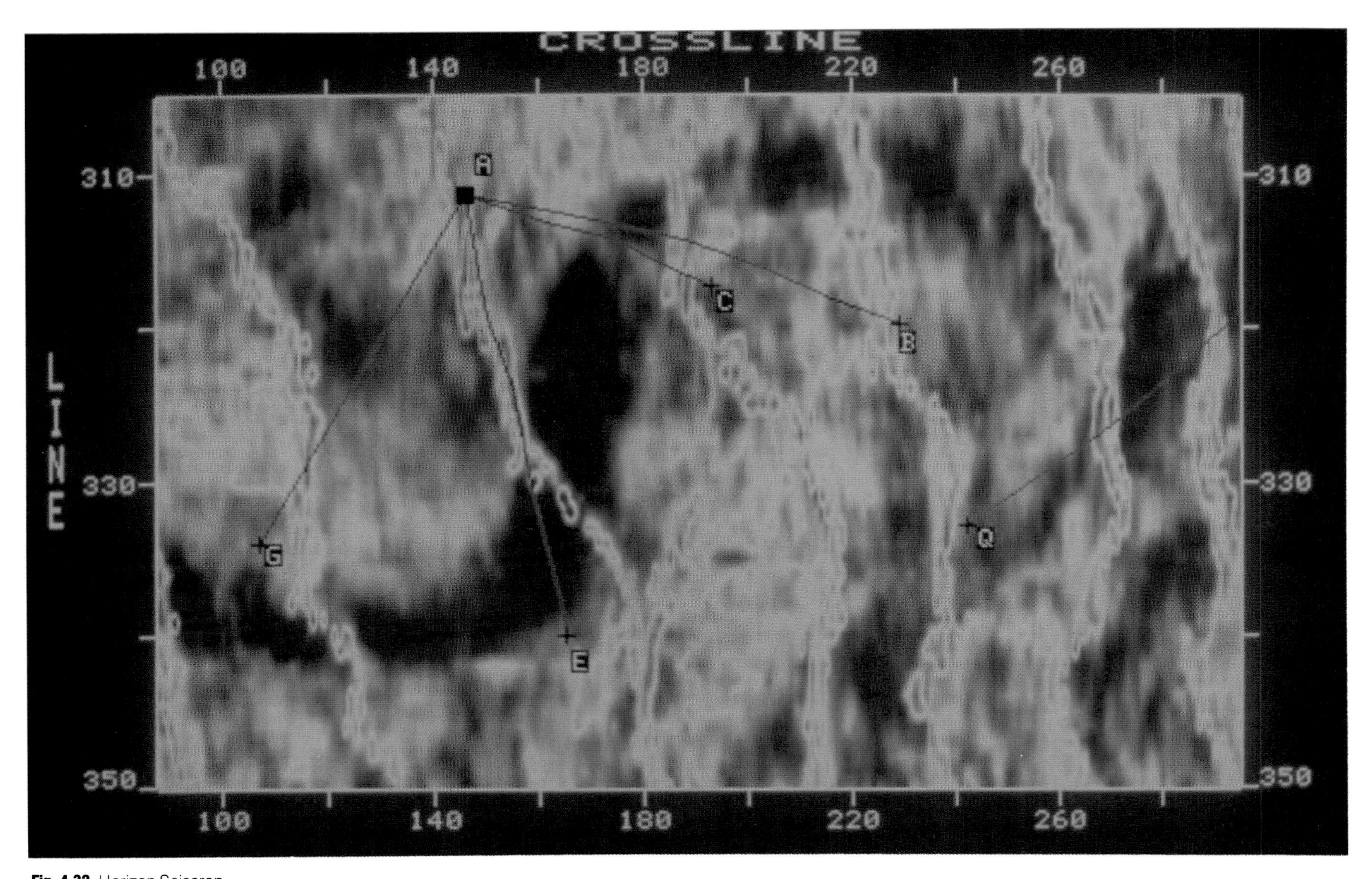

Fig. 4-32. Horizon Seiscrop section showing spatial distribution of amplitude over the horizon mapped in Figure 4-31. Gas production has been established in the meandering channel. (Courtesy Unocal Thailand Inc.)

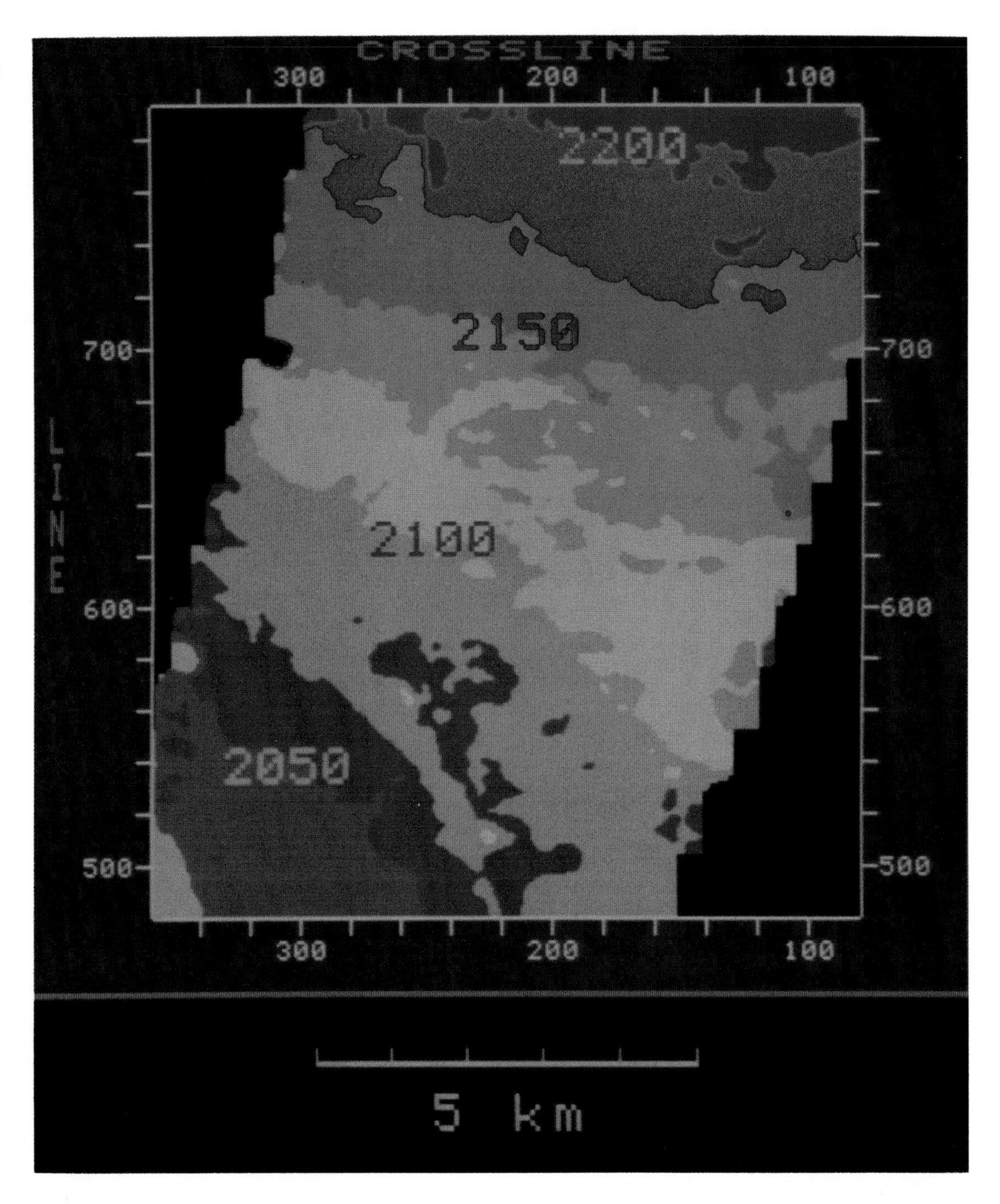

Fig. 4-33. Time structure map for a horizon interpreted from a 3-D survey offshore Louisiana. The large numbers are contour designations in milliseconds.

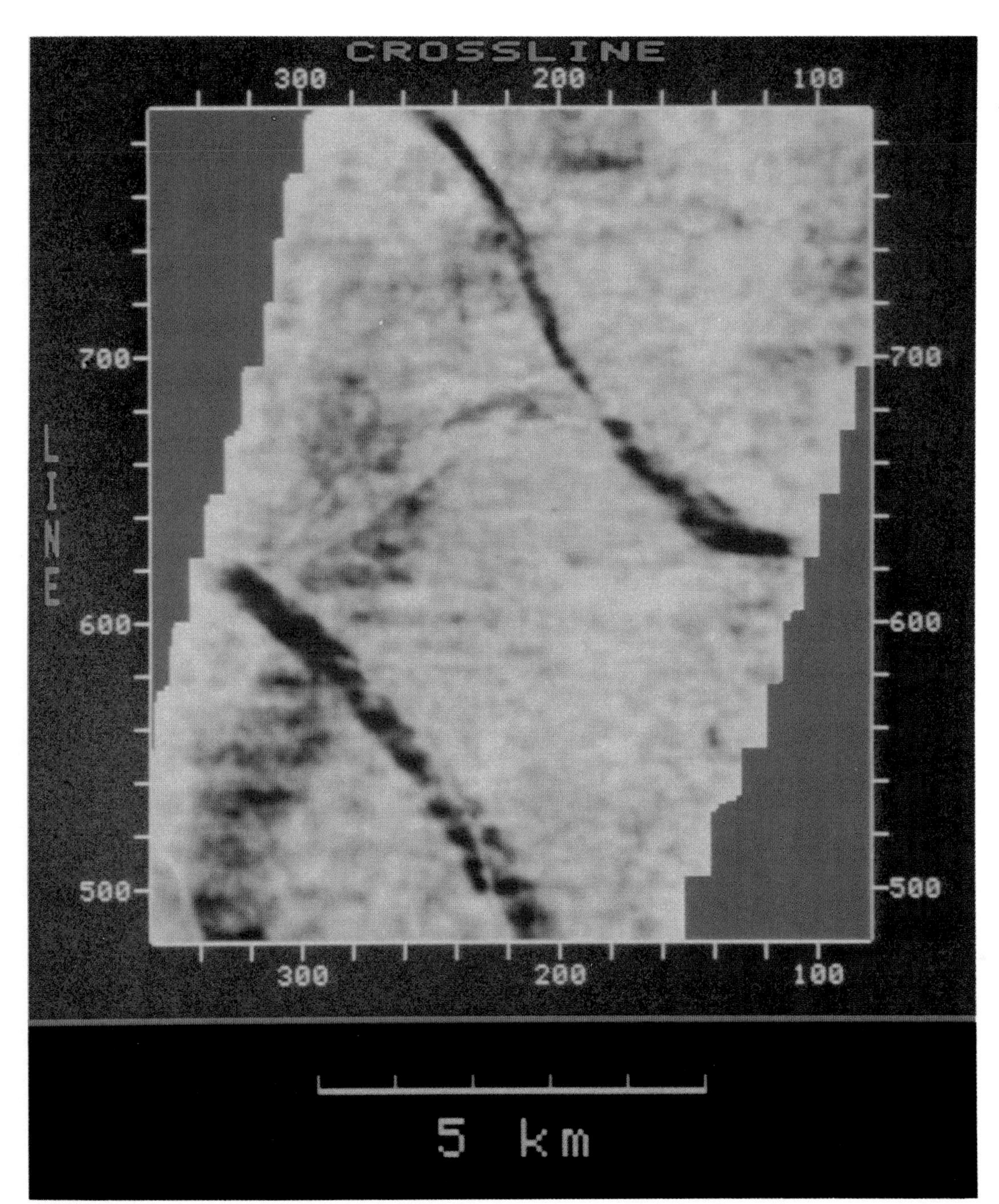

Fig. 4-34. Horizon Seiscrop section showing spatial distribution of amplitude over the horizon mapped in Figure 4-33 and indicating two channels.

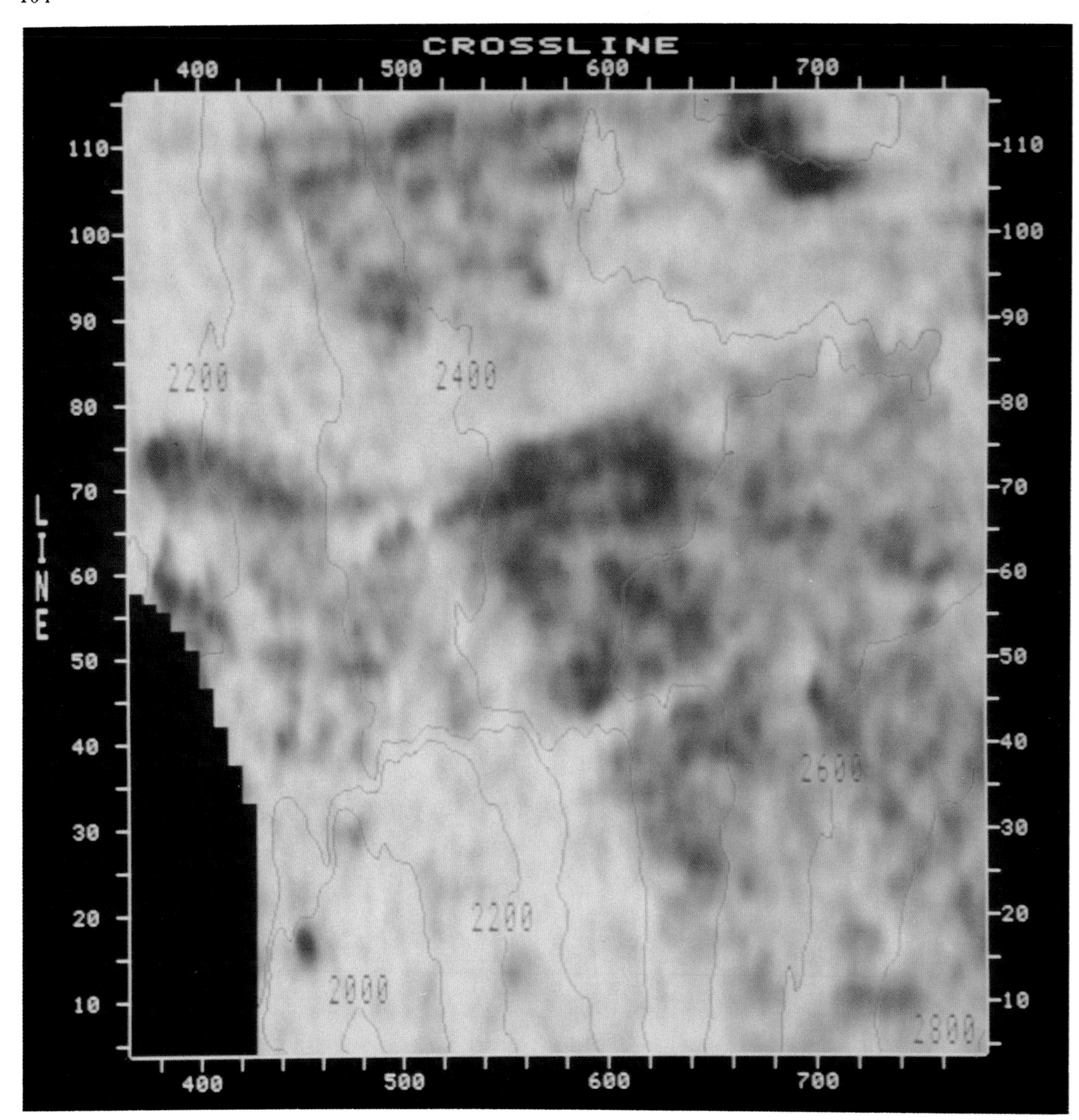

Fig. 4-35. Horizon Seiscrop section and superimposed structural contours from a Gulf of Mexico 3-D survey. The amplitude lineations indicate faults and bed truncations. (Courtesy Chevron U.S.A. Inc.)

BRIGHT SPOT VALIDATION

Bright Spots As They Used To Be

Figure 5-1 shows a bright spot presented by Tegland (1973). This was one of the early examples studied and was observable because amplitude had been preserved in seismic processing. In earlier years, when records were normally made with automatic gain control, there was little opportunity for studying amplitudes. The bright spot of Figure 5-1 is actually a very good one for its era because it also shows a flat spot, presumably a fluid contact reflection. The flat spot terminates laterally at the same points as does the bright spot; we would consider this a simple form of bright spot validation, increasing the interpreter's confidence that the anomaly indicates the presence of hydrocarbons.

With the improvements in seismic processing over the last decade, we can now consider polarity and phase as well as amplitude and spatial extent. Frequency, velocity, amplitude/offset and shear wave information can also help greatly in the positive identification of hydrocarbon indicators. These are all subjects of this chapter.

Most direct hydrocarbon indication relates to gas rather than oil reservoirs as the effect on acoustic properties of gas in the pore space is significantly greater than oil. Figure 5-2 (derived from Gardner, Gardner, and Gregory, 1974) summarizes the different effects of gas and oil and shows that the effect of either diminishes with depth. Much of the detailed studies in this book, therefore, concern gas sands at depths less than about 7,000 ft (2,100 m).

Backus and Chen (1975) were very thorough in their discussion of the diagnostic benefits of flat spots, and Figure 5-3 shows a flat spot at 1.47 seconds that they discussed. Figure 5-4 is interpreted sufficiently to highlight the various hydrocarbon indicators on the section. The flat spot is easily identified by its flatness, and because it is unconformable with adjacent reflections. Hence it is a good indicator of the hydrocarbon/water contact.

The reflection from the top of the reservoir (Figure 5-4) changes from a peak to a trough across the fluid contact and this again implies a significant change in acoustic properties between the gas sand above the hydrocarbon/water contact and the water sand beneath it. This phase change, or a polarity reversal, will be discussed in more detail in the next section.

The Character of a Modern Bright Spot

If the seismic data under interpretation have been processed to zero phase (see Chapter 2), then the detailed character of the bright spots, flat spots and other hydrocarbon indicators can be very diagnostic. Figure 5-5 shows diagrammatically the hydrocarbon indicators which may be associated with different relative acoustic impedances of gas sand, water sand and embedding medium. The polarity convention is the same as that explained in Chapter 2, namely that a decrease in acoustic impedance is expressed as a peak and an increase as a trough. Peaks and troughs are symmetrical if they are the zero-phase expressions of single interfaces.

The top diagram of Figure 5-5 illustrates the most common situation: the water sand has an acoustic impedance lower than the embedding medium and the impedance of the gas sand is further reduced. For this situation the signature of the sand is peak-over-trough and, for the gas-filled portion, the amplitude is greater. This is the classical zero-phase bright spot. If the sand

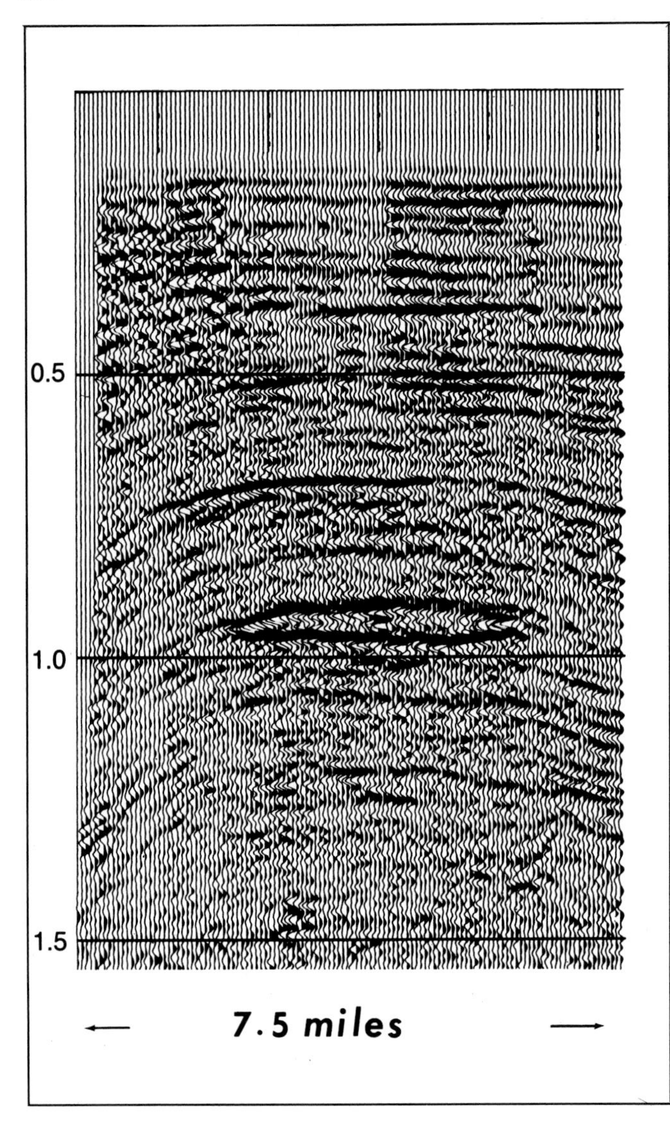

Fig. 5-1. A Gulf of Mexico bright spot and flat spot from the early 1970's.

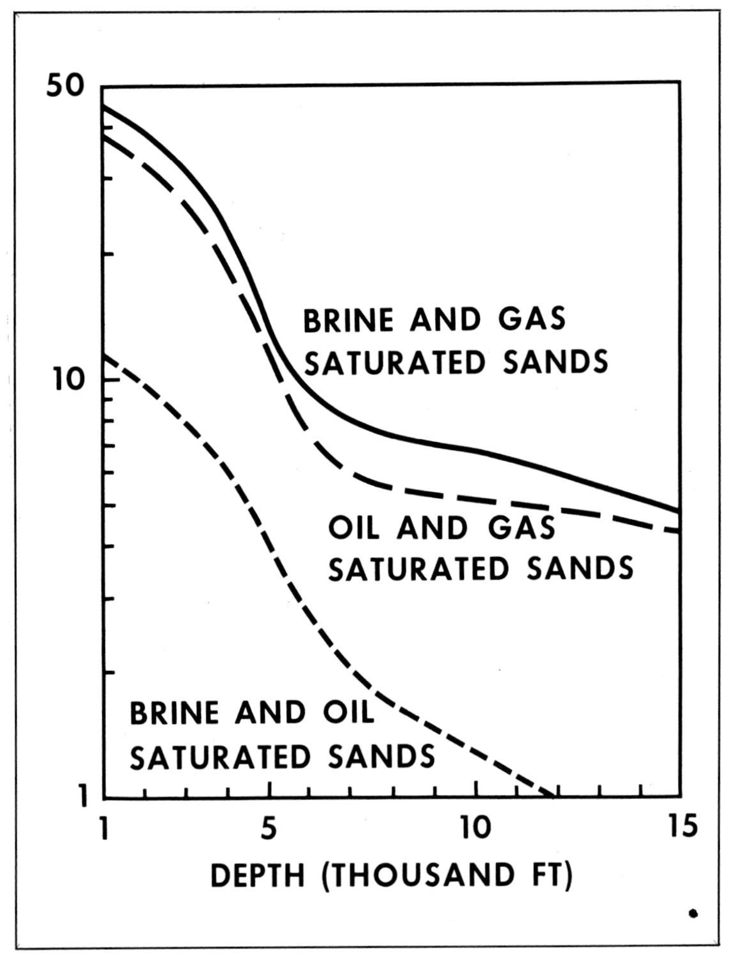

Fig. 5-2. Velocity difference between sands saturated with different fluids (derived from Gardner, Gardner, and Gregory, 1974).

is thick enough for the top and base reflections to be resolved, then a flat spot or fluid contact reflection should be visible between the gas sand and the water sand, that is at the point where brightening occurs. The flat spot reflection will be a trough because it must be an increase in acoustic impedance.

In the second diagram the situation is reversed; the water sand has a higher acoustic impedance than the embedding medium and hence has a signature of trough-over-peak. When gas replaces some of the water in the pores of the sand, the acoustic impedance is reduced, the contrast is reduced at the upper and lower boundaries, and the reservoir is seen as a dim spot. Again, if the sand is thick enough, a flat spot can be expected at the point where the dimming occurs and this again will be a trough.

In the third diagram the reduction in acoustic impedance of the sand, because of gas saturation, causes the acoustic impedance to change from a value higher than that of the embedding medium to one lower than that of the embedding medium. Hence the polarities of the reflections for the top and the base of the sand switch. The signature changes from trough-over-peak to peak-over-trough across the fluid contact. In order to observe such a phase change, or polarity reversal, in practice, the structural dip must be clearly determined from non-reservoir reflections just above and/or just below the sand under study. Again, if the sand is thick enough, a fluid contact reflection should be visible and it will be a trough.

Some Practical Examples

Figure 5-6 shows a Gulf of Mexico bright spot known to be a gas reservoir. The reservoir reflections have very high amplitude and hence the interference from other nearby reflections, multi-

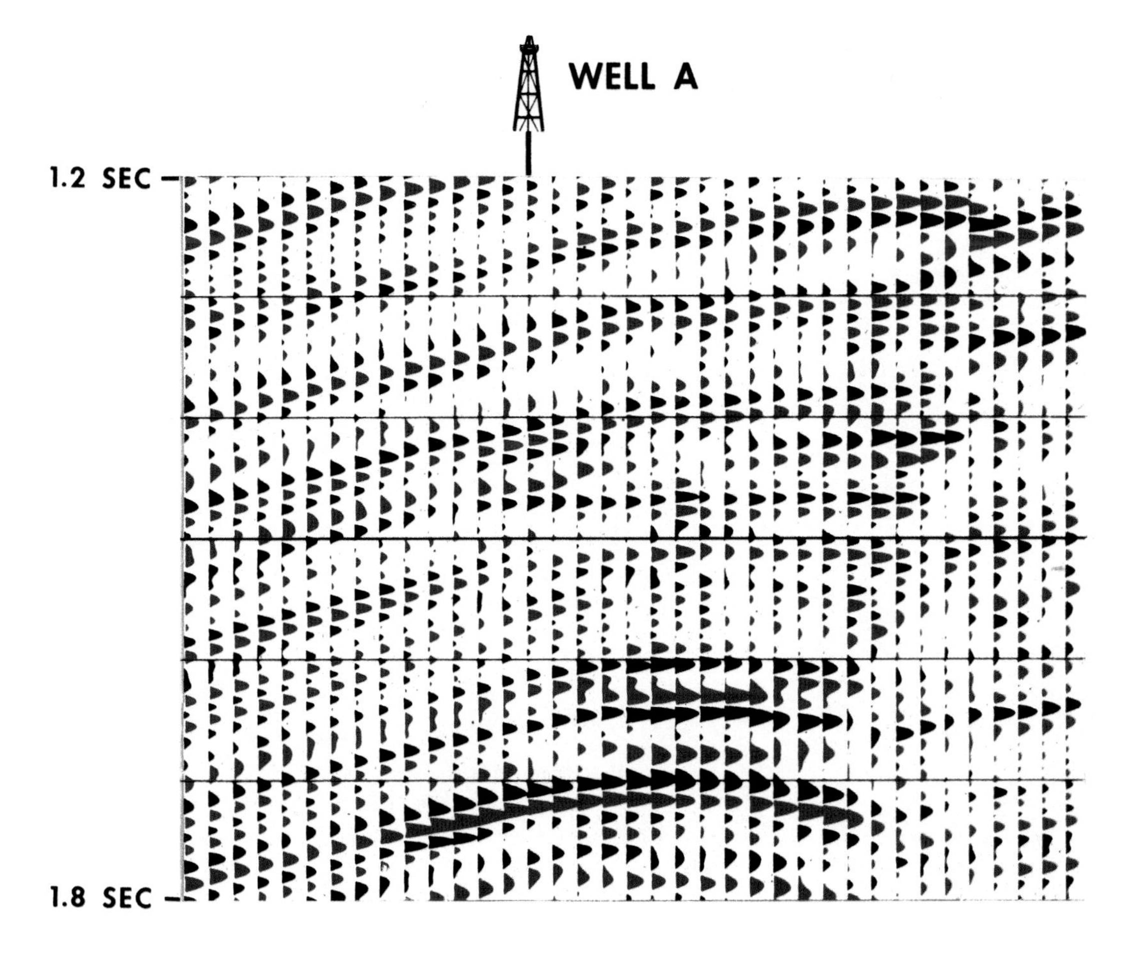

Fig. 5-3. Dual polarity section showing bright spot at 1.72 seconds and a flat spot at 1.47 seconds.

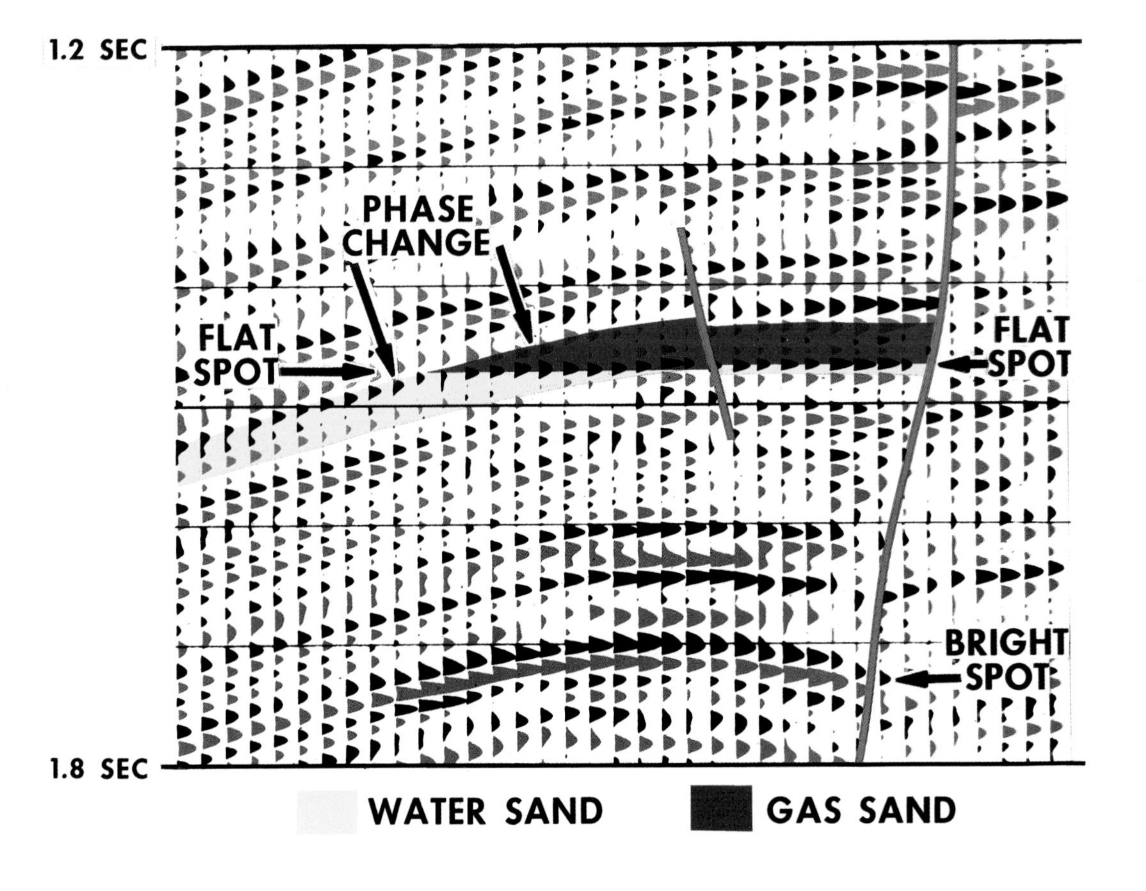

Fig. 5-4. Same section as Figure 5-3 showing the interpreted position of the gas reservoir and demonstrating a phase change between the reflections from the gas sand and the water sand.

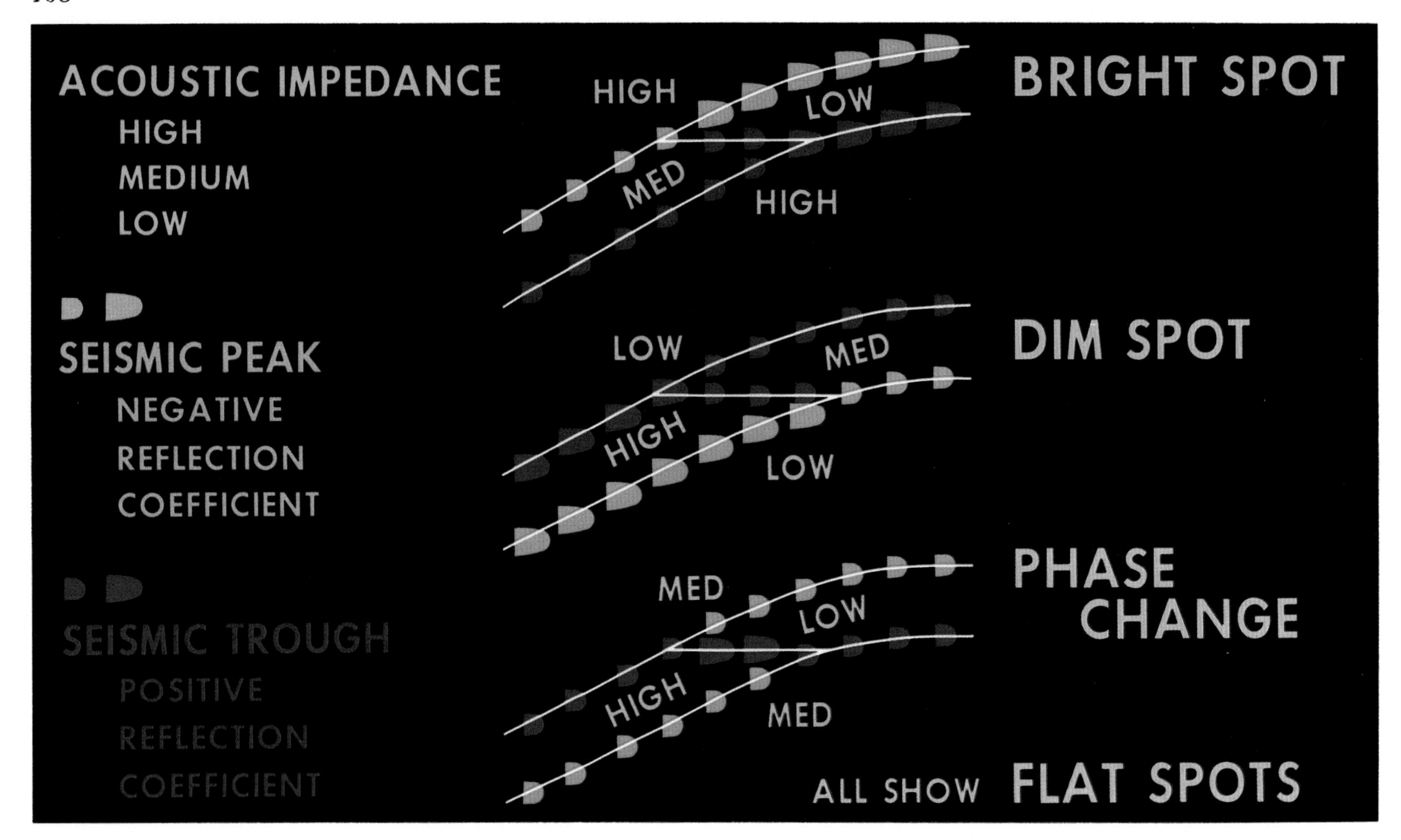

Fig. 5-5. Schematic diagram of the zero-phase response of reservoirs for different acoustic contrasts between the reservoir and the embedding medium.

ples or noise is small. The bright reflections show the zero-phase response of two reservoir sands, each peak-over-trough and located one on top of the other. The upper sand is fairly thin so there is only a hint of a flat spot reflection at the downdip limit of brightness. The lower sand is much thicker and the flat spot reflection is very clear.

Flat spot reflections are highly diagnostic indicators of gas but the interpreter should make several validity checks before drawing a conclusion. In Figure 5-6 the flat spot reflection is flat, bright and shows one symmetrical trough. It occurs at the downdip limit of the bright events and is unconformable with them. Figure 5-7 is a structure map of the base of the gas sand; it shows structural consistency for the flat spot reflection in the extent of the purple color.

Figure 5-8 is a practical example of a dim spot. The discovery well penetrates a gas column of about 400 ft (130 m) but the acoustic contrast of the gas sand with its embedding medium is small. Outside the reservoir the contrast between the sand and the embedding medium is much greater, as the amplitudes indicate.

Figures 5-9, 5-10, 5-11 and 5-12 illustrate a phase change; all four figures are exactly the same piece of data displayed with different colors and gains. Figure 5-9 uses the standard blue and red gradational scheme and the amplitude anomaly is clearly visible. Its visibility is perhaps enhanced further by the yellow, green and gray color scheme of Figure 5-10. In order to check for a phase change, or polarity reversal, it is necessary to judge the structural continuity from the bright reflections to their non-bright equivalents downdip. There is a very great difference in amplitude between these, causing a great difference in color intensity. Figures 5-11 and 5-12 use the same colors respectively as Figures 5-9 and 5-10 but with a higher gain applied to the data. This makes it easier to judge the downdip continuity on the left of the bright spot and hence to observe that red correlates with blue (Figure 5-11) and green correlates with yellow (Figure 5-12). In this way a polarity reversal is established.

Figure 5-13 shows a horizon Seiscrop section (for definition and derivation, see Chapter 4) indicating a channel. To the northeast the channel is bright, to the southwest it is not. The structural contours for this horizon have been superimposed and they demonstrate that the bright part of the channel is structurally above the dim part. This combination of structural and strati-

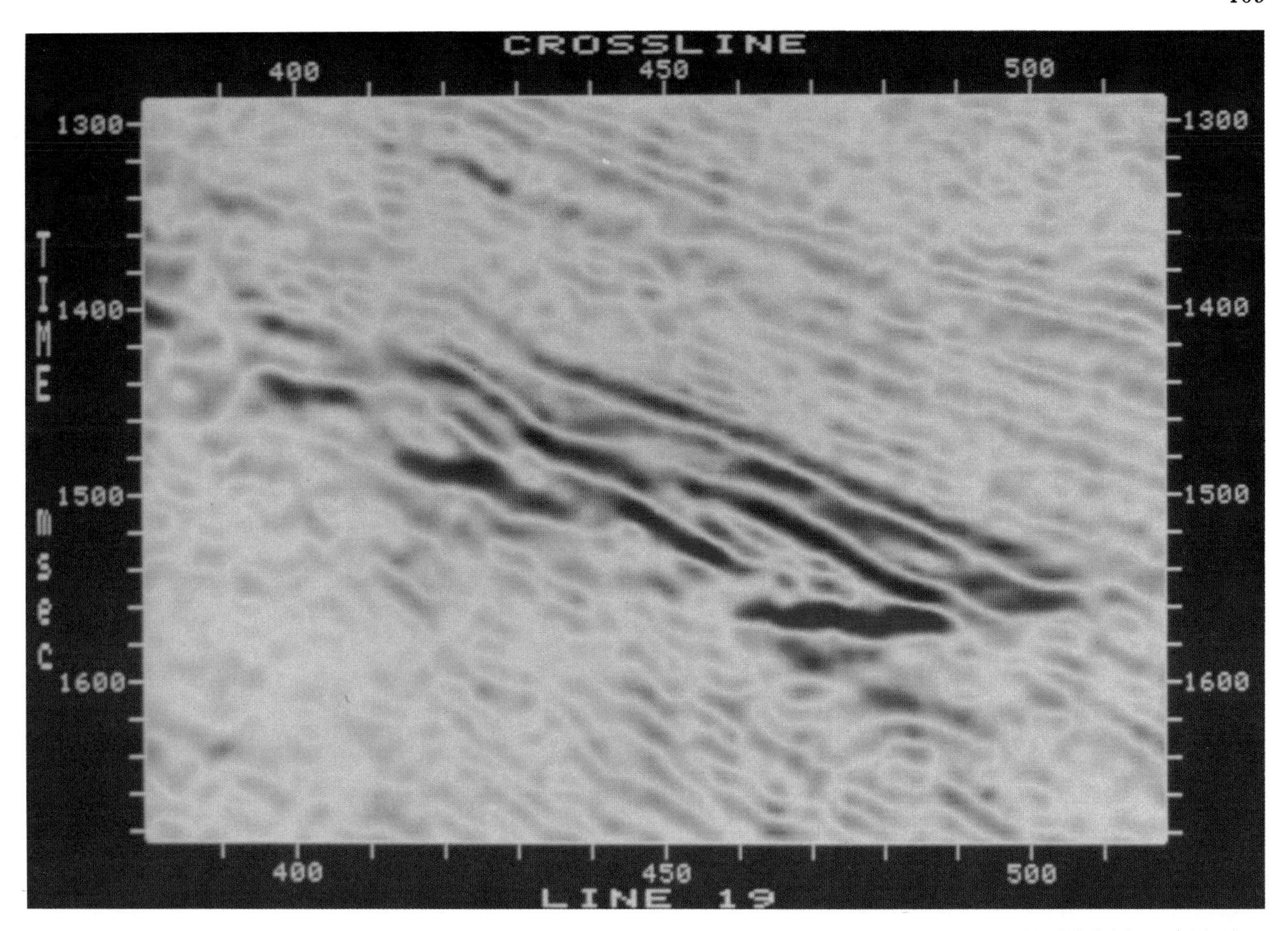

Fig. 5-6. Bright and flat reflections from a Gulf of Mexico gas reservoir known to be subdivided into upper and lower sand units. (Courtesy Chevron U.S.A. Inc.)

graphic information helps validate gas content. Figure 5-14 is another way of graphically illustrating the same relationship; the representation of the channel in amplitude is superimposed on the structural configuration of the horizon surface.

Figure 5-15 demonstrates gas velocity sag on a flat spot reflection. The trough (red event) dipping west between 1560 and 1600 ms should presumably be flat in depth but is depressed in time by the increased travel through the low velocity, wedge-shaped gas sand. Figure 5-16 is another example of gas velocity sag. Here the high amplitudes are still in blue and red but the lower amplitudes are expressed in gradational gray tones. This provides the double benefit of highlighting the bright reflections and also helping establish fault definition by increasing the visibility of low amplitude event terminations. This section also demonstrates another phenomenon: there are bright events *within* the reservoir which have little expression outside. This subject is discussed further in Chapter 7.

Figure 5-17 shows, at the position of the arrow, a bright event known from the well penetrating it to contain gas. By tracking the event and presenting the amplitude as a horizon slice (Figure 5-18) it was possible to relate the spatial distribution of horizon amplitude to gas content. Well 7X penetrated gas at the level of this horizon; well 10X did not.

Use of Frequency, Amplitude Variations With Offset and Shear Waves

Gas reservoirs attenuate high frequencies more than rocks without gas saturation. Following this principle, Taner, Koehler, and Sheriff (1979) have shown that low instantaneous frequency immediately below a suspected reservoir can be a good indicator of gas. The author has found this to be a rather unreliable indicator; several gas reservoirs studied with good data have yielded ambiguous results in instantaneous frequency.

Interval velocity is reduced if a low velocity gas sand is included in the interval studied. For

Continued on page 115

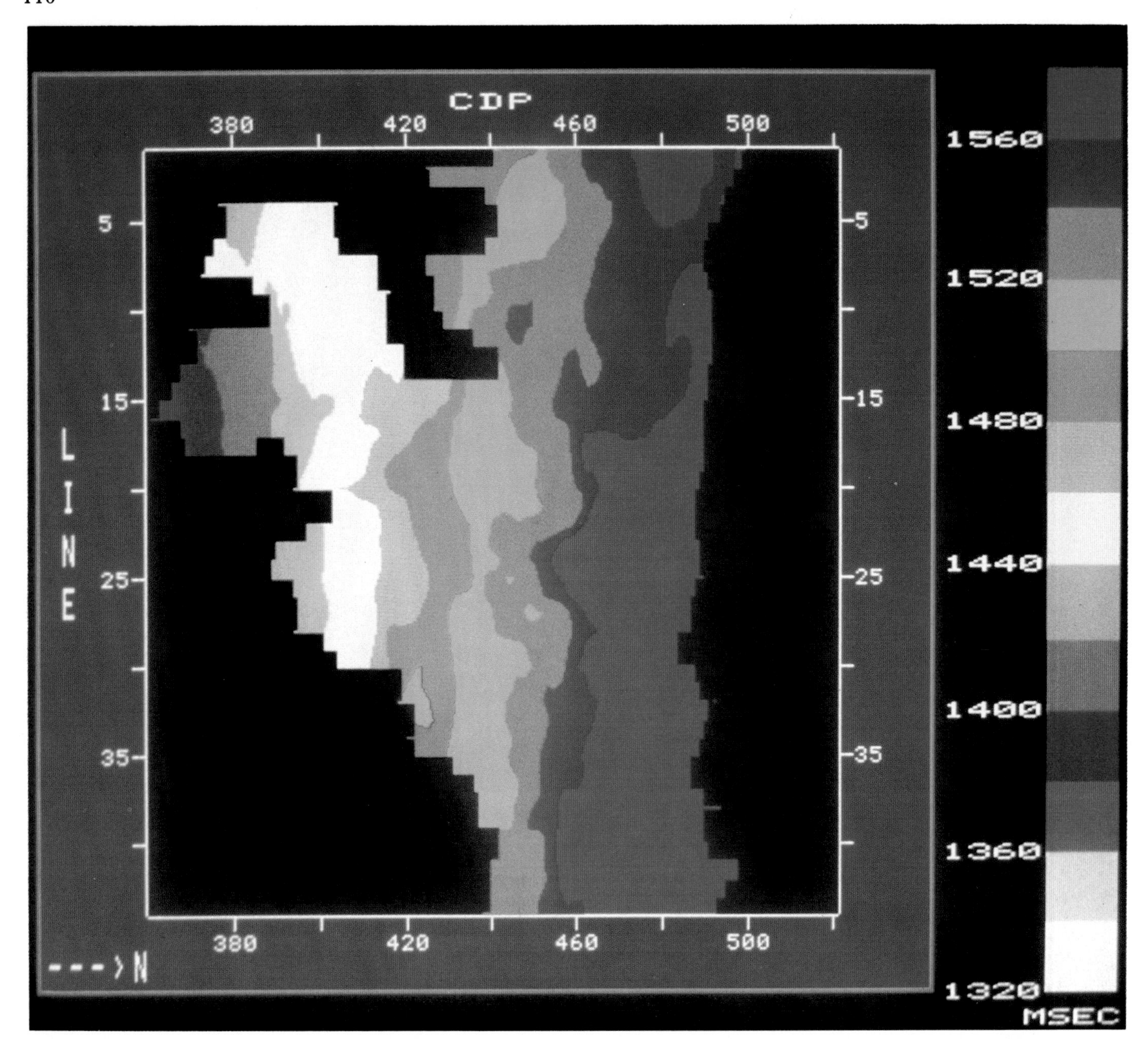

Fig. 5-7. Structure map of the base of the lower gas sand showing the areal extent of the flat spot seen in Figure 5-6. (Courtesy Chevron U.S.A. Inc.)

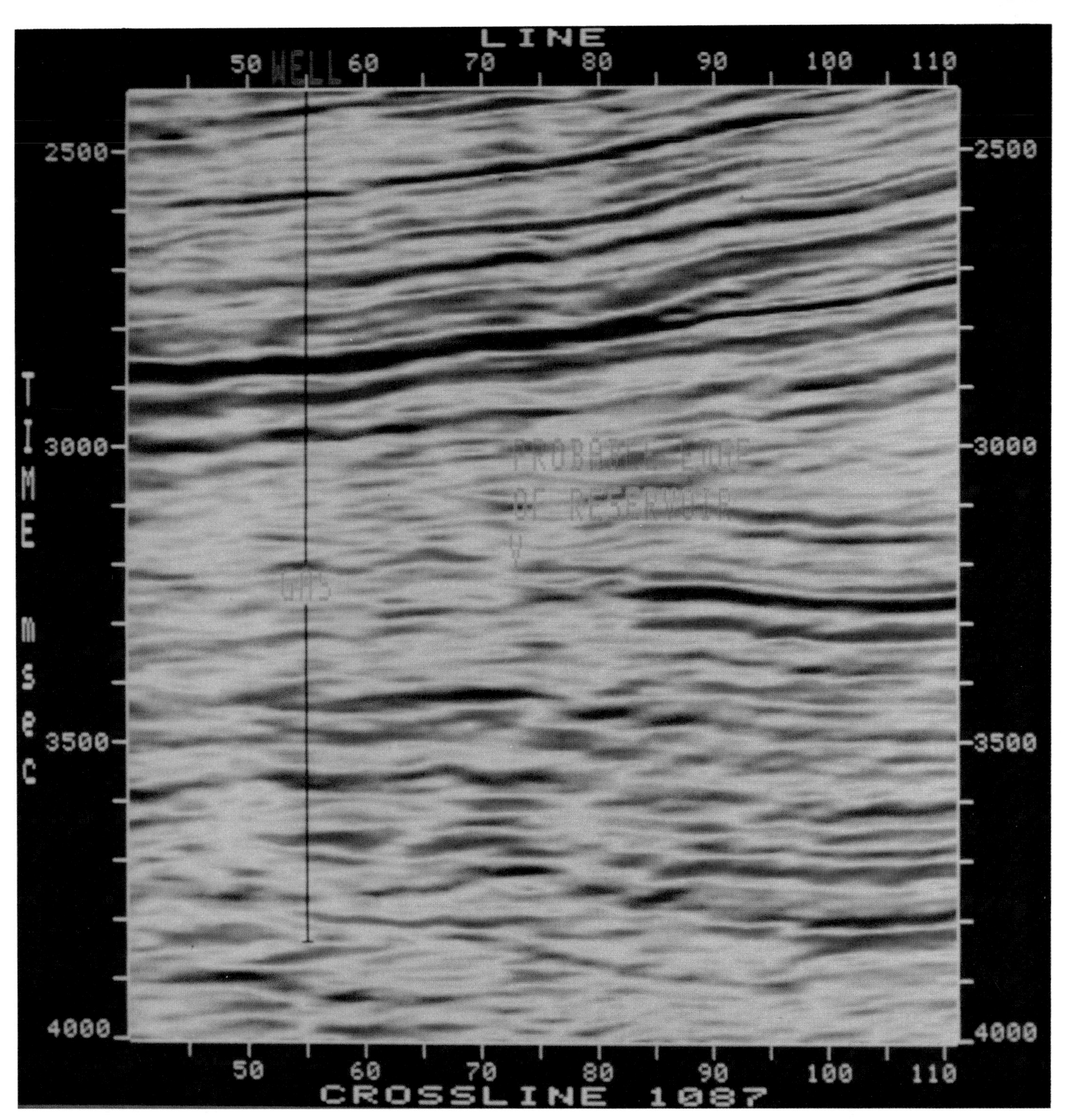

Fig. 5-8. A dim spot from a known gas reservoir offshore Trinidad. (Courtesy Texaco Trinidad Inc.)

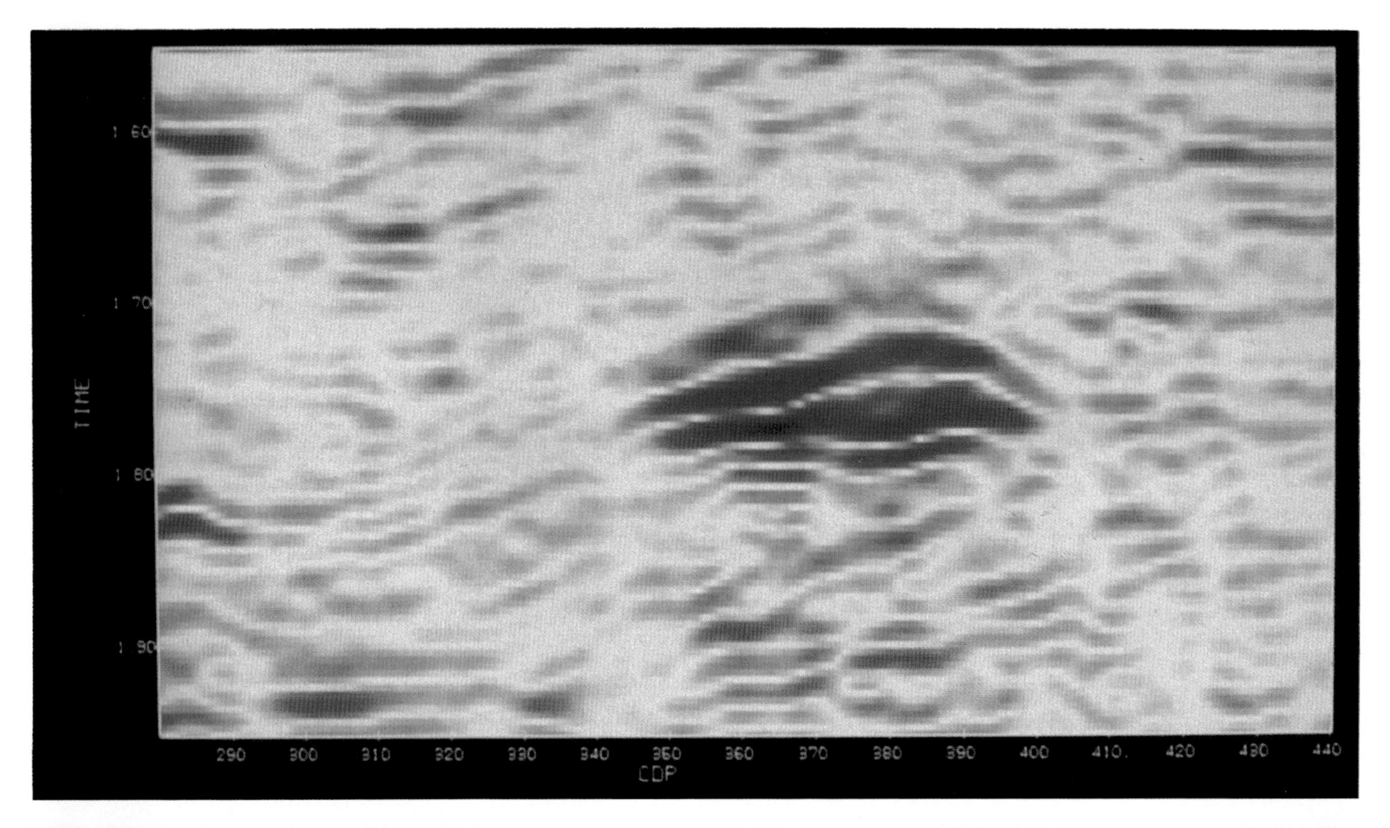

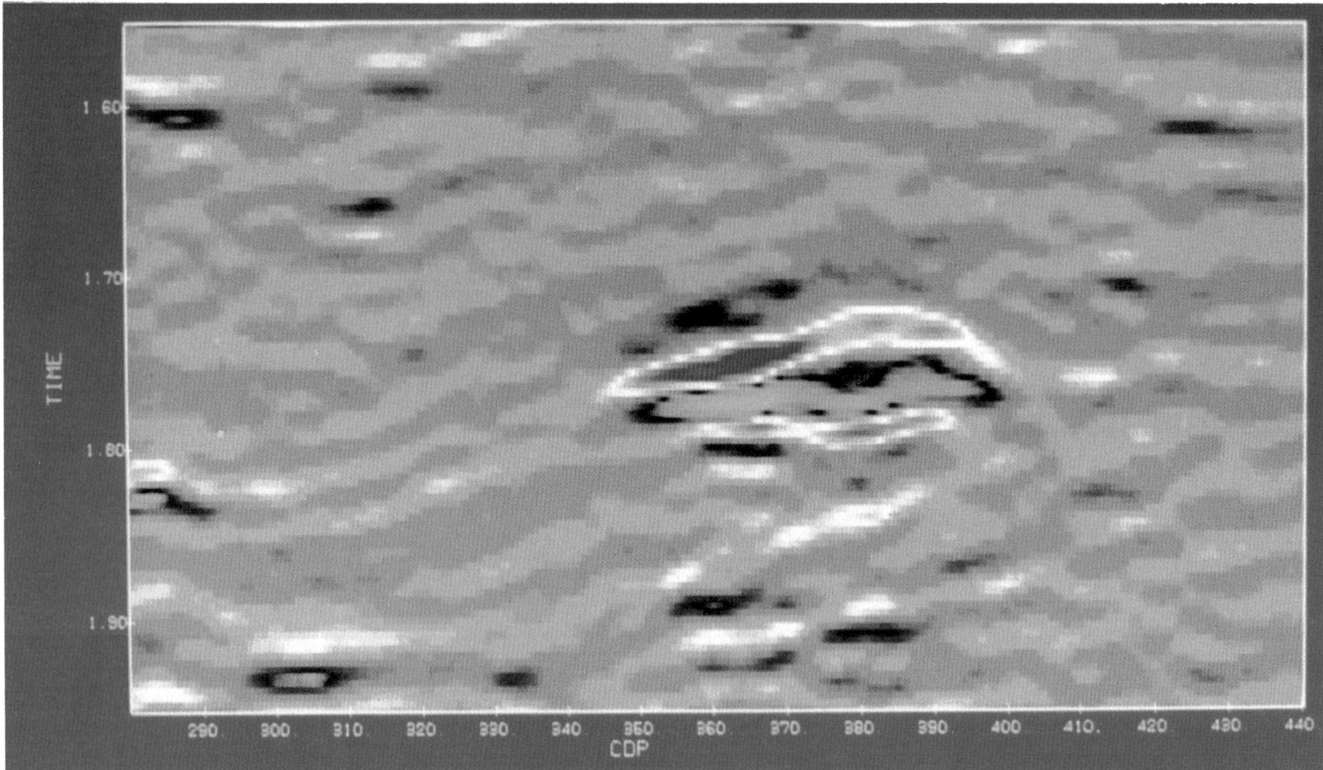

Fig. 5-9. (Top) Gulf of Mexico bright spot displayed in gradational blue and red with the gain set to maximize visual dynamic range and hence increase prominence of the amplitude anomaly. (Courtesy Chevron U.S.A. Inc.)

Fig. 5-10. (Bottom) Same bright spot as Figure 5-9 displayed in yellow, green and gray also in order to increase the prominence of the amplitude anomaly. (Courtesy Chevron U.S.A. Inc.)

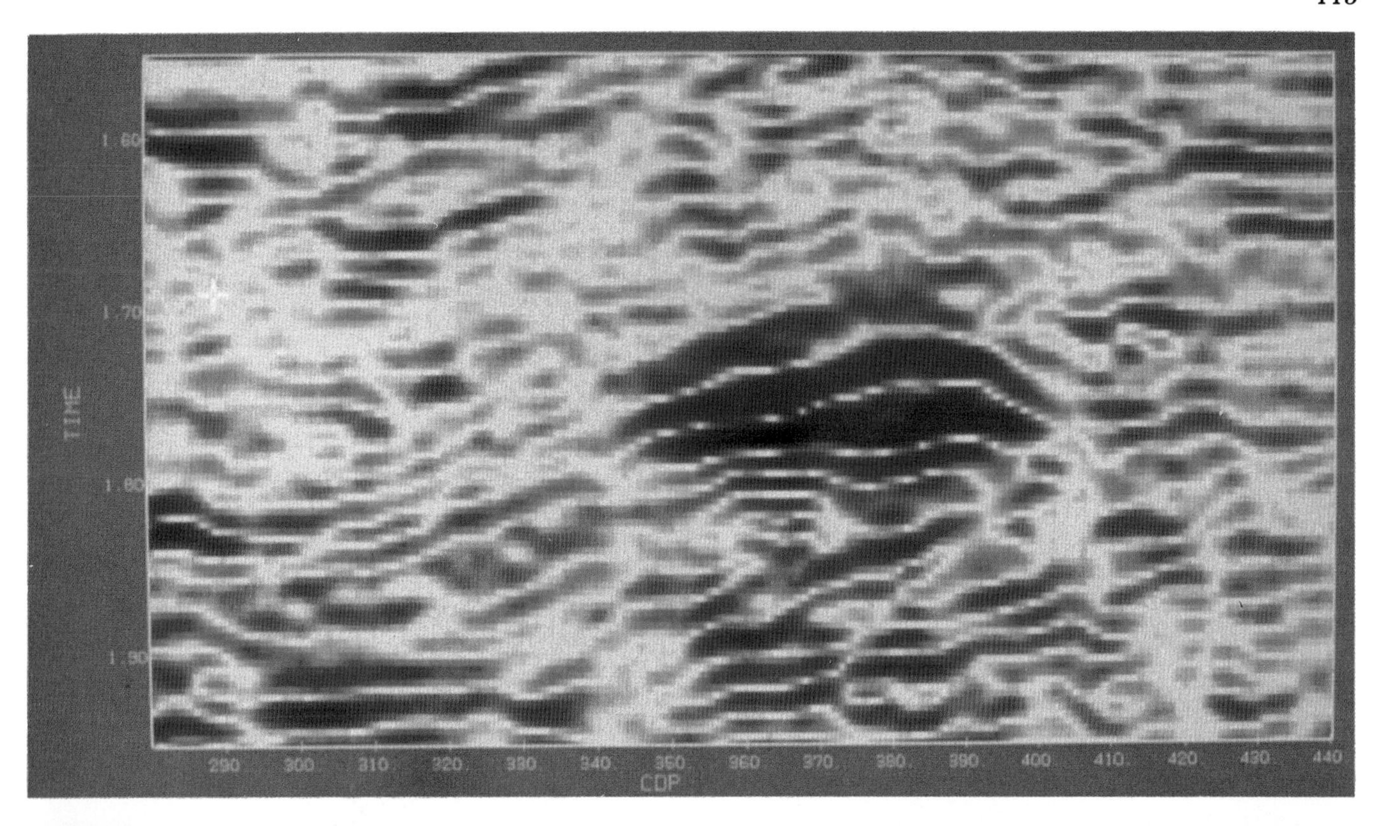

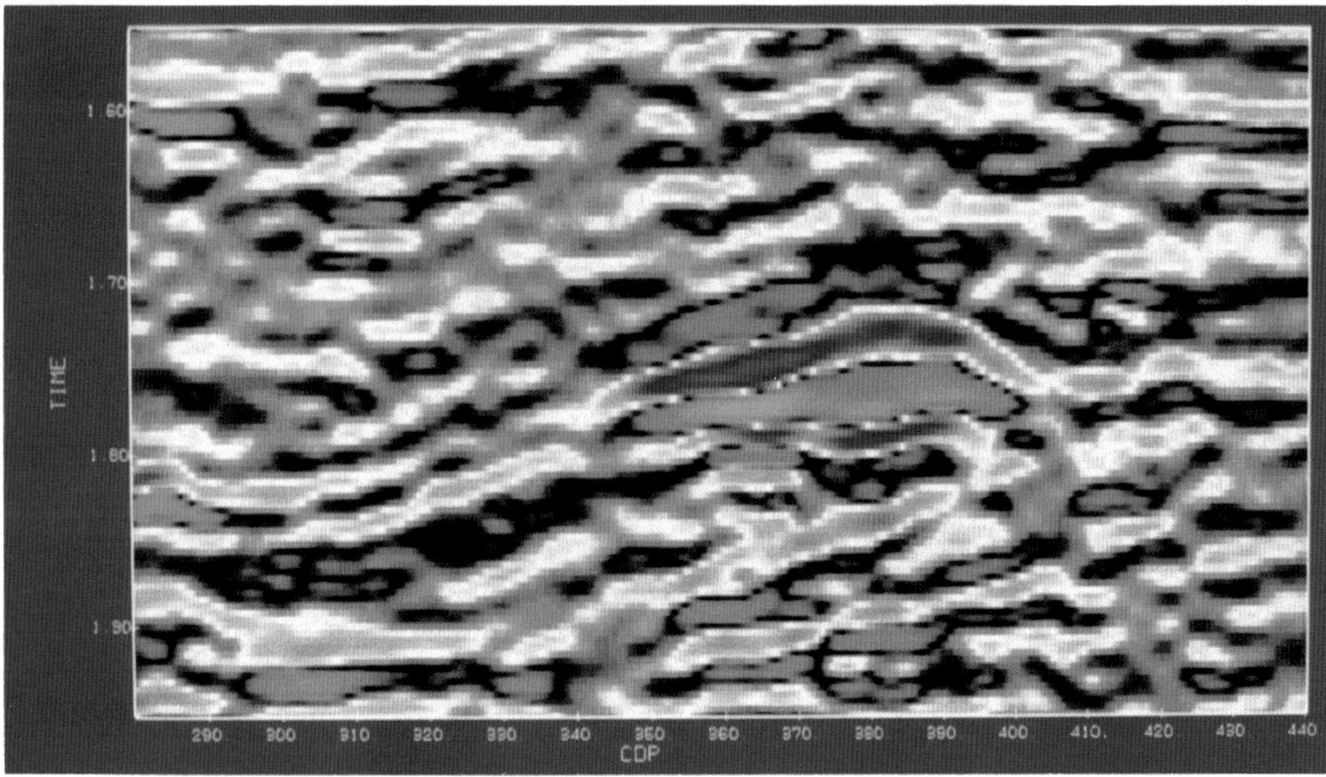

Fig. 5-11. (Top) Same bright spot and color scheme as Figure 5-9 but with the gain increased to study the continuity of reflections off the flank of the bright spot. Blue correlates with red and vice versa downdip indicating a phase change or polarity reversal at the edge of the bright spot. (Courtesy Chevron U.S.A Inc.)

Fig. 5-12. (Bottom) Same bright spot as Figure 5-9, same color scheme as Figure 5-10 and same gain as Figure 5-11. The correlation of reflections downdip from the bright spot again indicates a phase change at the edge of the reservoir. (Courtesy Chevron U.S.A. Inc.)

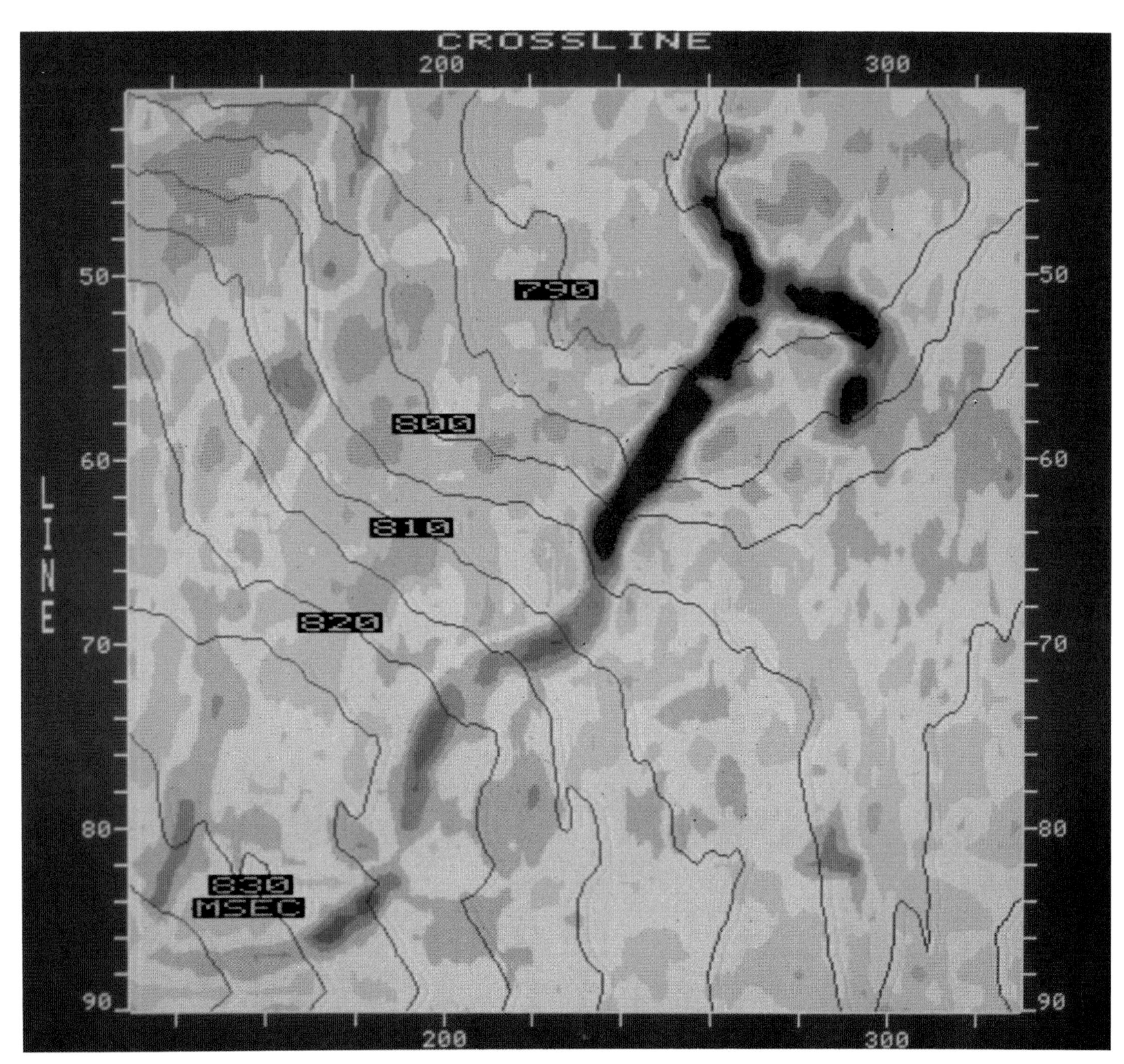

Fig. 5-13. Horizon Seiscrop section showing Gulf of Mexico channel discussed in Chapter 4. The superimposed structural contours indicate that the bright part of the channel is shallower than the dim part. (Courtesy Chevron U.S.A Inc.)

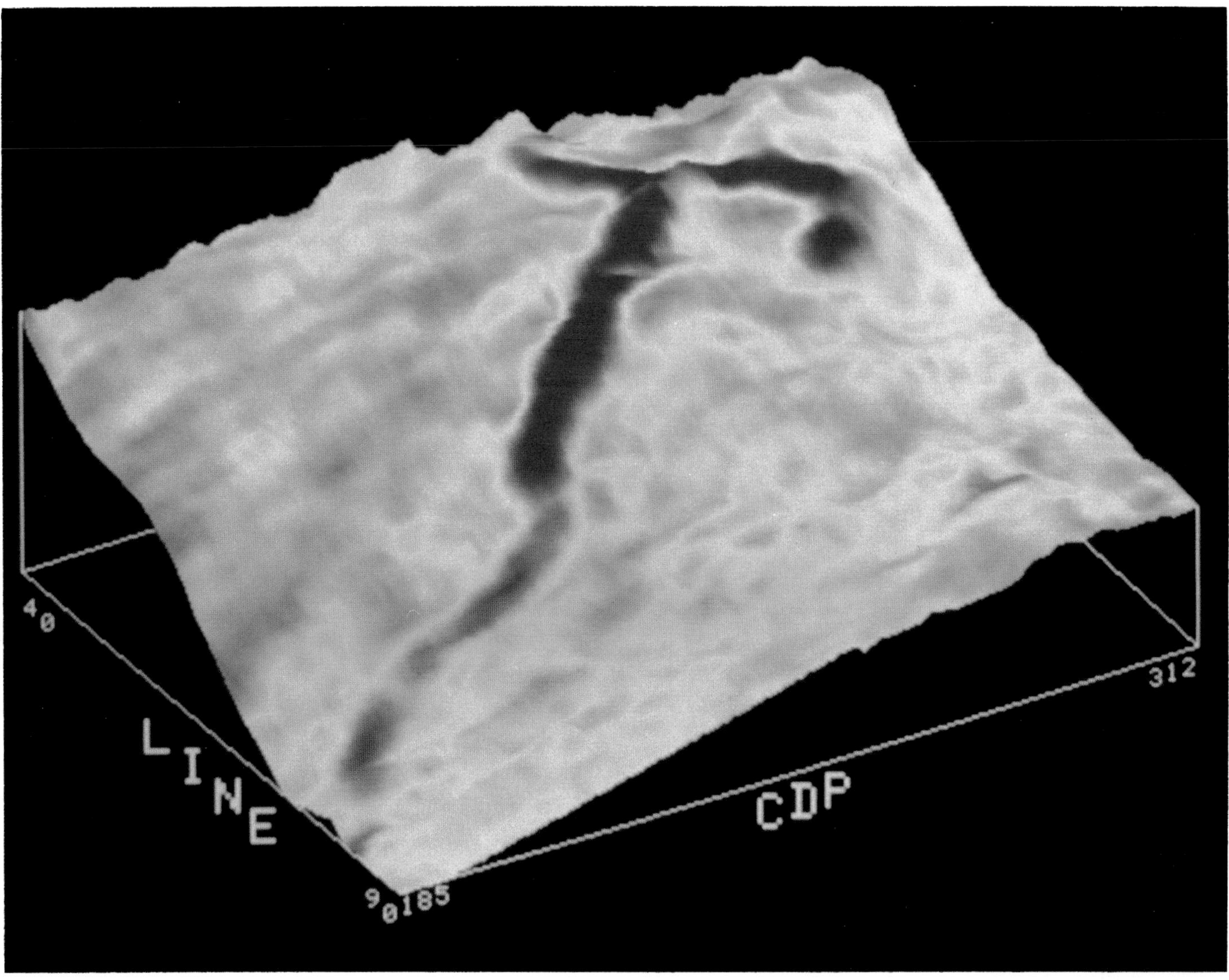

Fig. 5-14. Combination of the same horizon amplitude and structural information as Figure 5-13 using different colors and a three-dimensional perspective surface. (Courtesy Chevron U.S.A. Inc.)

many years RMS velocities derived from normal moveout have been used to compute interval velocities, and for gross effects and trends this is valuable. However, the stability of interval velocities gets progressively worse for greater depths and also for thinner beds. This generally means that interval velocities are not sufficiently accurate to play a useful role in bright spot validation.

The variation of amplitude with recording offset has recently become a popular subject because of the possibility of extracting a significant amount of lithologic information from this kind of data. However, there are many difficulties both of a theoretical and practical nature (Backus and Goins, 1984). Among the practical issues, the data are prestack and hence have a lower signal-to-noise ratio, and, being multidimensional, there are many possible modes of display.

Ostrander (1984) demonstrated that in many practical cases gas sands show an increase of amplitude with offset and that this can be used as a means of validating gas bright spots. He studied the data in the form of common-depth-point gathers, normally stacking together common and adjacent offsets to improve signal-to-noise ratio. Common-depth-point gathers corrected for normal moveout but without any stacking are shown in Figure 5-19. The somewhat bright reflections at and below the black arrow are from a known Gulf of Mexico gas sand. Increase of amplitude with offset is just visible.

The application of the horizon slice concept has significantly increased the visibility of amplitude/offset effects for one horizon. Consider a volume of one line of prestack seismic data

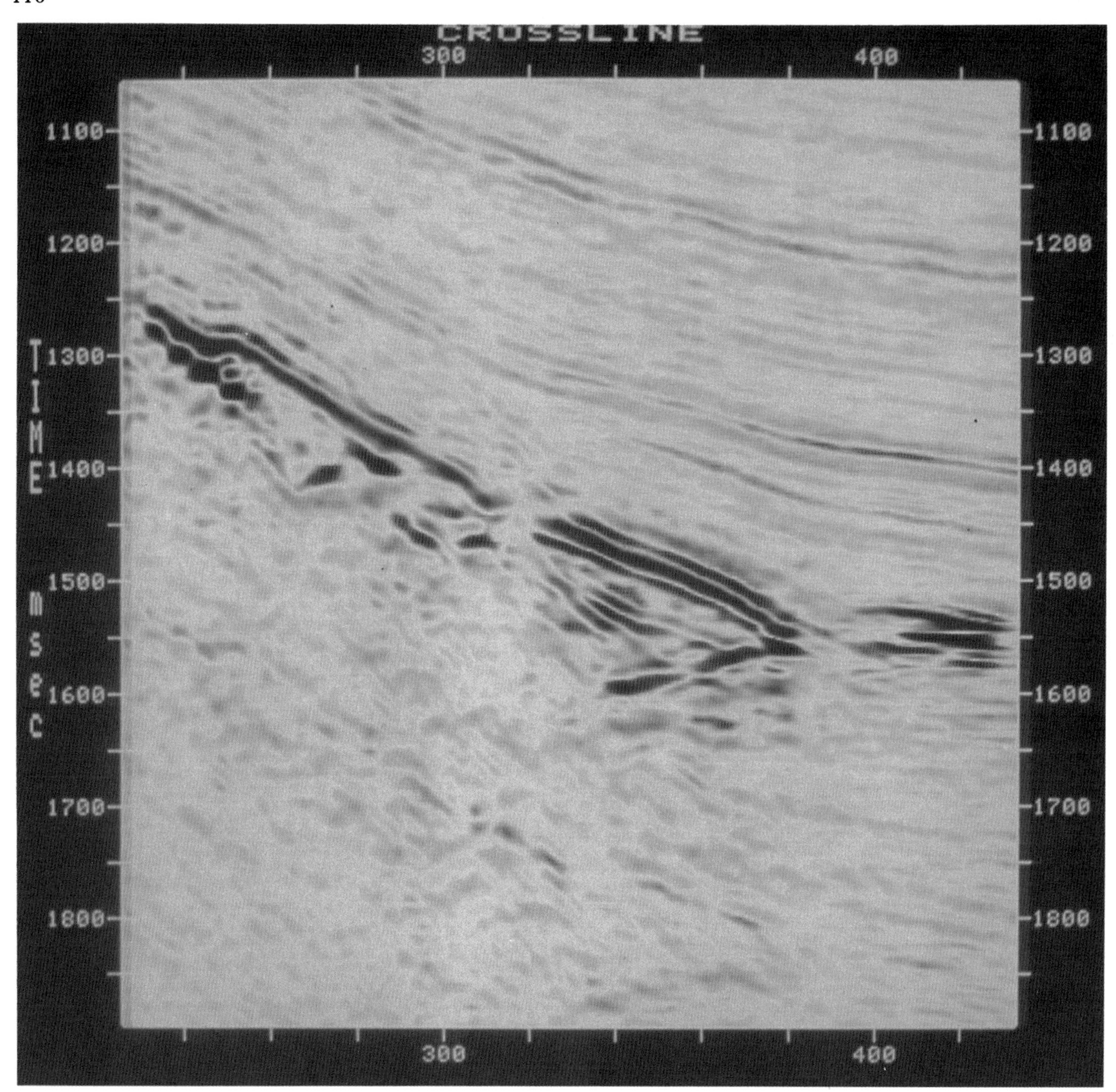

Fig. 5-15. Bright spot from a rather thick and complex gas sand. The red event dipping right-to-left is a flat spot displaying gas velocity sag. (Courtesy Chevron U.S.A Inc.)

(Figure 5-20). The three dimensions are (1) CDP position along the line, (2) traveltime and (3) recording offset. The shape of one reflection without normal moveout corrections is a cylindrical hyperbola as shown. By tracking this horizon and displaying the resultant amplitudes as a horizon slice, a **horizon offset section** is obtained.

A horizon offset section prepared in this way is shown in Figure 5-21. The variation in amplitude with CDP position and with offset (approximately converted to incident angle) is shown for the trough immediately below the black arrow in Figure 5-19. The horizon offset section has been spatially smoothed, as an alternative to partial stacking, for increase of signal-to-noise ratio. The interpreter can observe, on this one section, the variation of amplitude with offset over many depth points for this horizon of interest. The amplitude increases with offset for most of the depth points and is hence consistent with gas content.

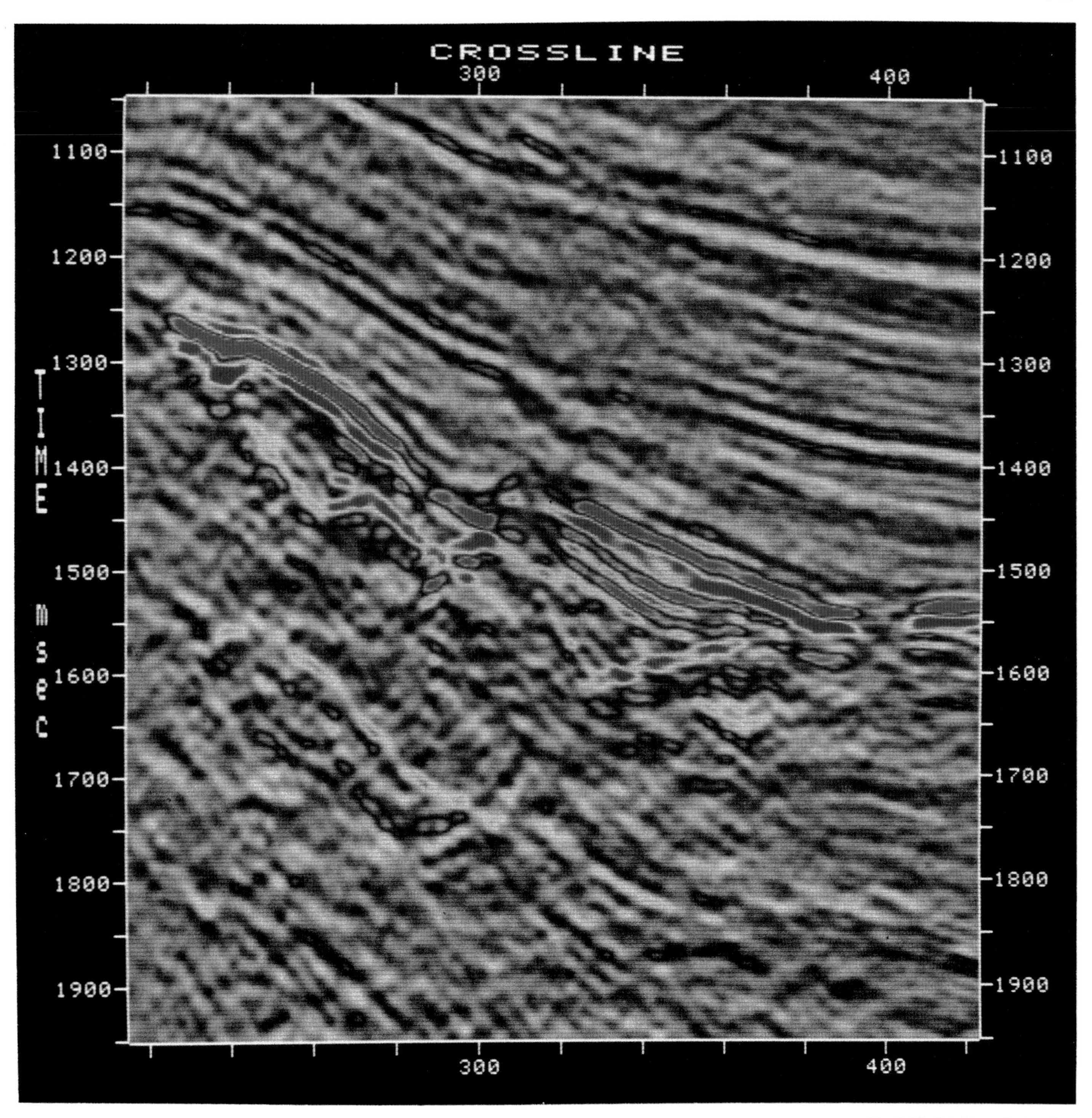

Fig. 5-16. Bright spot showing similar phenomena to Figure 5-15; the lower amplitudes are here displayed in gray tones. (Courtesy Chevron U.S.A. Inc.)

This method of validation requires that the gas sand and the embedding medium have very different Poisson's ratios. Because this is not always the case this method lacks certainty, even on theoretical grounds. On practical grounds poor signal-to-noise ratio is a common problem. Chiburis (1984) used amplitude ratio between the target horizon and a reference horizon and, because he used 3-D data, extensive smoothing was possible. In this way he delineated a gas cap from the amplitude/offset data and found reasonable agreement with engineering results.

Onstott, Backus, Wilson, and Phillips (1984) have used color in a novel way to study amplitude/offset information. They made vertical section substacks of near, mid and far traces and assigned to each of these one of the additive primary colors — red, green and blue. After

Continued on page 122

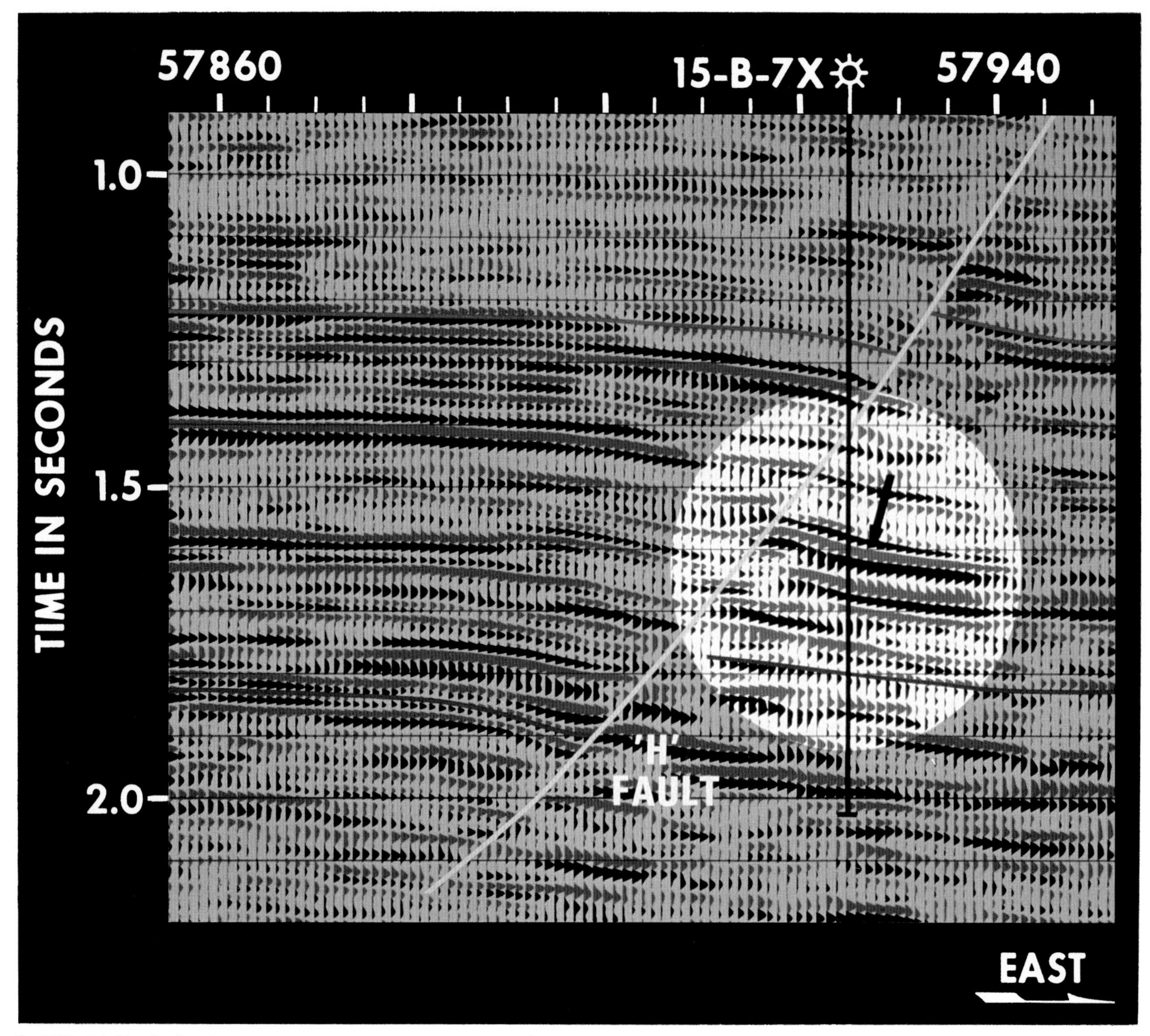

Fig. 5-17. Bright reflection in upthrown fault block known to result from gas. (Courtesy Texas Pacific Oil Company Inc.)

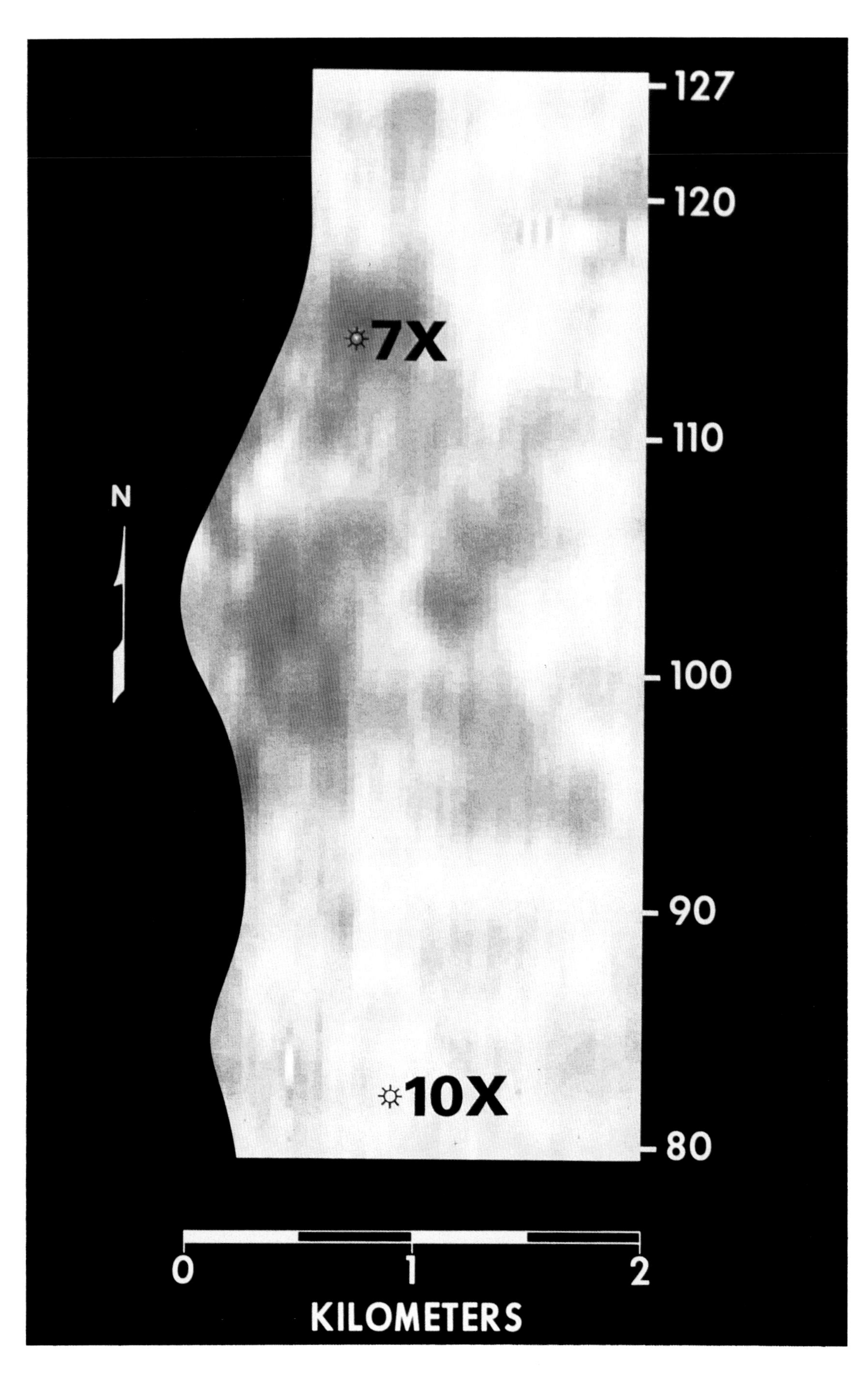

Fig. 5-18. Horizon Seiscrop section for event indicated on Figure 5-17 showing spatial extent of the high amplitude associated with gas. (Courtesy Texas Pacific Oil Company Inc.)

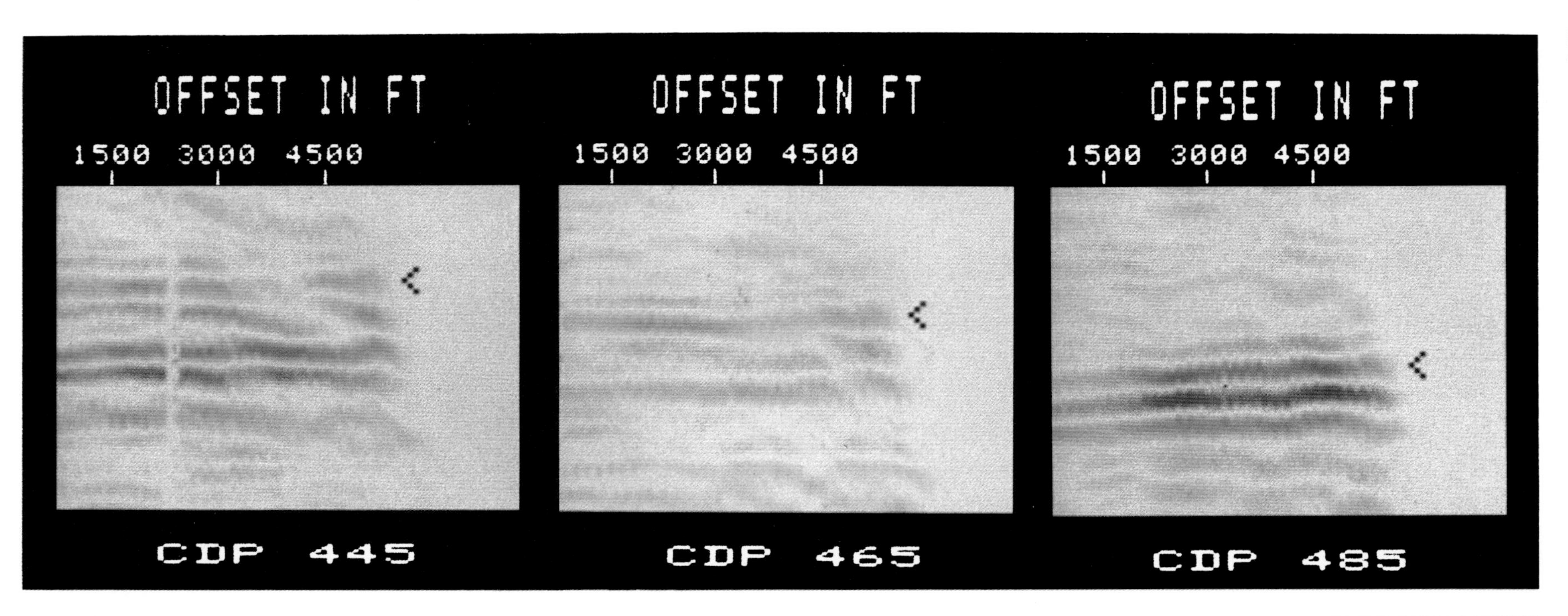

Fig. 5-19. Common-depth-point gathers corrected for moveout showing variation of amplitude with offset for several horizons.

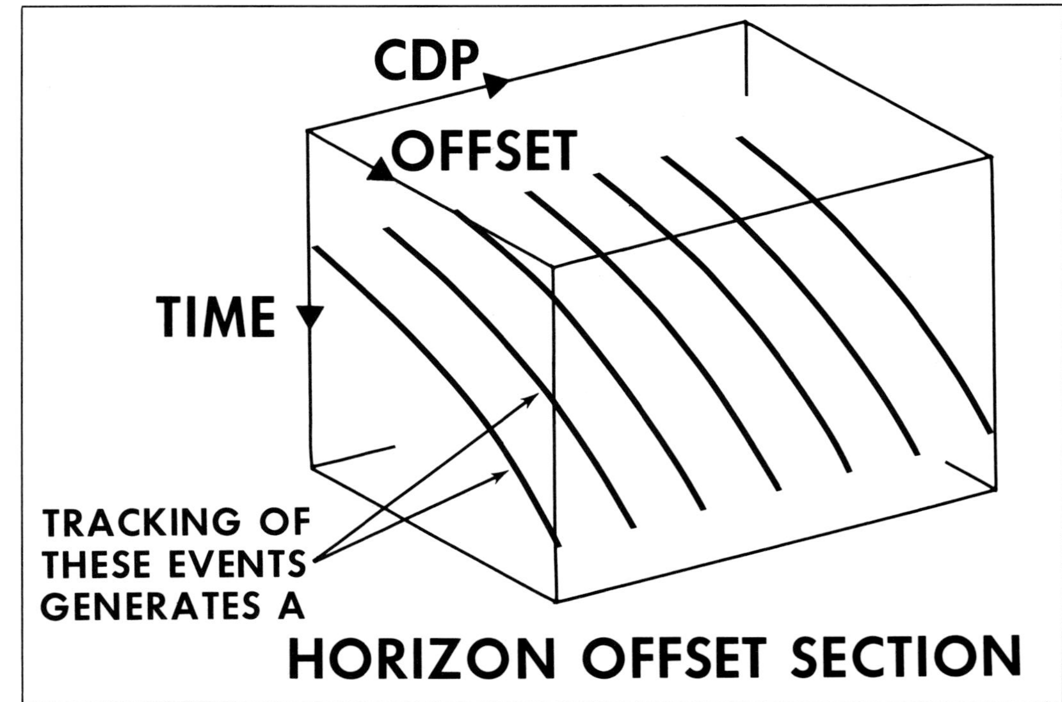

Fig. 5-20. The concept of a horizon offset section generated by tracking a horizon on a sequence of common-depth-point gathers.

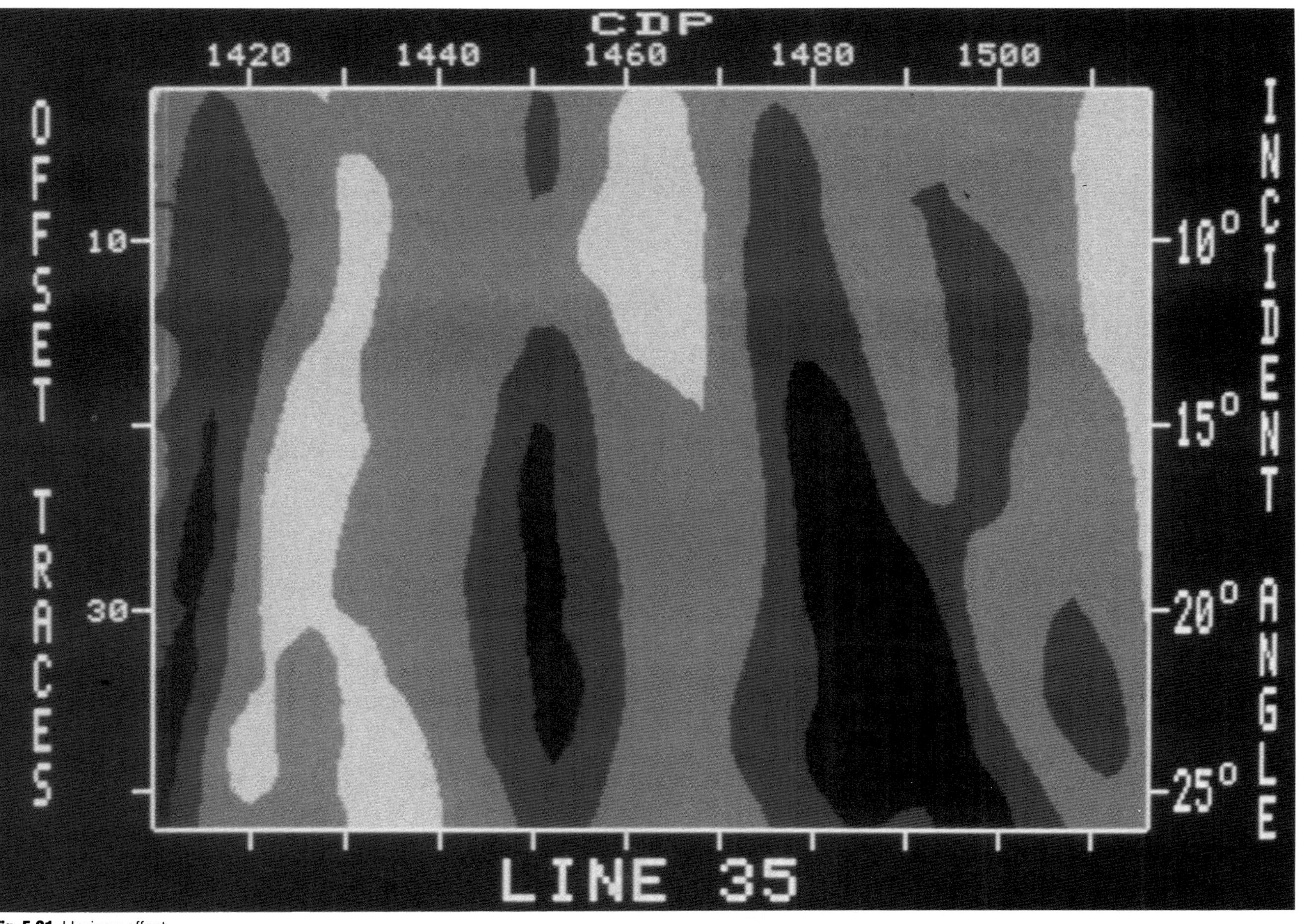

Fig. 5-21. Horizon offset section for a Gulf of Mexico bright spot showing increase of amplitude with offset for most depth points.

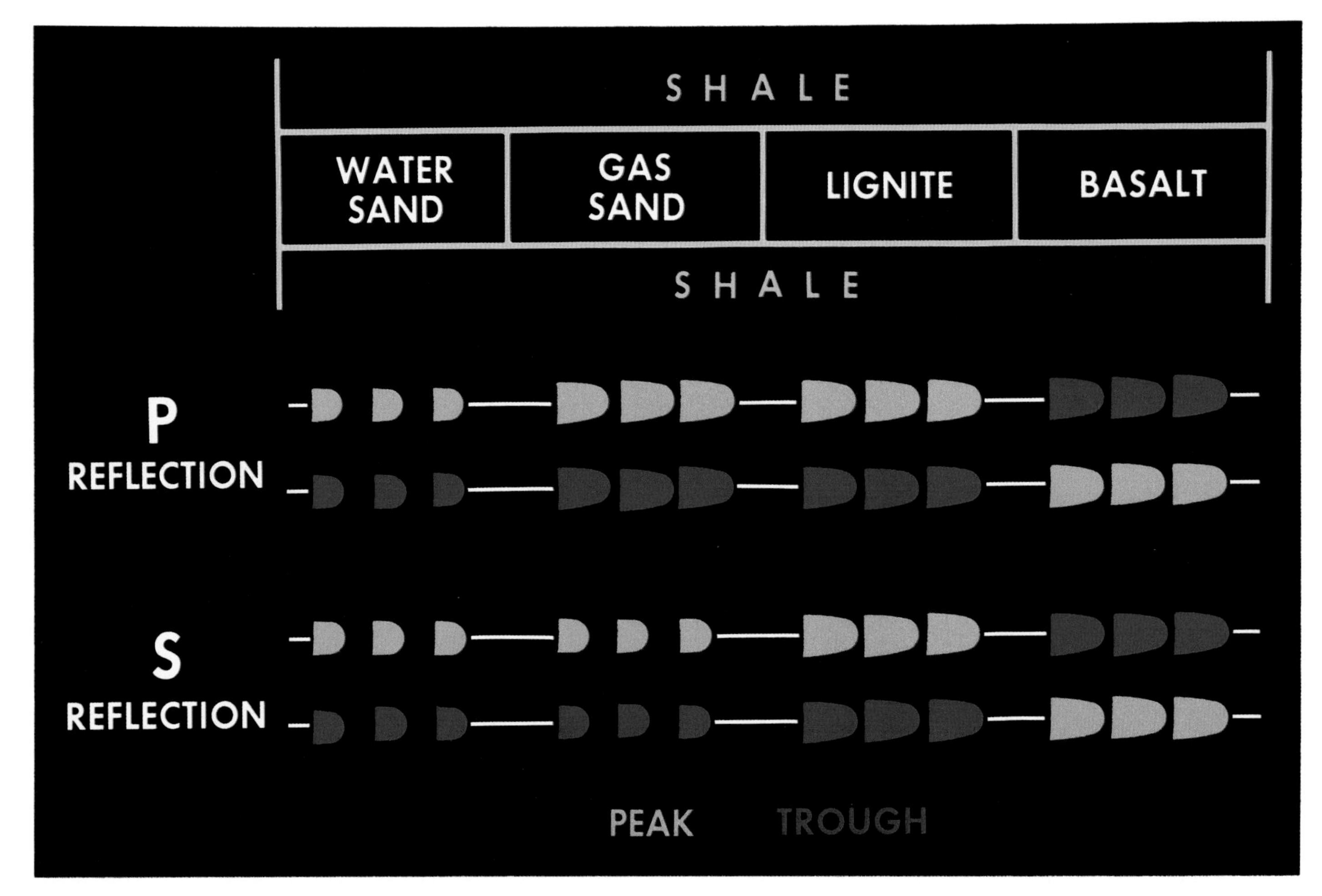

Fig. 5-22. Schematic diagram of the P-wave and S-wave zero-phase response for different beds encased in shale.

combining these colored sections, the final section carried amplitude/offset information encoded in the color of every event.

Interpretation of shear wave amplitudes in conjunction with conventional compressional wave amplitudes can provide a powerful method of bright spot validation. On land, S-wave data have generally been collected in a separate operation. S-waves are not transmitted through water so, at sea, it is necessary to use PSSP waves, mode converted at the water bottom. Tatham and Goolsbee (1984) have separated mode-converted S-wave data from P-wave data collected where the water bottom was hard.

Figure 5-22 summarizes the response of a water sand, a gas sand, a lignite bed, and a basalt bed to P- and S-wave energy; it should be studied in conjunction with Figure 5-5. Lignite has very low velocity and can be confused with a gas sand on the basis of P-wave response alone. Basalt, although high velocity, may also show a similar response if the polarity and phase of the data are not well understood.

The diagnostic comparison between P- and S-wave sections for a reflection from a gas sand is the presence of a P-wave bright spot and the absence of an amplitude anomaly for the correlative S-wave event. Figures 5-23 and 5-24 show comparative P-wave and S-wave sections reprinted from Robertson and Pritchett (1985). The 3rd Starkey reflection is bright on the P-section but not on the S-section, indicating gas.

In fact, the fundamental underlying principle is that compressional waves are sensitive to the type of pore fluid within rocks, whereas shear waves are only slightly affected. Hence the S-wave response of a reservoir sand will change little from below to above the gas/water contact, while the P-wave response normally changes greatly. Referring to Figure 5-5, it is clear that the P-wave dim spot would correlate on an S-section with a higher amplitude reflection. Where a phase change occurs across the gas/water contact on the P-section, the correlative P-wave and S-wave reflections from the gas sand will have opposite polarity. This is the situation interpreted by Ensley (1984).

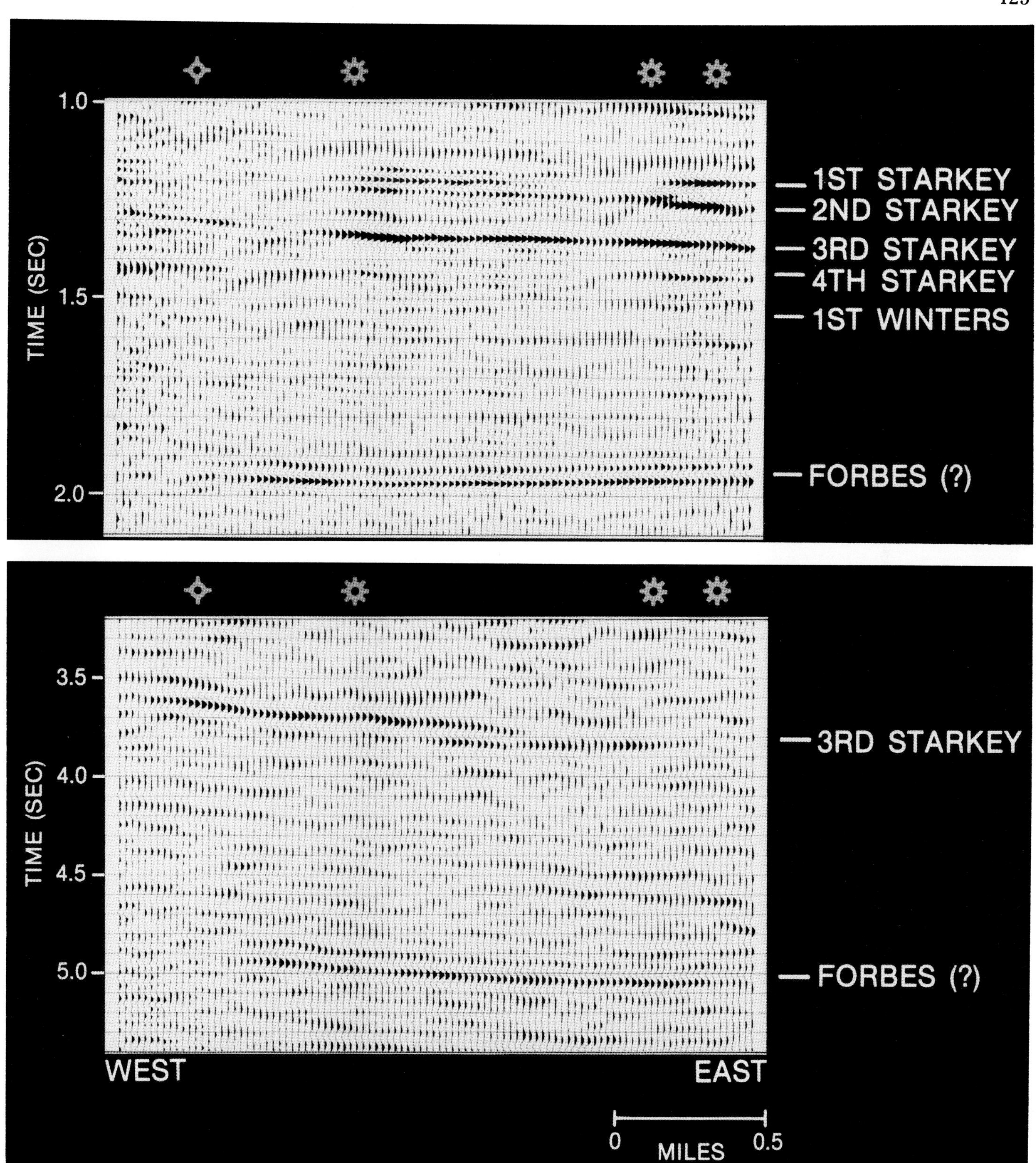

Fig. 5-23. (Top) P-wave section over California bright spot (from Robertson and Pritchett, 1985).

Fig 5-24. (Bottom) S-wave section from same location as Figure 5-23 showing no amplitude anomaly for correlative event (from Robertson and Pritchett, 1985).

Questions an Interpreter Should Ask in an Attempt to Validate Hydrocarbon Indicators

(1) Is the reflection from the suspected reservoir anomalous in amplitude, probably bright?
(2) Is the amplitude anomaly structurally consistent?
(3) If bright, is there one reflection from the top of the reservoir and one from the base?
(4) Do the amplitudes of the top and base reflections vary in unison, dimming at the same point at the limit of the reservoir?
(5) Are the data zero phase?
(6) Is a flat spot visible?
(7) Is the flat spot flat or dipping consistent with gas velocity sag?
(8) Is the flat spot unconformable with the structure but consistent with it?
(9) Does the flat spot have the correct zero-phase character?
(10) Is the flat spot located at the downdip limit of brightness (or dimness)?
(11) Is a phase change visible?
(12) Is the phase change structurally consistent and at the same level as the flat spot?
(13) Do bright spot, dim spot, or phase change show the appropriate zero-phase character?
(14) Is there a low-frequency shadow below the suspected reservoir?
(15) Is there an anomaly in moveout-derived interval velocity?
(16) Is a study of amplitude versus offset on the unstacked data likely to yield further validation evidence?
(17) Are any shear wave data available for further validation evidence?

In practice, any one indication can be spurious. Hydrocarbon validation on seismic data necessarily involves accumulation of evidence. The more questions to which you can answer "yes", the greater is your confidence in the identification of hydrocarbons.

Figure 5-25 contains several suspected hydrocarbon indicators. Try asking the above questions for this data. You should find affirmative answers for questions 1 through 14 and conclude there are three separate reservoirs.

Figure 5-26 labels most of the hydrocarbon indicators on this section. A horizon tracked above the hydrocarbons has been redrawn twice lower on the section to draw attention, firstly, to the phase change associated with the uppermost reservoir and, secondly, to the gas sag caused by the total accumulation of hydrocarbons. Figure 5-27 presents the same section with a different color scheme; gradational cyan and yellow are used for the highest levels of amplitude leaving the gradational blue and red to represent only the moderate and lower amplitudes.

References

Backus, M. M., and R. L. Chen, 1975, Flat spot exploration: Geophysical Prospecting, v. 23, p. 533-577.

——— and N. Goins, 1984, Change in reflectivity with offset, Research Workshop report: Geophysics, v. 49, p. 838-839.

Chiburis, E. F., 1984, Analysis of amplitude versus offset to detect gas/oil contacts in the Arabian Gulf: Proceedings, SEG 54th Annual Meeting, p. 669-670.

Ensley, R. A., 1984, Comparison of P- and S-wave seismic data: a new method for detecting gas reservoirs: Geophysics, v. 49, p. 1420-1431.

Gardner, G. H. F., L. W. Gardner, and A. R. Gregory, 1974, Formation velocity and density–the diagnostic basics for stratigraphic traps: Geophysics, v. 39, p. 770-780.

Onstott, G. E., M. M. Backus, C. R. Wilson, and J. D. Phillips, 1984, Color display of offset dependent reflectivity in seismic data: Proceedings, SEG 54th Annual Meeting, p. 674-675.

Ostrander, W. J., 1984, Plane-wave reflection coefficients for gas sands at non-normal angles of incidence: Geophysics, v. 49, p. 1637-1648.

Robertson, J. D., and W. C. Pritchett, 1985, Direct hydrocarbon detection using comparative P-wave and S-wave seismic sections: Geophysics, v. 50, p. 383-393.

Taner, M. T., F. Koehler, and R. E. Sheriff, 1979, Complex seismic trace analysis: Geophysics, v. 44, p. 1041-1063.

Tatham, R. H., and D. V. Goolsbee, 1984, Separation of S-wave and P-wave reflections offshore western Florida: Geophysics, v. 49, p. 493-508.

Tegland, E. R., 1973, Utilization of computer-derived seismic parameters in direct hydrocarbon exploration and development, *in* Lithology and direct detection of hydrocarbons using geophysical methods: Dallas Geophysical and Geological Societies symposium.

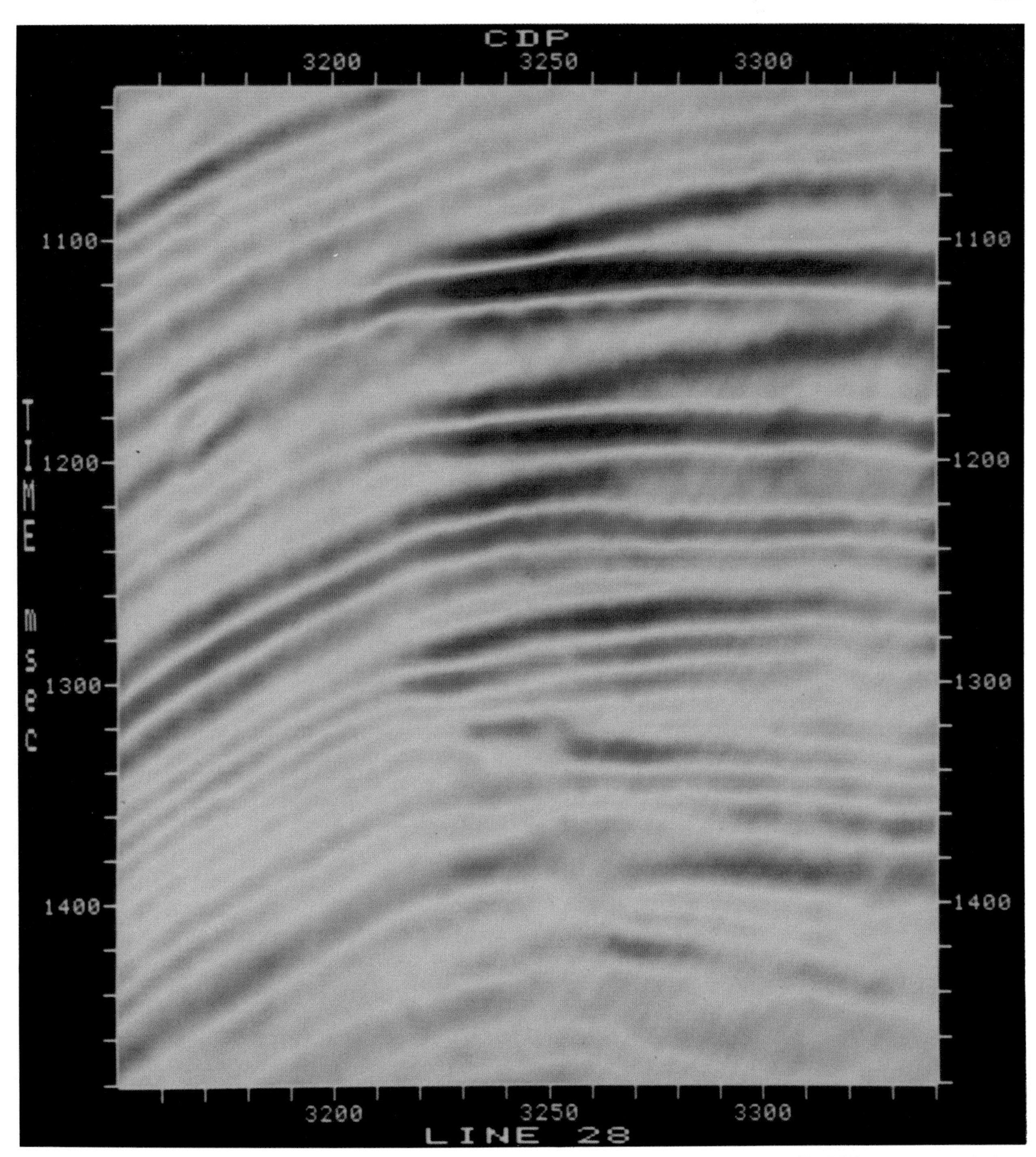

Fig. 5-25. Hydrocarbon indicators offshore California. It is intended that the reader interrogates this section with the "Questions an interpreter should ask in an attempt to validate hydrocarbon indicators".

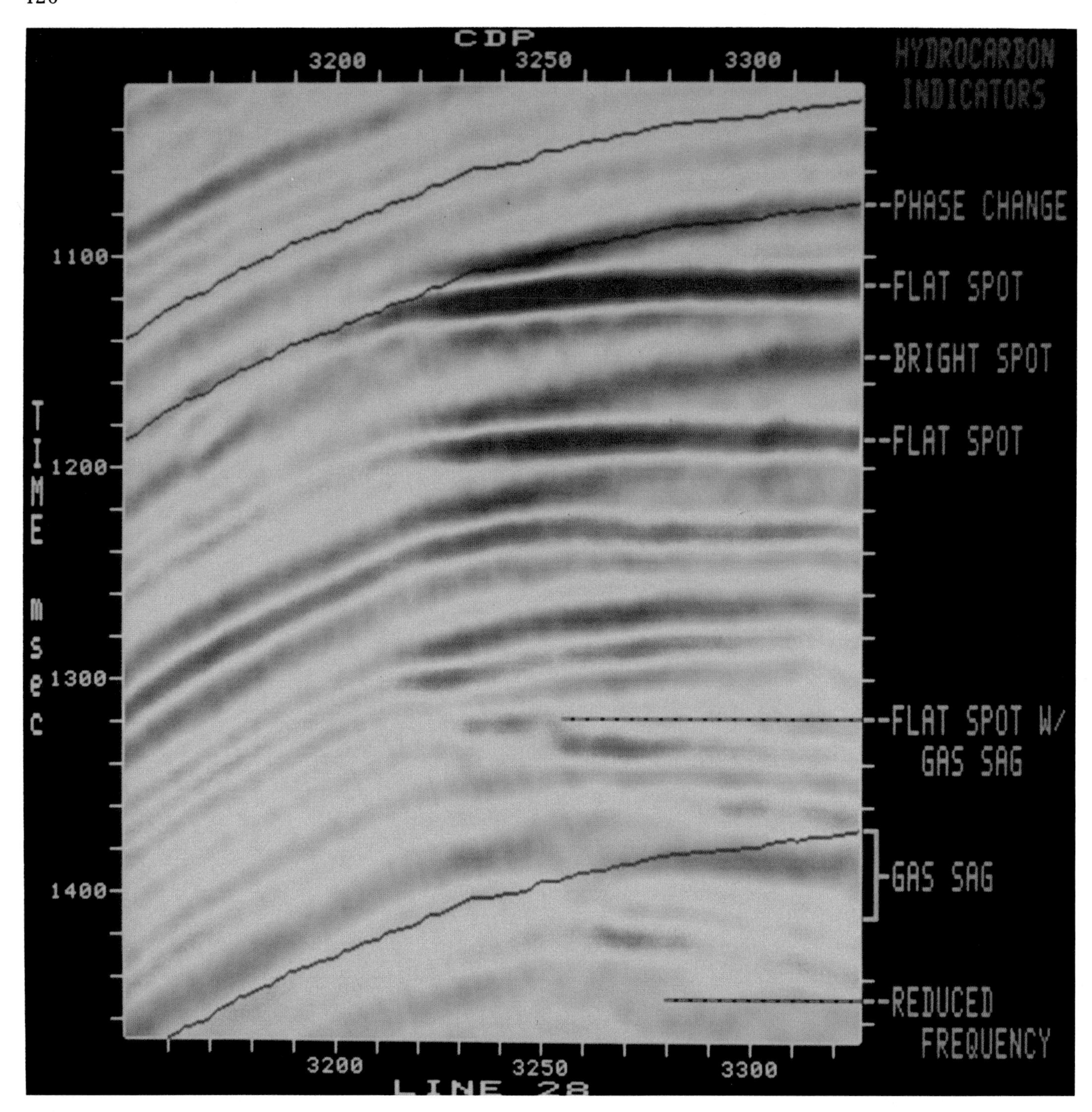

Fig. 5-26. Hydrocarbon indicators offshore California with most of the indicators labeled.

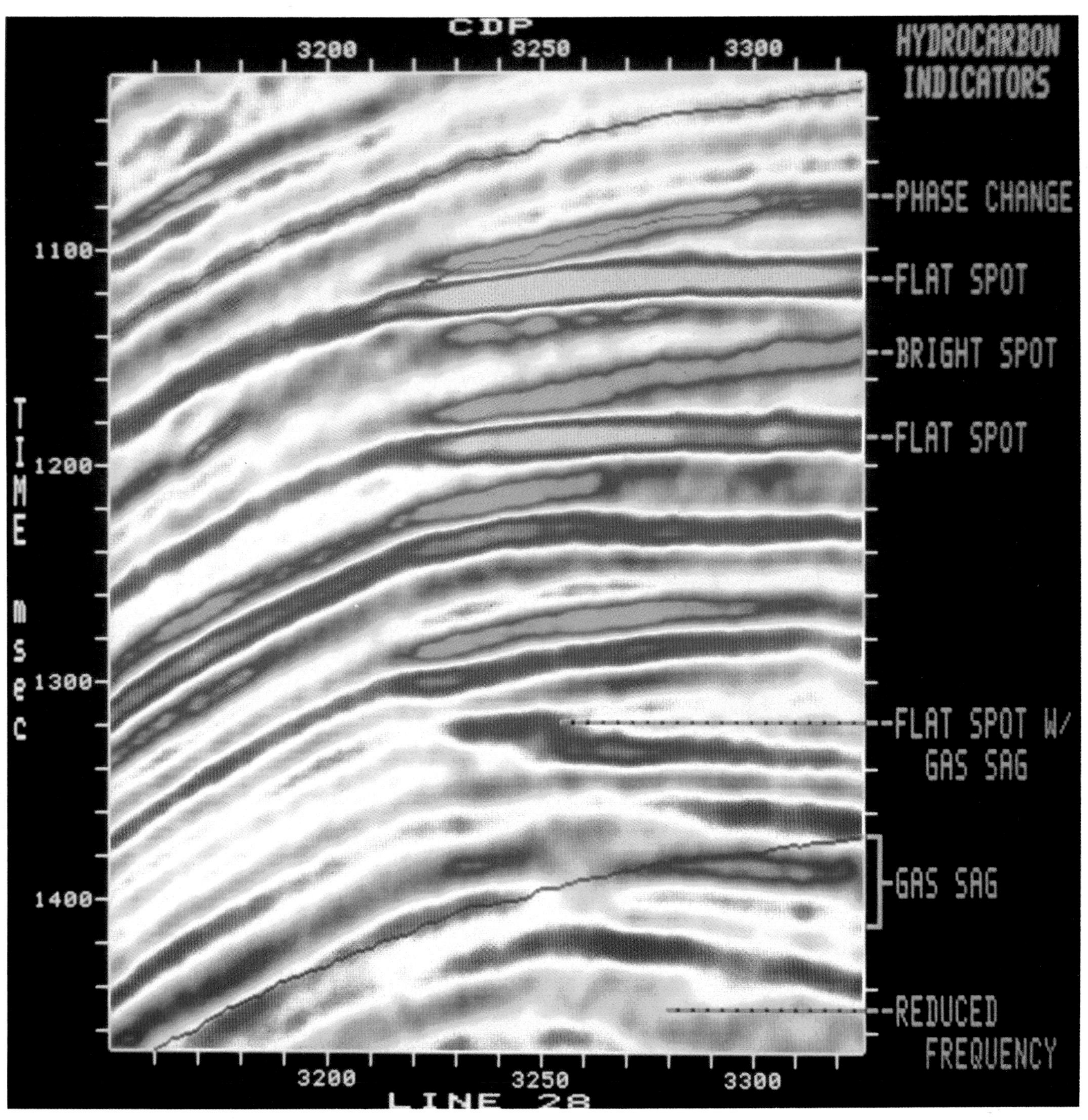

Fig. 5-27. Hydrocarbon indicators offshore California with color scheme designed to further increase visual dynamic range.

TUNING PHENOMENA

Effect of Tuning on Stratigraphic Interpretation

Widess (1973) demonstrated the interaction of closely-spaced reflections. In his classic paper, "How thin is a thin bed?", he discussed the effect of bed thickness on seismic signature. For a bed thickness of the order of a seismic wavelength or greater there is little or no interference between the wavelets from the top and the bottom of the bed and each is recorded without modification. For thinner beds these wavelets interfere both constructively and destructively. Considering wavelets of opposite polarity, the amplitude of the composite wavelet reaches a maximum for a bed thickness of one-quarter wavelength (one-half period) and this is known as the tuning thickness. For beds thinner than this the shape of the composite wavelet stays the same but its amplitude decreases. Clearly, the bed thicknesses at which these phenomena occur depend on the shape of the wavelet in the data and hence on its frequency content.

These tuning phenomena are of considerable importance to the stratigraphic interpreter. They must be recognized as effects of bed geometry as opposed to variations in the acoustic properties of the medium. Figure 6-1 shows a sedimentary pod. As the reflections from the top and the base come together (within the black square) the amplitude abruptly increases; this is interpreted as tuning between the top and base reflections.

Figure 6-2 shows some bright spots which are reflections from the top and base of gas sands of variable thickness. The base of the gas sands (the bright red events) are fluid contacts at most of the downdip limits. Hence the top and base reflections in many places constitute thinning wedges. Close inspection of Figure 6-2 reveals several local amplitude maxima close to the downdip limits of brightness. At these points the apparent dip also changes. The interfering wavelets are unable to approach each other more closely than a half period. Therefore, the composite bed signature for each of these thin beds assumes a dip attitude which is the mean of the real dips of the top and the base of the bed. Because the base gas sand is flat at the downdip pinchout, it is easy to see the dip of the composite wavelet turning to assume this intermediate value.

Tuning effects are not always a nuisance; in fact, they can be used to increase the visibility of thin beds. Amplitude tuning occurs for a layer thickness of one-half period of the dominant seismic energy, as already discussed. Frequency tuning, on the other hand, occurs for layer thicknesses of one-quarter period or less. Robertson and Nogami (1984) used instantaneous frequency sections to study thin, porous sandstone lenses based on this phenomenon.

Deterministic Tuning Curves

Tuning phenomena are usually described by graphs such as those of Figure 6-3. In this simple form the principles of tuning are well understood and widely published (for example, see Meckel and Nath, 1977). Figure 6-3 shows that measured thickness, indicated by the time separation of the reflections from the top and base of a bed, is only an acceptable measure of the true thickness of the bed for thicknesses above the tuning thickness. Also at tuning thickness the amplitude of the reflections reaches a maximum due to constructive interference between the reflected energy from the top and bottom of the bed.

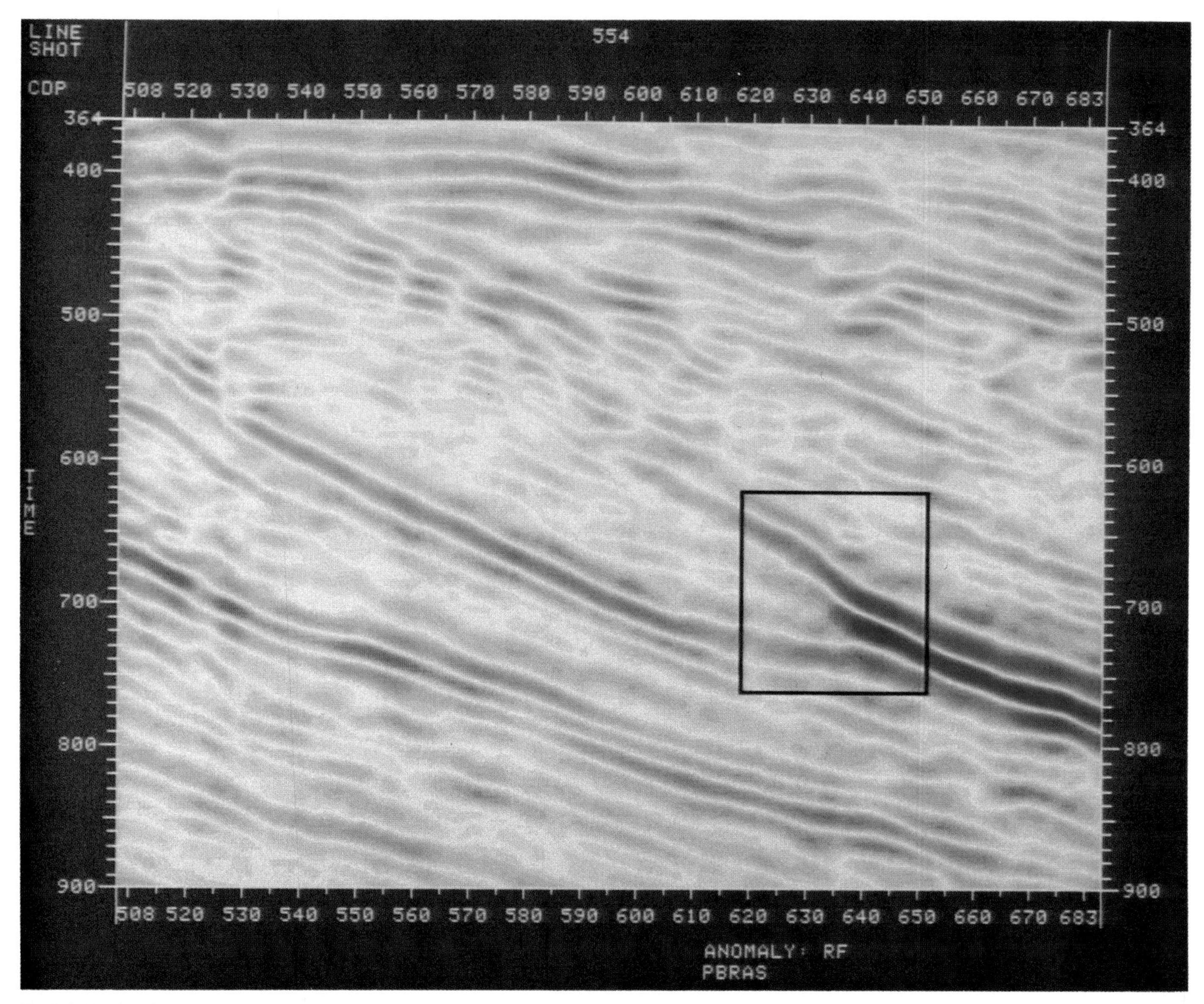

Fig. 6-1. Amplitude increases as reflections converge because of tuning. (Courtesy Petroleo Brasileiro.)

The upper diagram of Figure 6-4 shows how the wavelets from the top and the base of a sand bed must be aligned to produce the principal tuning amplitude maximum; here it is assumed that the reflection coefficients are equal in magnitude and opposite in polarity. It is apparent that the shape of the tuning curve is dependent on the shape of the side lobes of the wavelet. Constructive interference occurs when the central peak of the wavelet from the base of the sand is aligned with the *first* negative side lobe of the wavelet from the top of the sand.

The lower diagram of Figure 6-4 shows how a second tuning maximum is caused. In this case the central peak of the wavelet from the base of the sand is aligned with the *second* negative side lobe of the wavelet from the top. Hence multiple wavelet side lobes generate multiple maxima in the tuning curve. Kallweit and Wood (1982) studied the resolving power of zero-phase wavelets and reported multiple maxima in their tuning curves (Figure 6-5).

Figure 6-6 illustrates deterministic tuning curves derived from four different wavelets. At the top of the page the Ricker wavelet has no side lobes beyond the first and consequently the tuning curve determined from it has only one maximum. This is the classical type of tuning curve, similar to that illustrated in Figure 6-3 and to that published by Meckel and Nath (1977).

The second wavelet in Figure 6-6 is a zero-phase wavelet derived from four corner frequencies defining a band-pass filter. It has, as can be seen, the same width of central peak as the Ricker wavelet but otherwise was randomly selected. This wavelet simply illustrates that multiple side

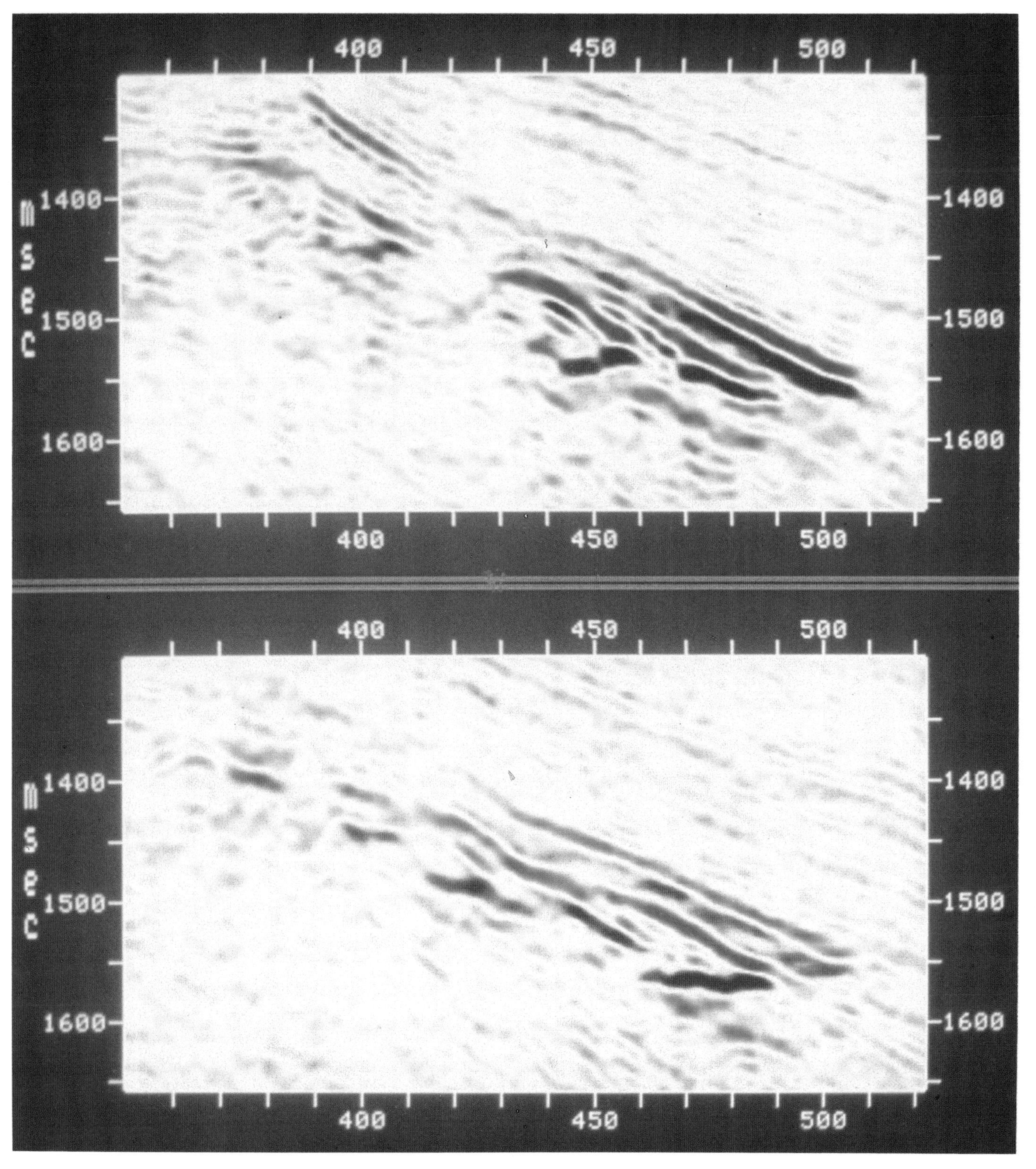

Fig. 6-2. Bright reflections from gas sands of variable thickness shows tuning effects in both amplitude and dip as events converge. (Courtesy Chevron U.S.A. Inc.)

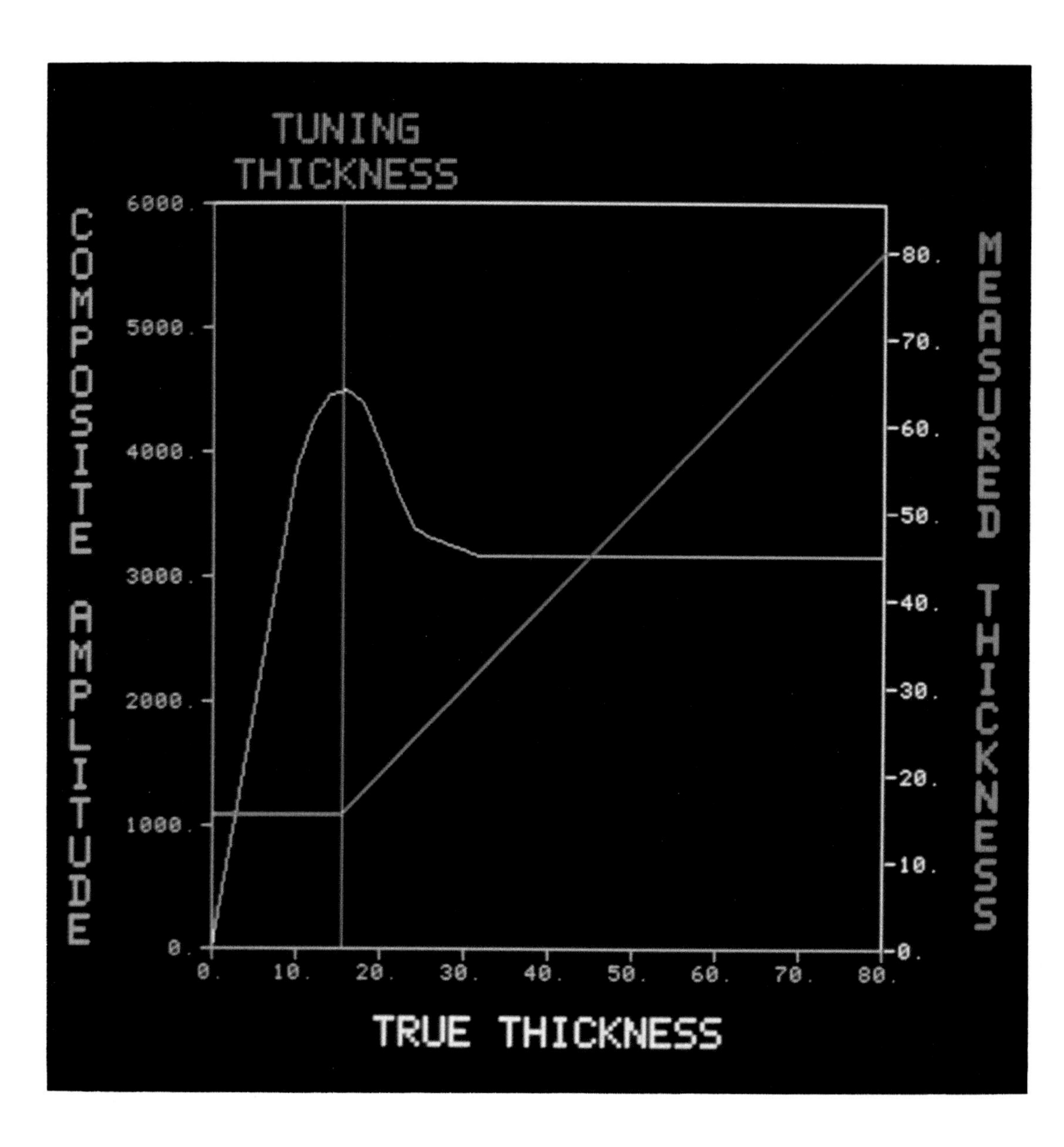

Fig. 6-3. Basic concepts of tuning for thin beds.

lobes in the wavelet generate multiple maxima in the deterministic tuning curve. In fact it is interesting to note the similarity in shape between the tuning curve and half of the wavelet upside-down.

The third wavelet is again zero phase. Its simpler shape generated only two maxima in the tuning curve.

The fourth wavelet in Figure 6-6 was extracted from zero-phase data by a cross-correlation technique between the processed seismic trace and the synthetic seismogram at a well. The wavelet is seen to be almost, but not quite, zero phase. The deterministic tuning curve derived from the extracted wavelet shows some complexity but principally two maxima.

An amplitude spectrum was generated from this tuning curve. By interpreting this spectrum in terms of four corner frequencies it was possible to compute an ideal zero-phase equivalent wavelet and its tuning curve. For the extracted wavelet at the bottom of Figure 6-6 the ideal zero-phase equivalent wavelet is shown directly above as the third wavelet on the page.

In a practical situation the interpreter may be striving for a tuning curve applicable to the zone of interest over some broad area of a prospect. Inevitably, the interpreter will wonder whether a deviation from zero-phaseness such as that shown by the extracted wavelet at the bottom of Figure 6-6 is applicable to the whole area. He may reasonably consider the ideal zero-phase equivalent wavelet and its tuning curve to be more universal.

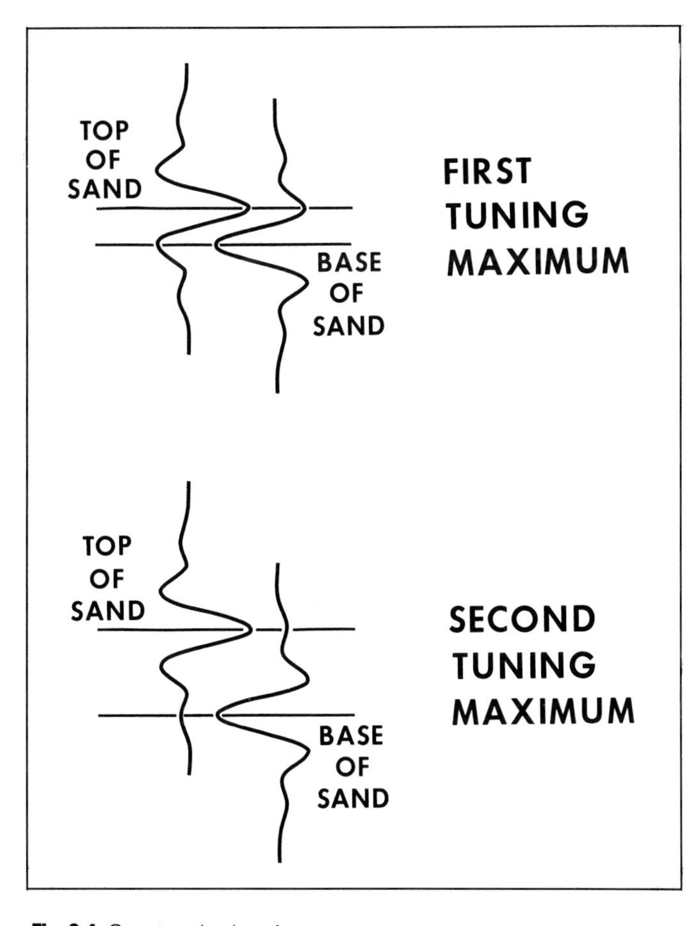

Fig. 6-4. Constructive interference of zero-phase wavelets to produce tuning maxima.

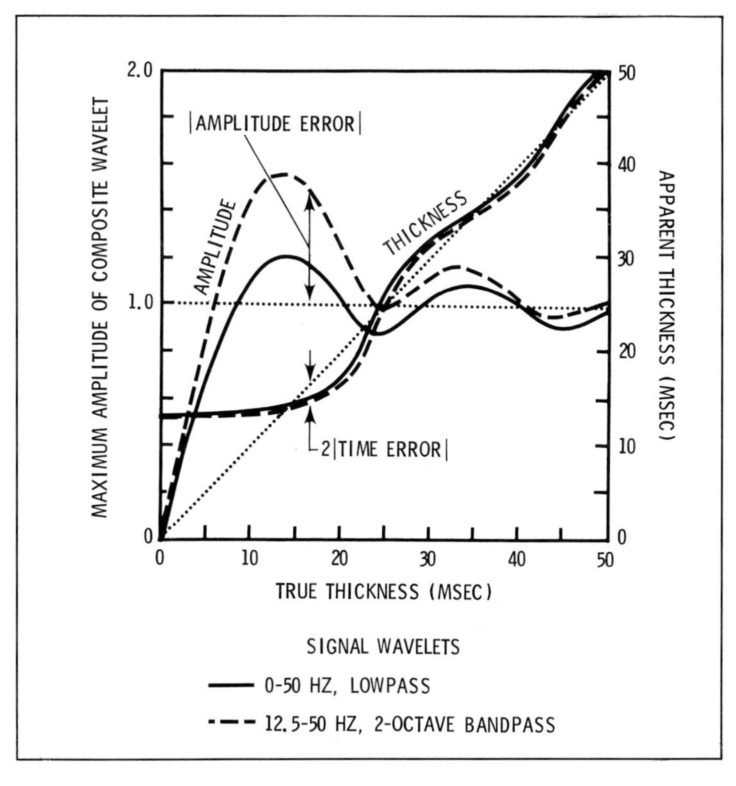

Fig. 6-5. Tuning curves for two zero-phase wavelets showing multiple amplitude maxima (after Kallweit and Wood, 1982).

Statistical Tuning Curves

Tracked horizon data in time, amplitude and other attributes are normally mapped before drawing conclusions from them. It is also possible to crossplot one attribute against another from the same subsurface position. Crossplotting operates within a user-specified subsurface area. With this capability statistical analysis of horizon data can be an important part of interactive interpretation.

In studying the detailed character of bright spots and the tuning phenomena therein, it may be desirable to make the simplifying assumption that lateral variations in amplitude are due to lithologic changes in the reservoir or to tuning effects, and *not* due to changes in the acoustic properties of the embedding media. Figure 6-7 shows a crossplot of the top sand amplitudes against the base sand amplitudes for a particular reservoir. The general proportionality between the two as indicated by the extension of the points along the diagonal yellow line indicates that, to a first approximation, the lateral changes in amplitude do result from lateral changes within the reservoir rather than in the encasing material.

In pursuing the quantitative study of reservoirs (Chapter 7), the absolute value summation of top and base reservoir amplitudes accentuates the properties of the reservoir, lithologic or geometric, relative to those of the encasing material. This absolute value summation is referred to as composite amplitude. Figure 6-8 shows a crossplot of composite amplitude against gross isochron (that is, measured thickness). These are the parameters for studying tuning (Figure 6-3). This crossplot incorporates many thousands of points, so it would be a daunting task to plot manually.

The principal maximum in composite amplitude (Figure 6-8) occurs at 16 ms, the tuning thickness. In addition there is a second maximum evident at about 35 ms. The meaning of these two maxima in terms of wavelet interaction was explained schematically in Figure 6-4. The first interpretation of a statistical tuning curve from this crossplot is then the envelope of the plotted

Fig. 6-6. Various wavelets and their corresponding deterministic tuning curves.

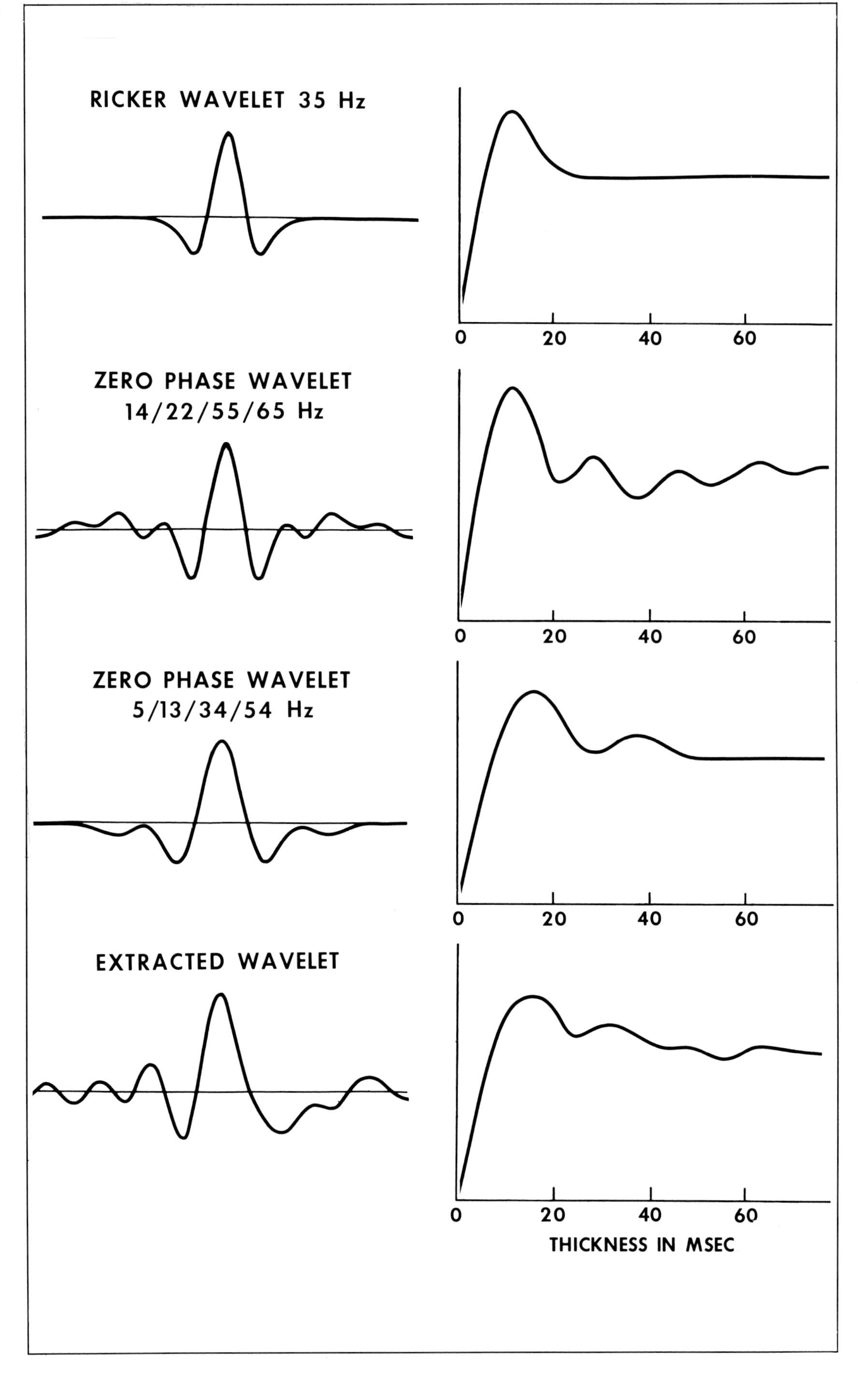

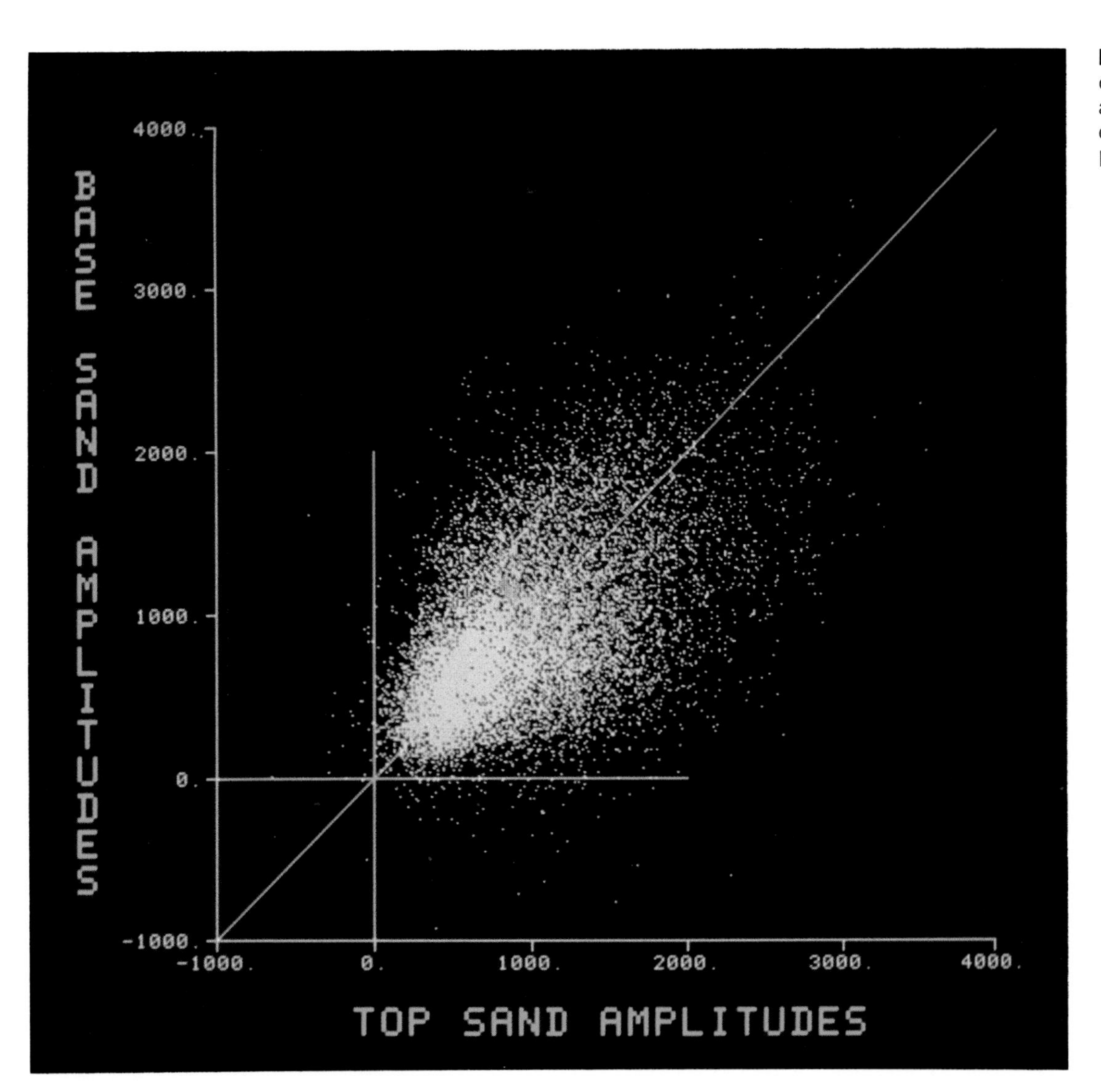

Fig. 6-7. Interactive crossplot of base sand amplitudes against top sand amplitudes demonstrating approximate proportionality.

points (Figure 6-9). This is based on the assumption that the highest amplitude points all indicate the maximum acoustic response of the interval under study and therefore that the variable shape of this envelope with isochron indicates geometric effects alone.

The horizontal blue line to the right of Figure 6-9 is the baseline and indicates the maximum untuned amplitude. The ratio of the amplitude of the tuning maximum to this baseline value is controlled by the side lobe levels of the interfering wavelets. On this basis the tuning maxima as drawn in Figure 6-9 are too high. Considering the very large number of points plotted for isochrons in the 10-40 ms range, it is reasonable that some amplitudes are spuriously high because of constructive interference of the already-tuned reflections with nonreservoir interfaces, multiples or noise. In Figure 6-10, 99th percentile points computed over 2 ms isochron gates are plotted as blue asterisks. They fall at more reasonable levels relative to the untuned baseline.

Hence the existence of two maxima in the tuning curve was indicated by the raw crossplot, but a statistical analysis of the points guided by the knowledge of the deterministic tuning curve was required to establish the shape of the final curve. This final interpreted curve is shown in yellow in Figure 6-10.

Figure 6-11 shows the deterministic tuning curve points and the final interpreted curve superimposed on the same crossplot. Deterministic tuning curves have arbitrary vertical scales. Hence it was necessary to interpretively judge the factor by which the deterministic points must be scaled so that they could be plotted on the same composite amplitude axis as the crossplot points. This was done by matching the deterministic points to the crossplot envelope at the

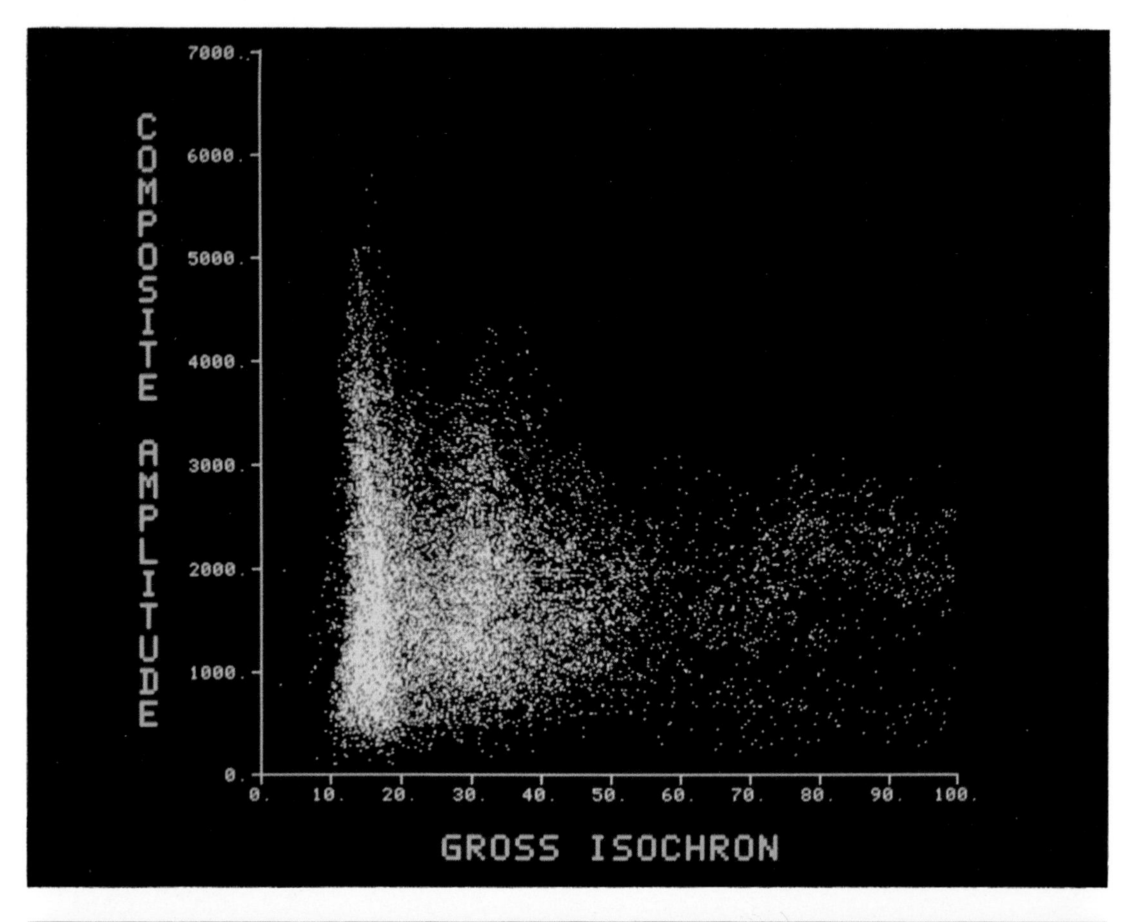

Fig. 6-8. Interactive crossplot of composite amplitude against gross thickness of a reservoir interval for all the interpreted data points in a prospect.

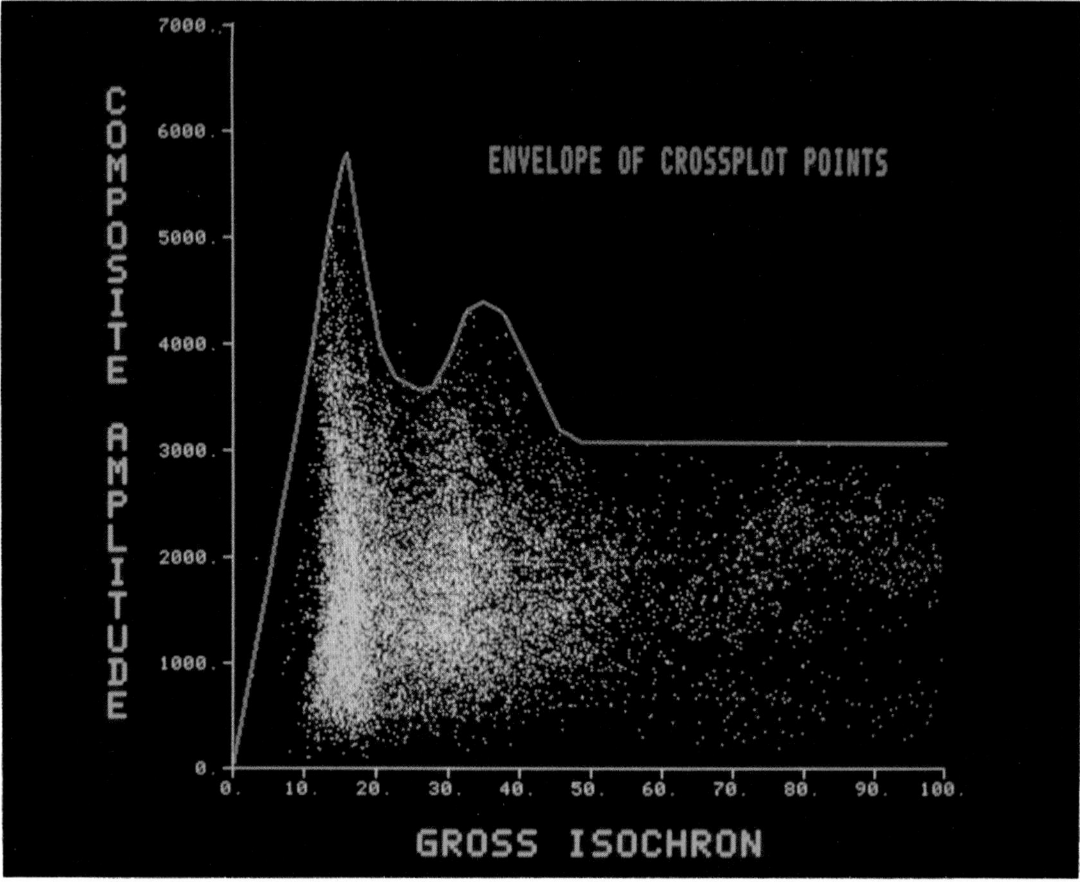

Fig. 6-9. Same crossplot as Figure 6-8 with upper envelope drawn as a first interpretation of a statistical tuning curve.

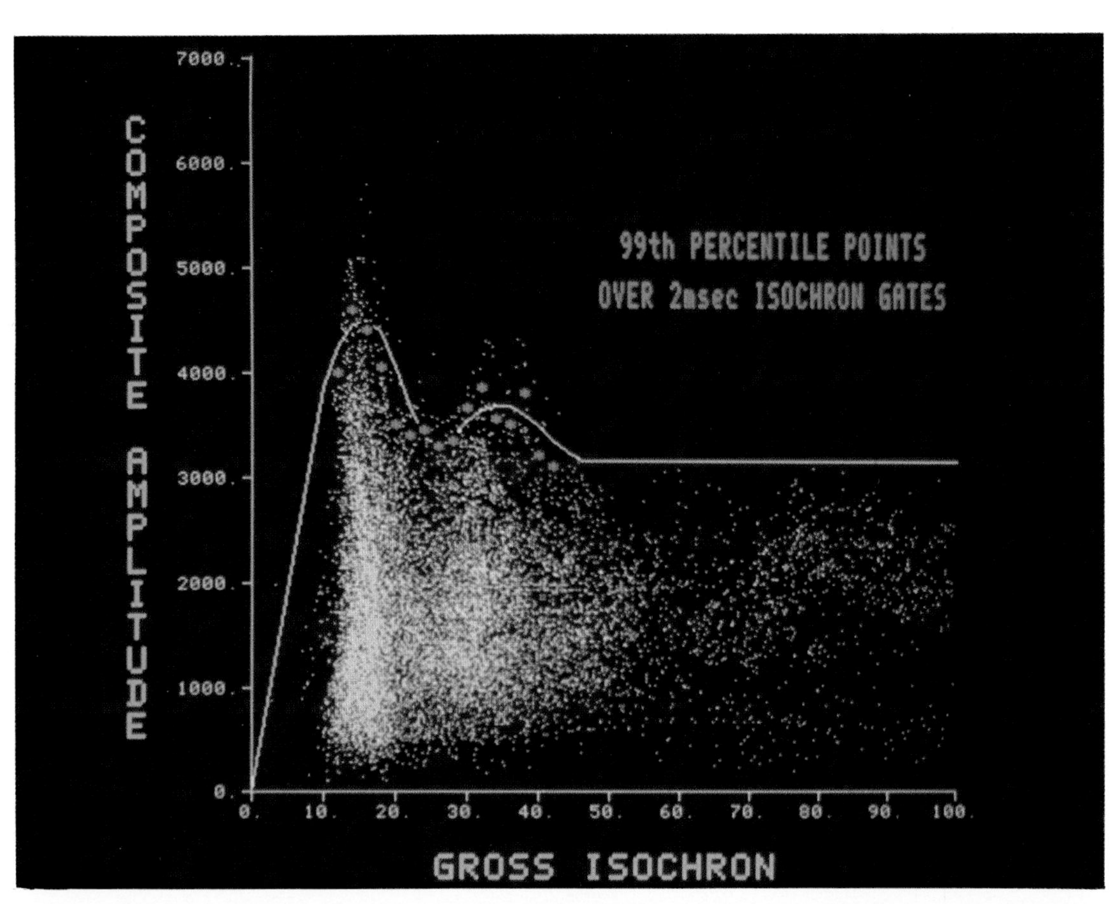

Fig. 6-10. Same crossplot as Figure 6-8 analyzed to yield 99th percentile points which show a more realistic peak-to-baseline ratio as required by deterministic studies.

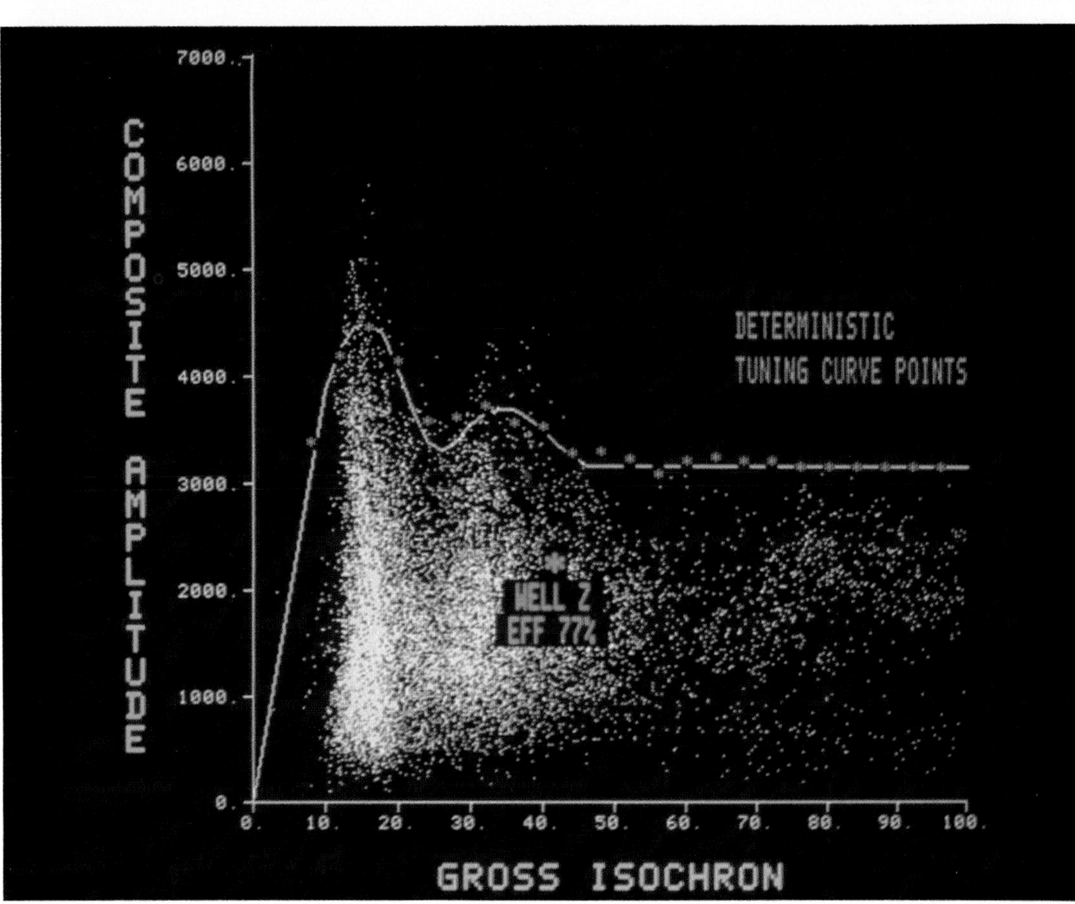

Fig. 6-11. Same crossplot as Figure 6-8 with deterministic tuning curve points computed from an extracted wavelet. The yellow final interpreted tuning curve is the same as in Figure 6-10 to provide comparison between statistical and deterministic tuning points.

greater thicknesses where little or no tuning effect exists, and was confirmed by plotting the model response at a control well. The yellow curve in Figures 6-10 and 6-11 is the same. It is repeated to demonstrate how the final interpreted tuning curve for the area under study tied both the statistical and deterministic inputs.

References

Kallweit, R. S., and L. C. Wood, 1982, The limits of resolution of zero-phase wavelets: Geophysics, v. 47, p.1035-1046.

Meckel, L. D., Jr., and A. K. Nath, 1977, Geologic considerations for stratigraphic modeling and interpretation, *in* C. E. Payton, ed., Seismic stratigraphy–applications to hydrocarbon exploration: AAPG Memoir 26, p. 417-438.

Robertson, J. D. , and H. H. Nogami, 1984, Complex seismic trace analysis of thin beds: Geophysics, v. 49, p. 344-352.

Widess, M. B., 1973, How thin is a thin bed?: Geophysics, v.38, p.1176-1180.

RESERVOIR EVALUATION

Reservoir Properties Affecting Amplitude

This chapter is an extension of the last two. "Bright spot validation" (Chapter 5) discussed the *identification* of hydrocarbon accumulations on the basis of seismic amplitude, an objective we would like to achieve reliably *before* drilling. "Tuning phenomena" (Chapter 6) discussed amplitude *distortions* affecting hydrocarbon indicators. This chapter considers the extraction of more detailed and quantitative reservoir information from seismic amplitudes. This endeavor is part of field development *after* the initial discovery and is often considered "production geophysics".

The principal reservoir properties which affect seismic amplitude can be divided into two groups:

GROUP A	GROUP B
nature of fluid	hydrocarbon saturation
lithology	porosity
pressure	net pay thickness

The properties in Group A are those which, to a first approximation, affect the reservoir as a whole. The difference between gas and oil was discussed in Chapter 5. The lithology of a reservoir rock generally does not change much within one reservoir; other related properties such as age, compaction and depth also will remain fairly constant. Anomalous reservoir pressure can boost seismic amplitude considerably but again this will generally affect the whole reservoir rather than only a part of it.

The properties in Group B are the ones which can vary laterally over short distances and therefore significantly affect the reserve estimates of a reservoir penetrated by only a small number of wells. A major objective of development and production geophysics is to map these spatially-varying reservoir properties so that wells and platforms can be located optimally and reserve estimates can be made with greater precision.

Lateral changes in amplitude of reservoir reflections can be caused by changes in any one or more of these Group B properties, so there is an inherent ambiguity. Other independent parameters offer little help. Frequency, although theoretically affected by gas, rarely if ever provides quantitative information. Interval velocity, derived from normal moveout, cannot normally be determined for sufficiently small intervals to be useful for reservoir studies. Derived attributes, such as inversion velocity, instantaneous amplitude, etc., may aid interpretation but do not add new information. Shear wave data can sometimes provide independent porosity estimates and have been used by Robertson (1983) to study carbonates. Hence the determination of reservoir properties using seismic amplitudes is normally an underspecified problem. Research efforts are underway to provide best fit solutions.

Today's interpretive approach to reservoir evaluation thus requires that simplifying assumptions be made. Conveniently, the amplitude of a seismic bright spot is higher where hydro-

Continued on page 144

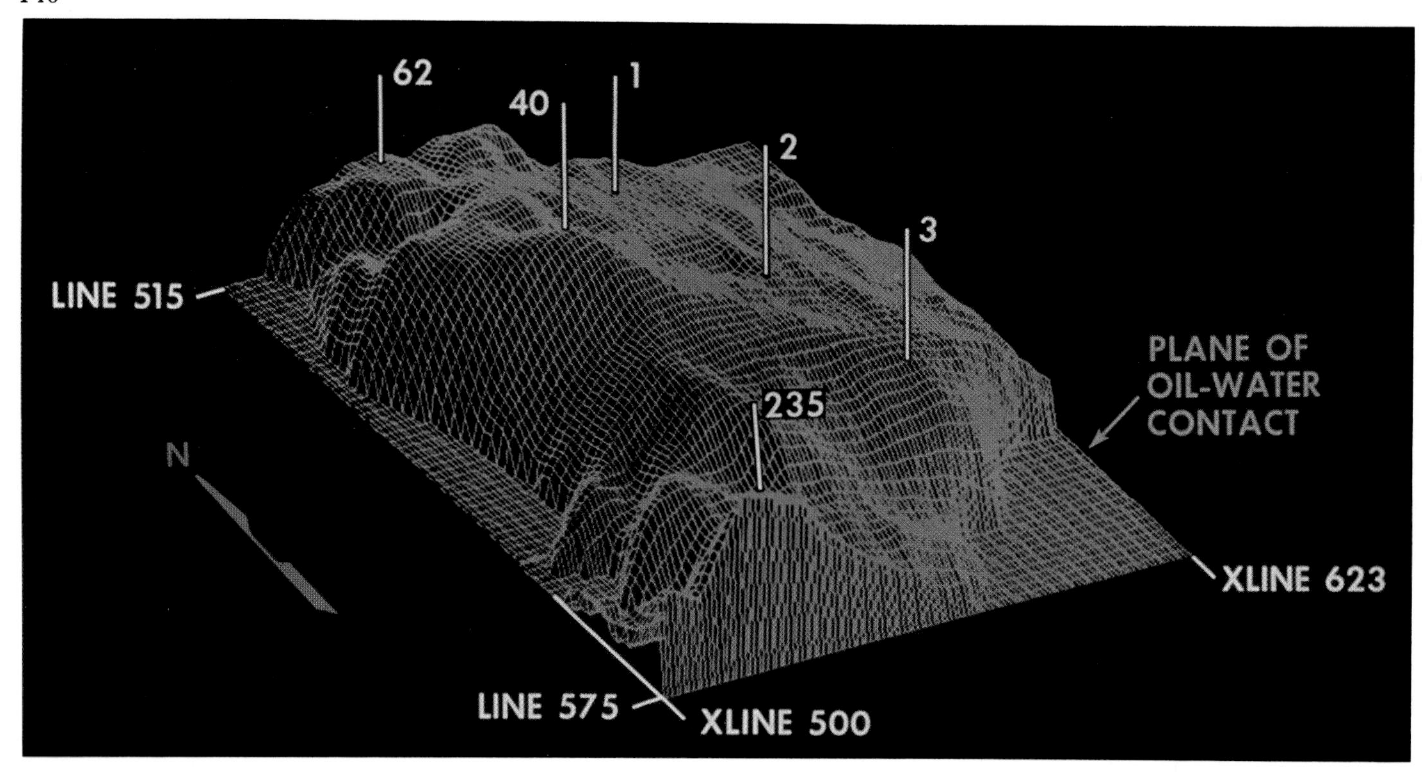

Fig. 7-1. Structural shape of the Macae calcarenite reservoir, Pampo oil field, offshore Brazil. (Courtesy Petroleo Brasileiro.)

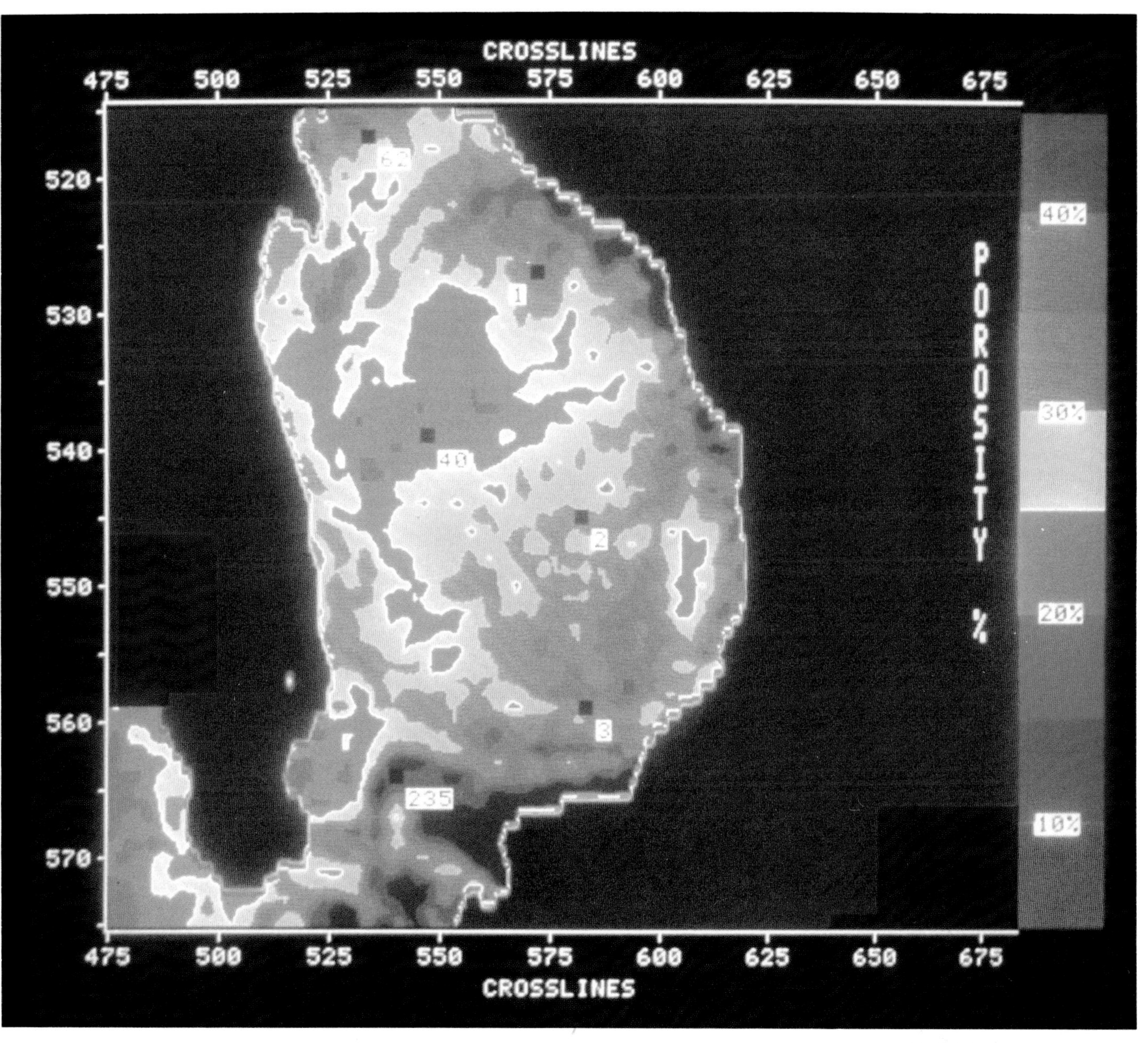

Fig. 7-2. Seiscrop porosity section through the Macae calcarenite reservoir at 1708 ms. (Courtesy Petroleo Brasileiro.)

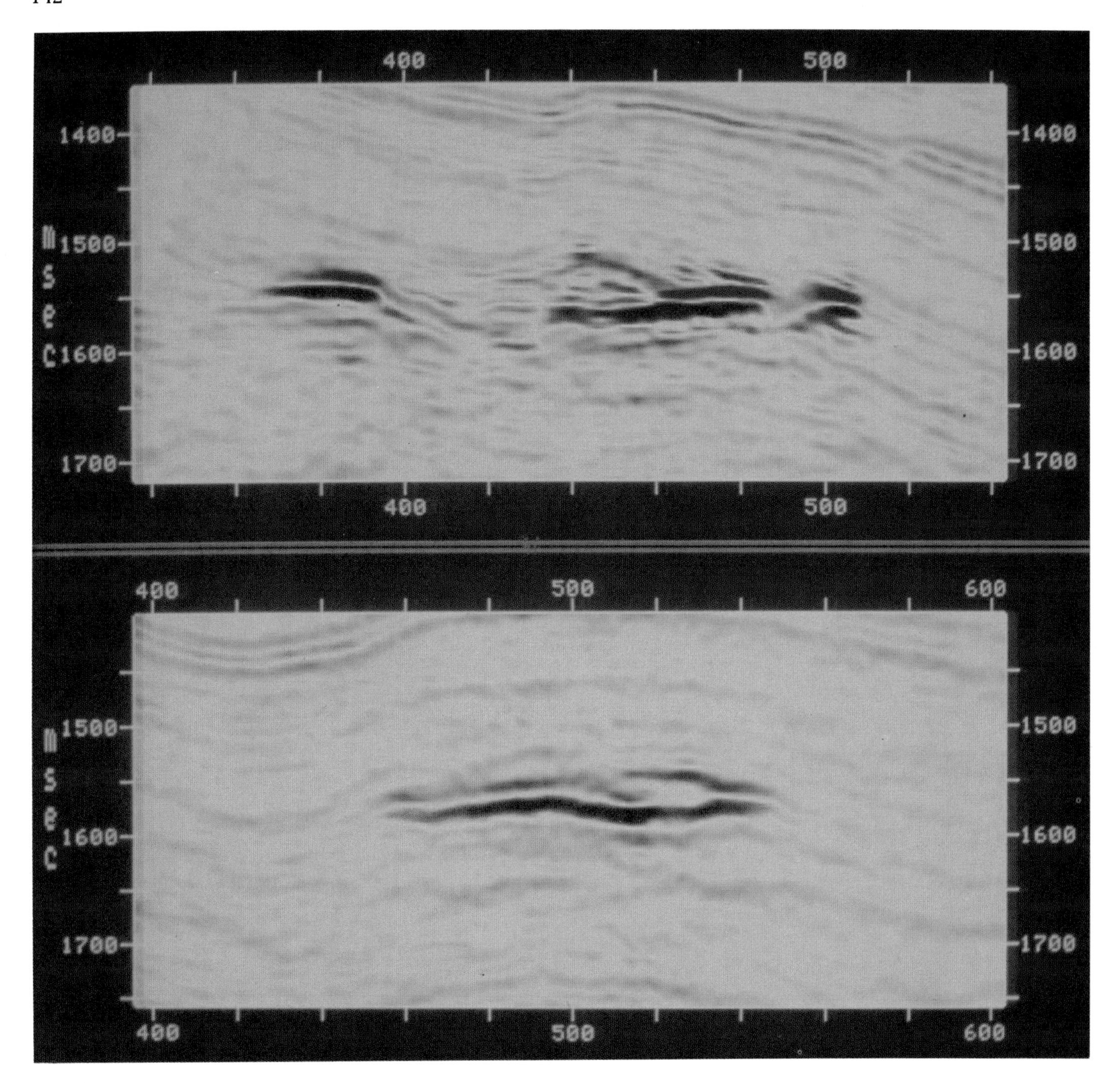

Fig. 7-3. Bright zero-phase reflections from the top and the base of one reservoir sand. (Courtesy Chevron U.S.A. Inc.)

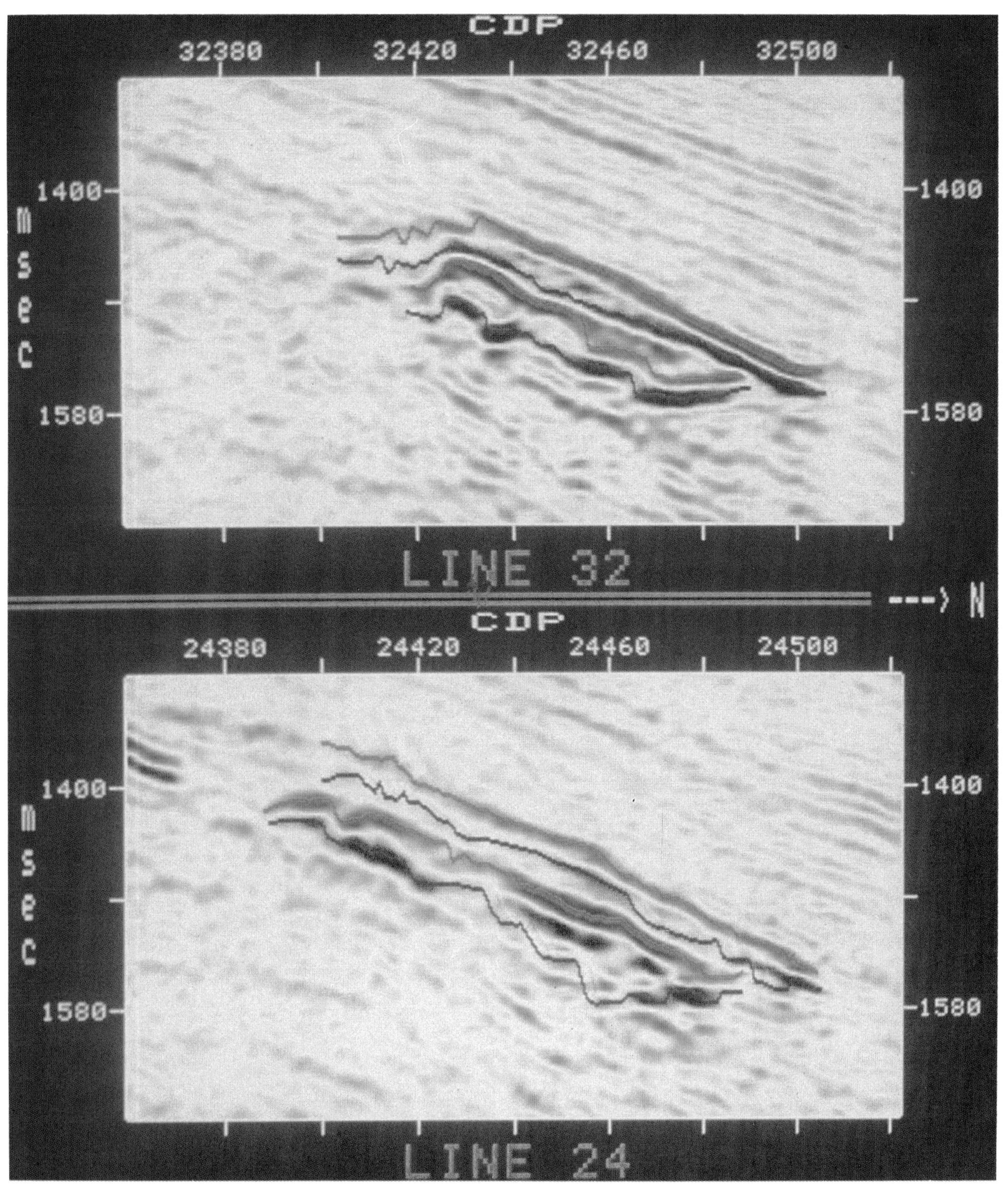

Fig. 7-4. Bright zero-phase reflections from the top and the base of two reservoir sands showing automatic tracks. (Courtesy Chevron U.S.A. Inc.)

carbon saturation is higher (although non-linearly; Domenico, 1974), where porosity is higher, and where net pay thickness is greater (with some complications due to tuning). It normally follows, therefore, that the brighter the bright spot, the better the prospect. In a particular prospect under study the wells may tell the interpreter that one of the reservoir properties is varying more than the others and hence that variations in amplitude can be ascribed principally to variations in that property. We will use case history examples to demonstrate the study of some different properties and also some other aspects of reservoir detail.

Porosity

Figure 7-1 shows the structural configuration of the Macae calcarenite reservoir in the Pampo oil field offshore Brazil (Curtis, Martinez, Possato, Saito, 1983). Amplitude variations of the calcarenite event were considered to result primarily from porosity changes within the reservoir.

The 3-D data volume was processed through recursive seismic inversion. The low-frequency interval velocity field originated from a 3-D inverse normal-incidence ray tracing procedure. The resultant velocities in the reservoir were then converted to apparent porosity using Wyllie's equation (Wyllie, Gregory, and Gardner, 1958). Figure 7-2 shows a horizontal slice through the Macae calcarenite displaying apparent porosity variations within the reservoir. A decrease in porosity toward the reservoir core is evident and is confirmed by well data.

Net Pay Thickness

In an area of Pleistocene sediments offshore Louisiana the wells indicated that each reservoir sand interval was composed of several thin productive lobes and that the position of these lobes within the sands and their thickness varied laterally over a short distance (Brown, Wright, Burkart, Abriel, 1984). The top and base of the gross sand intervals generate the seismic reflections and the nonproductive segments within them are caused by the sands becoming tight and shaly. The aggregate thickness of the productive lobes is what matters economically. Therefore, the overall objective is to use amplitude measurements, coupled with time thickness measurements, to determine the spatial distribution of net producible gas sand from the seismic data.

The use of seismic amplitude to measure the proportion of sand within a sand/shale interval was demonstrated by Meckel and Nath (1977) for beds less than tuning thickness. Here the principle has been extended to thicker beds on the assumption that the individual lobes of producible gas sand are each below tuning thickness and that here producible gas sand is a material of uniform acoustic properties.

Figure 7-3 shows bright reflections from one reservoir sand. The single peak-over-trough signature indicates zero-phaseness (see Chapter 2). Figure 7-4 shows two examples of two reservoir sands. Automatic tracking on an interactive interpretation system was used to track the bright reflection at the top and at the base of each reservoir interval. The tracker followed the maximum amplitude in the waveform while the interactive system stored the time and the amplitude of that pick in the digital database. Given that the data were zero phase, the time of the maximum amplitude is the correct time for the reservoir interface.

Figure 7-5 shows the interactive interpretive sequence which was then applied to the times and amplitudes provided by the horizon tracking. For any one sand, the horizon times provided structure maps for the top and the base reflections. Subtraction of these time maps yielded the gross isochron map for the sand. The horizon amplitudes provided horizon Seiscrop sections for the top and base sand reflections. These were then added together in absolute value to yield the composite amplitude response of the sand.

Tuning effects remained as a distortion in this composite amplitude response and had to be removed. The key is to understand the shape of the tuning curve in detail; this can be obtained deterministically from an extracted wavelet or statistically from a crossplot (see Chapter 6). In this offshore Louisiana example both methods were used to yield the yellow tuning curve of Figure 7-6. Editing was then required to change the response from that shown in yellow to that shown in orange, so that the amplitude as a function of gross reservoir thickness alone is constant above tuning thickness and linearly decaying to zero for decreasing thickness below tuning.

In order to conclude the interpretive sequence of Figure 7-5, the composite amplitude response was edited according to Figure 7-6, smoothed, and scaled to yield a map of net gas/gross sand ratio. Combining this by multiplication with the gross isochron map, a net gas isochron map was obtained. A constant gas sand interval velocity was then sufficient to convert this net gas isochron map to a net gas isopach map. In combining the gross isochron map with the net gas/gross sand ratio (derived by editing with the function in Figure 7-6), it should be remembered that there are no gross isochrons less than tuning thickness because of the tuning phenomenon itself (Figure 6-3). For actual gas sand thicknesses less than tuning, all the net gas sand information is encoded in the amplitude.

Figures 7-7 and 7-8 show comparable gross and net isochron maps, making clear the contribution of the net gas/gross sand ratio derived from the seismic amplitude. Note the two northwest-southeast thickness trends on the gross isochron map and then note that only one of them has survived in the net isochron map. Net gas sand maps derived in this way have been shown to tie well data acceptably. In practice, relative values are more accurate than absolute values because of the difficulty of determining the scale factor connecting edited amplitude to net gas/gross sand ratio. Interpolation between existing wells penetrating the reservoir is thus the soundest application of this approach.

When there is more than one mappable reservoir interval associated with the reservoir under study, each interval is treated separately and added together at net gas isopach stage. Figure 7-9 shows total net producible gas sand in color superimposed on the structural configuration of the top of the reservoir.

Integration of net gas isopach maps yields the volume of the reservoir. By integrating over chosen sub-areas, the reservoir volume over different lease blocks or areas of special interest can be readily determined.

McCarthy (1984) reported a different approach to mapping gas thickness. Using 2-D data he made several quantitative attribute measurements of detailed waveform character. Sets of these measurements were then related to pore fluid and gas thickness by statistical analysis.

Statistical Use of Tracked Horizon Data

Once the time and amplitude of the zero-phase reflections from the top and base of a reservoir are stored in a readily-accessible digital database, statistical studies of the horizon data are straightforward. The value of interactive crossplotting for the statistical analysis of tuning phenomena was explained in Chapter 6.

Figure 7-10 shows an interactive crossplot of gross isochron against top sand time; that is, vertical thickness (in time) against structural position. The general effect is triangular with several clearly visible lineations. Interpretation of these lineations is a statistical assessment of the many thousands of data points included in this crossplot.

Figures 7-11 and 7-12 are crossplots of sub-areas of the prospect, each accompanied by an exemplary data segment. Figure 7-11 makes it clear that the lineation along the bottom of the crossplot illustrates that thicknesses can only be measured down to just below tuning.

Sub-area C (Figure 7-11) includes many lines of data illustrating good flat spot reflections, one of which is illustrated in the data insert. The orange straight line labelled FLAT SPOT is a line of the form, $y = c - x$, that is, one representing equal increments of gross isochron and top sand time. Because top sand time plus gross isochron equals base sand time and because base sand time is approximately constant for a flat spot reflection, it is indeed expected that a flat spot would plot along such a diagonal line.

Sub-area B (Figure 7-12) is an area of thicker sands. An orange line represents the line of equal increments of gross isochron and top sand time as before. Most of the points fall to the right of this line, indicating varying degrees of gas velocity sag. Extreme sag is illustrated in the data insert and marked on the crossplot by a pink line.

An interpreted version of the total area crossplot of Figure 7-10 is shown in Figure 7-13. This interpretation incorporates the observations from several sub-areas including the two discussed here. The convergence of many of the lineations in the lower right corner and also the great concentration of crossplot points at the same location suggest a common gas/water contact over the majority of the reservoir. This in turn suggests that much of the reservoir is in communication, at least in terms of flow rates effective over geologic time. However, there is a faint suggestion of a lineation with the correct slope to be another flat spot; this lineation intersects the top sand axis at 1,440 ms. The concentration of points in a swath along the lower part of the triangular pattern suggests that many of the sands were preferentially deposited with thicknesses of 30 ms (approximately 25 m) or less.

Figure 7-14 shows an interpreted crossplot of gross isochron against base sand time. This

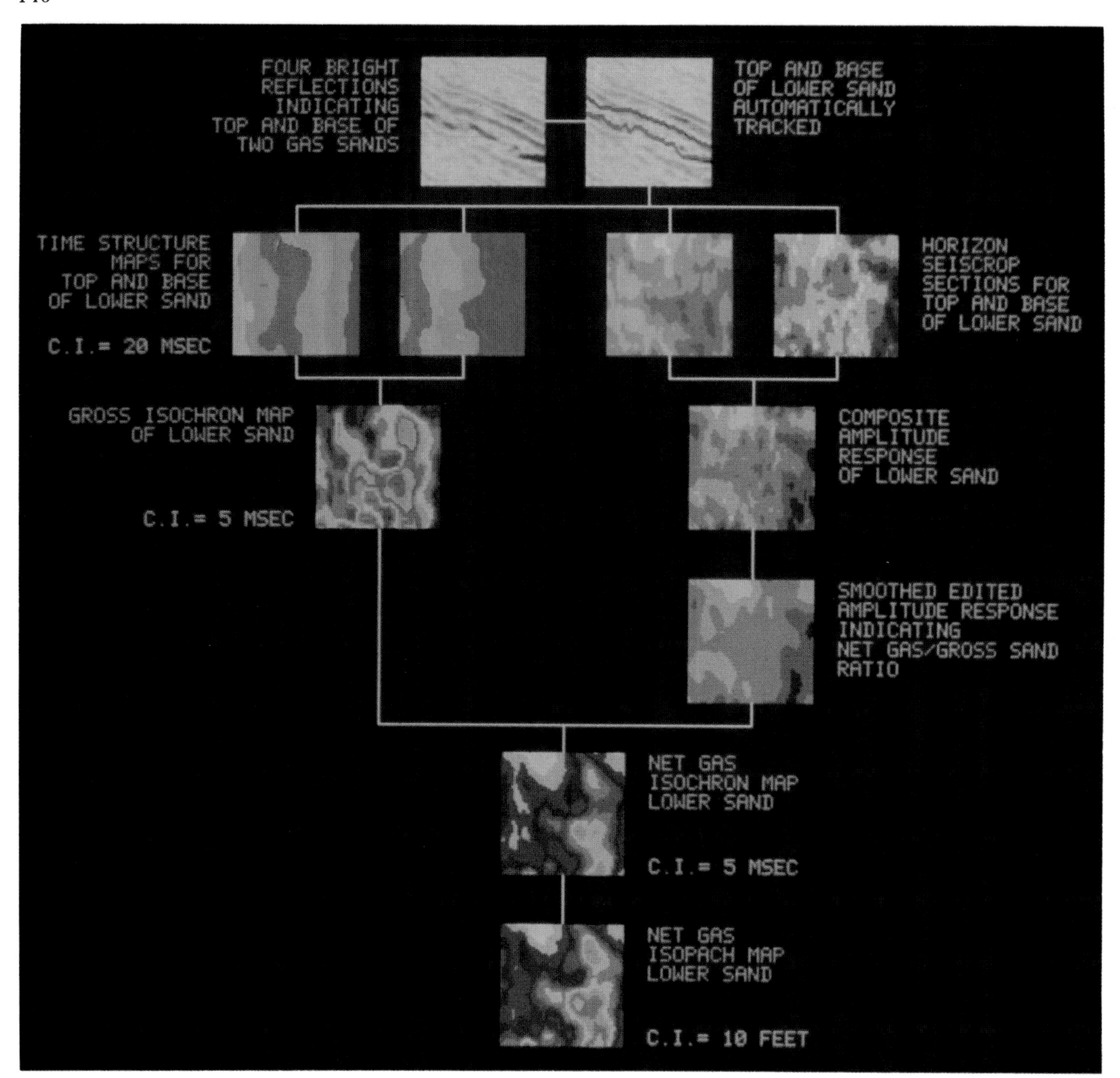

Fig. 7-5. Interpretive sequence used and intermediate products generated in the course of deriving net producible gas sand maps. C.I. = Contour Interval. Time structure maps show dip down to the right; the purple area is the flat spot at the base of the Lower Sand. The isochron and isopach maps have greens and blues indicating the thicker zones. All the four amplitude products have darker colors indicating the higher amplitudes. (Courtesy Chevron U.S.A Inc.)

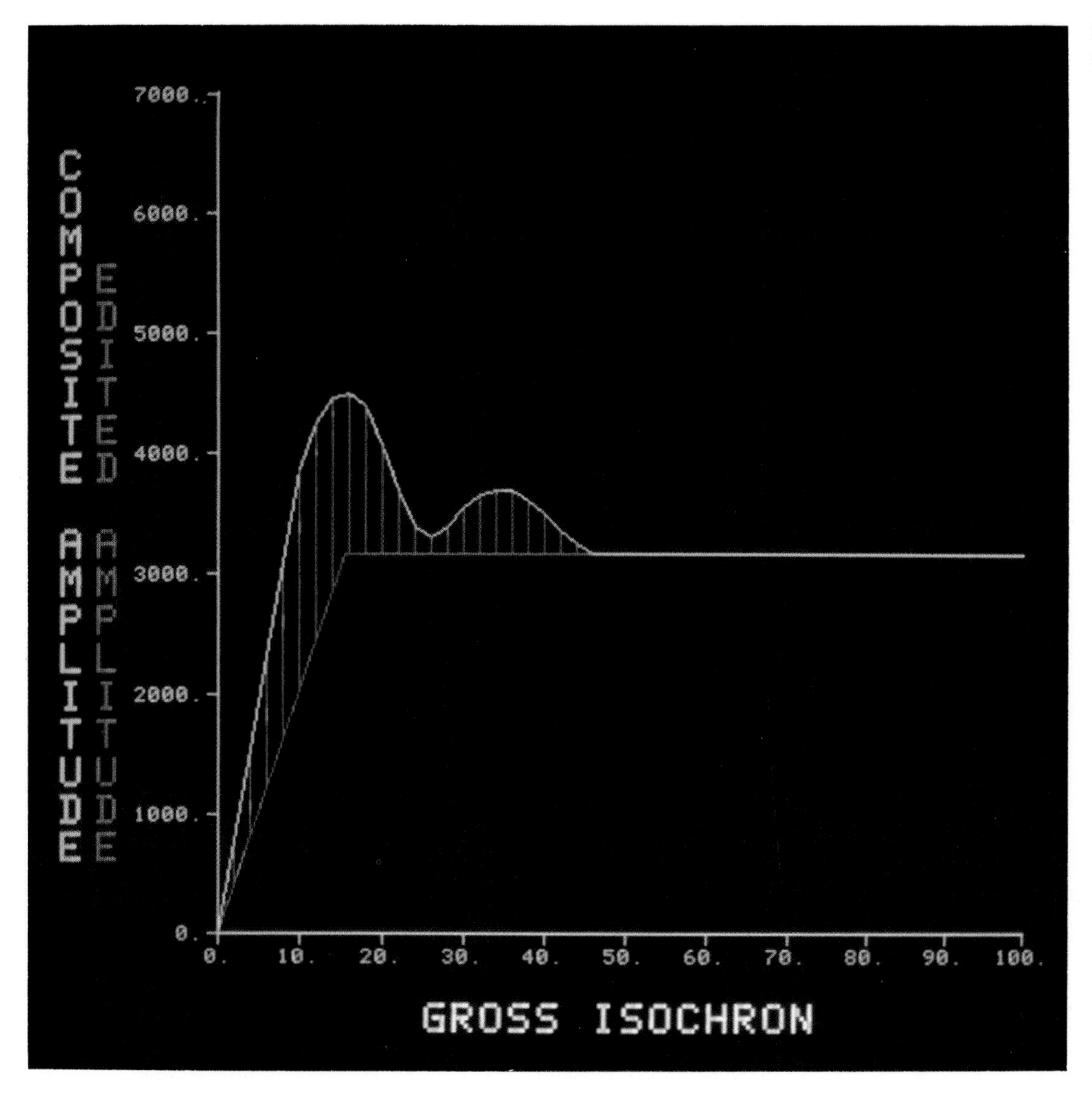

Fig. 7-6. The editing of tuning effects.

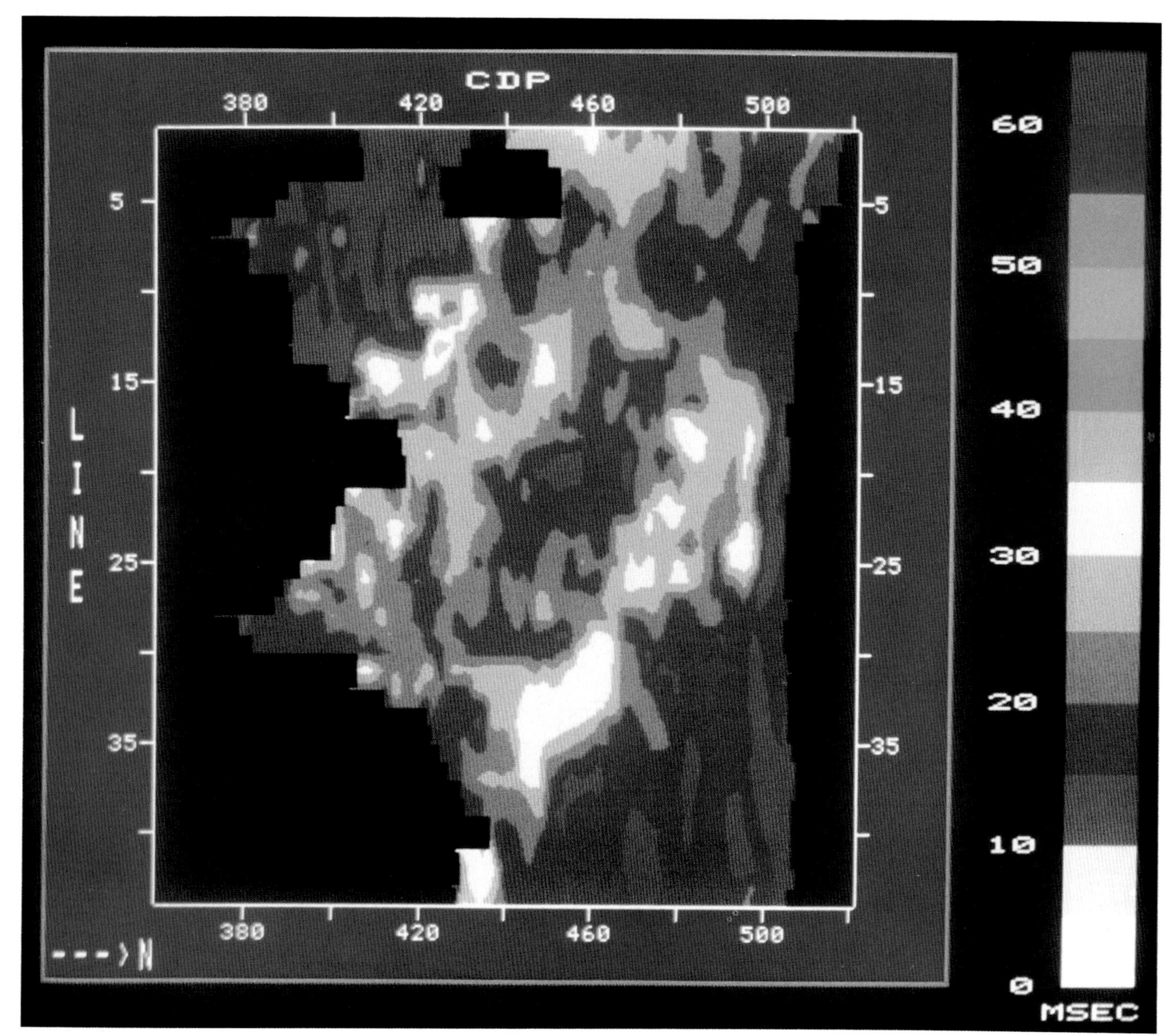

Fig. 7-7. Gross isochron map of Upper Sand showing two thickness trends. (Courtesy Chevron U.S.A. Inc.)

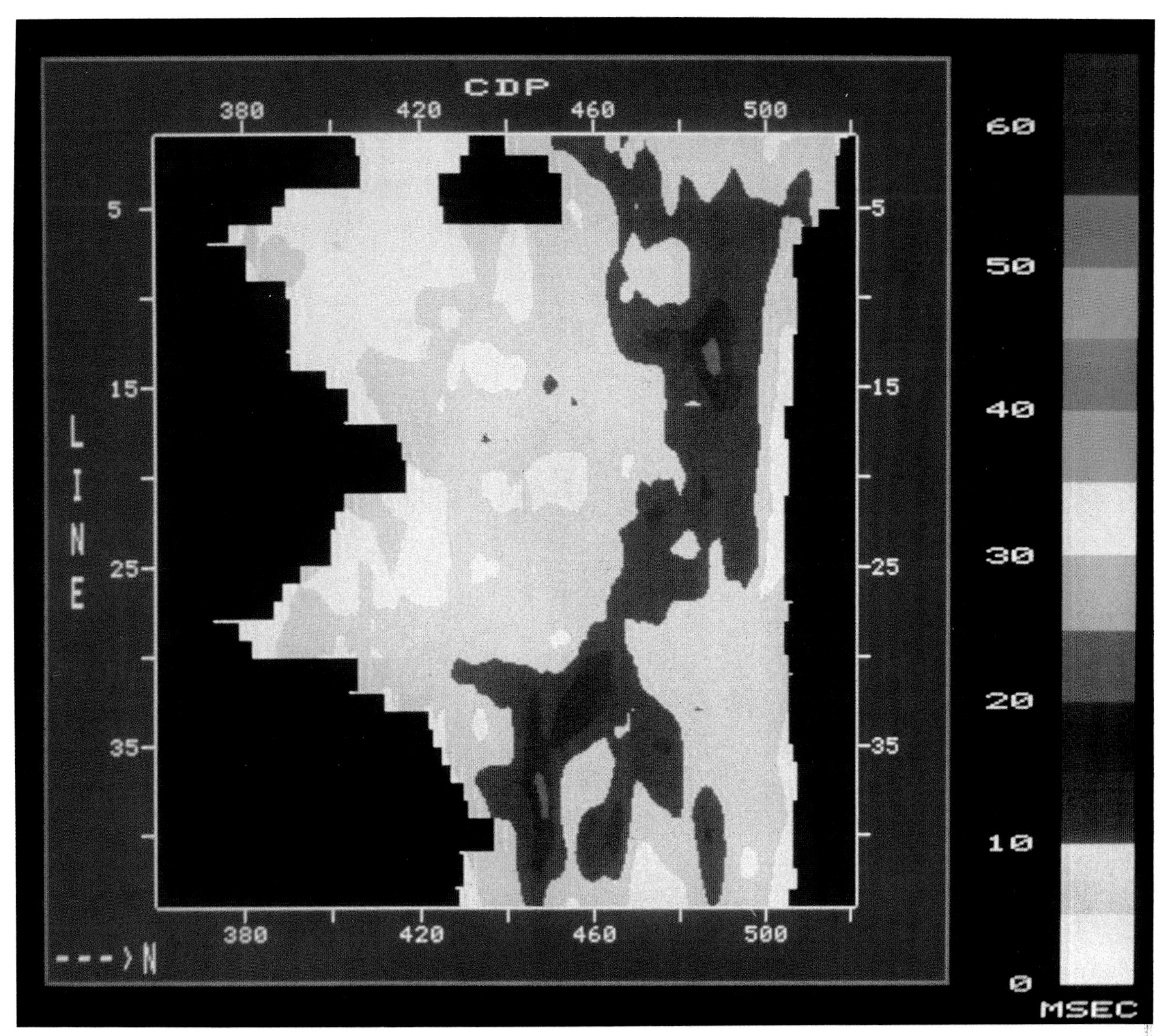

Fig. 7-8. Net gas isochron map of Upper Sand showing one thickness trend. (Courtesy Chevron U.S.A. Inc.)

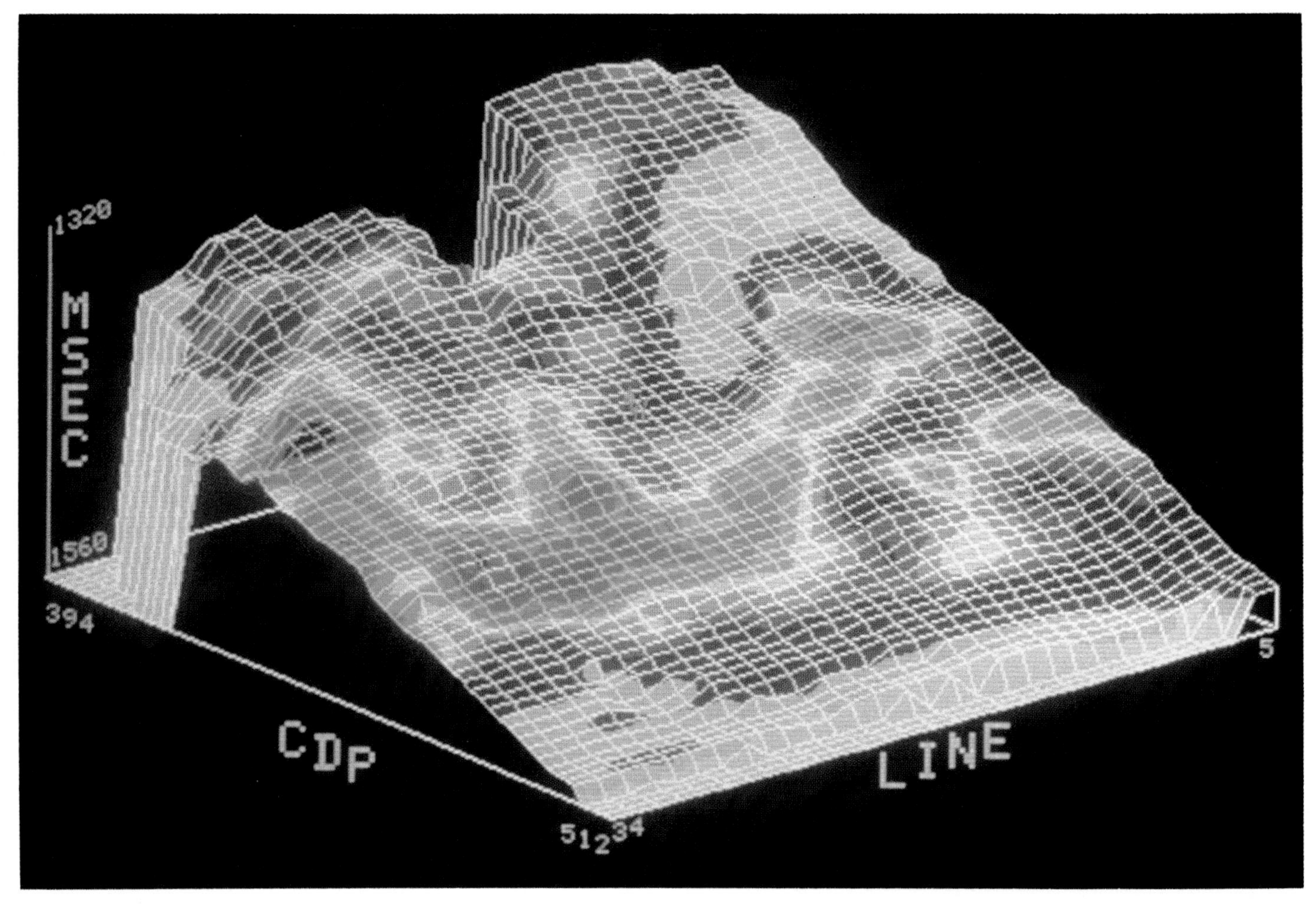

Fig. 7-9. Total net gas sand isopach map superimposed on the structure of the top of the reservoir. The greens and blues indicate the thicker net gas zones. (Courtesy Chevron U.S.A. Inc.)

provides the same information as Figure 7-13 simply with a different crossplot abscissa.

The lineation forming the left side of the triangular crossplot (the blue straight line in Figure 7-13) demonstrates that the sands thin updip. This regular relationship between maximum gross isochron and minimum top sand time indicates the combined effects of paleoslope at the time of deposition and structural movement since that time. The points lying on the edge lineation came from Line 478 illustrated in Figure 7-15. By flattening this section on the top reservoir reflection it is possible to hypothesize how these sands appeared at the time of deposition. The dip here is to the north and is probably caused by a deeper salt swell. These reservoir sands are believed to have been deposited from the north, slumping into the depo-trough contained by the salt-induced structure. On the assumption that the trough filled to yield a horizontal upper surface, Figure 7-15 then indicates that the magnitude of the maximum paleoslope encountered by the depositing sand was 13°.

Further Observations of Reservoir Detail

Figure 7-16 illustrates several aspects of a thick Gulf of Mexico gas reservoir. The upper right panel shows a vertical section indicating a wedge-shaped gas zone. The lower bright red reflection is from the fluid contact dipping to the left because of gas velocity sag. The structural dip is in the opposite direction. Layering within the reservoir thus crosses the fluid contact and, because of different properties from layer to layer, causes varying amplitude along the fluid contact reflection.

Automatic tracks on the top and base of the gas reservoir are shown in the lower left panel of Figure 7-16. The time structure map for the base of the reservoir in the upper left panel illustrates the zone of major gas velocity sag by the area of dark blue. The horizon slice showing the spatial variation in amplitude over the base reservoir reflection is seen in the lower right panel. Within

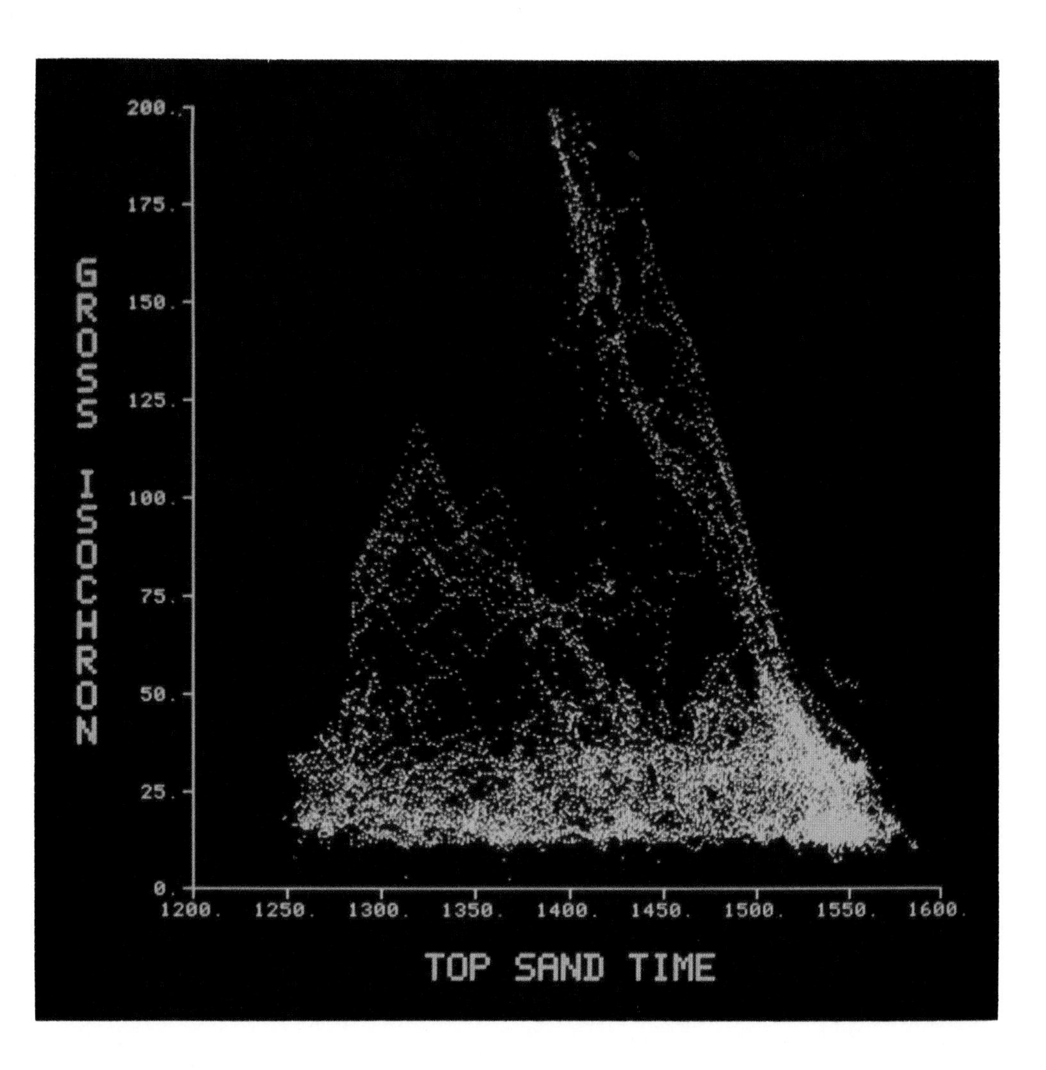

Fig. 7-10. Interactive crossplot of gross isochron against top sand time.

the zone of major gas sag, approximately north-south high amplitude streaks illustrate the areas where layers of superior reservoir quality intersect the fluid contact reflection.

Figures 7-17 and 7-18 were derived from 3-D data from Peru. The Vivian pay sand is a proven hydrocarbon reservoir but the reflection from it, as illustrated in the lower panel of Figure 7-17, is unspectacular. The data were considered to be zero phase and thus appropriate for seismic inversion. The velocity section after inversion is shown in the upper panel of Figure 7-17. The lower velocity in the Vivian sand on top of the structure can be clearly seen. This is a case where seismic inversion has significantly increased the visibility of a reservoir feature.

The whole 3-D volume was then inverted and the velocity low identified as the Vivian sand sliced in a horizon-consistent manner. The resulting horizon slice in velocity, in the right panel of Figure 7-18, shows an oval-shaped area of low velocity. This field has now been substantially developed and the left panel of Figure 7-18 shows that the producing wells all lie within the low velocity zone.

References

Brown, A. R., R. M. Wright, K. D. Burkart and W. L. Abriel, 1984, Interactive seismic mapping of net producible gas sand in the Gulf of Mexico: Geophysics, v. 49, p. 686-714.

Curtis, M. P., R. D. Martinez, S. Possato and M. Saito, 1983, 3-dimensional seismic attributes contribute to the stratigraphic interpretation of the Pampo oil field, Brazil: Proceedings, SEG 53rd Annual Meeting, p. 478-481.

Domenico, S. N., 1974, Effect of water saturation on seismic reflectivity of sand reservoirs encased in shale: Geophysics, v. 39, p. 759-769.

McCarthy, C. J., 1984, Seismic prediction of pore fluid and gas thickness: Proceedings, SEG 54th Annual Meeting, p. 326-328.

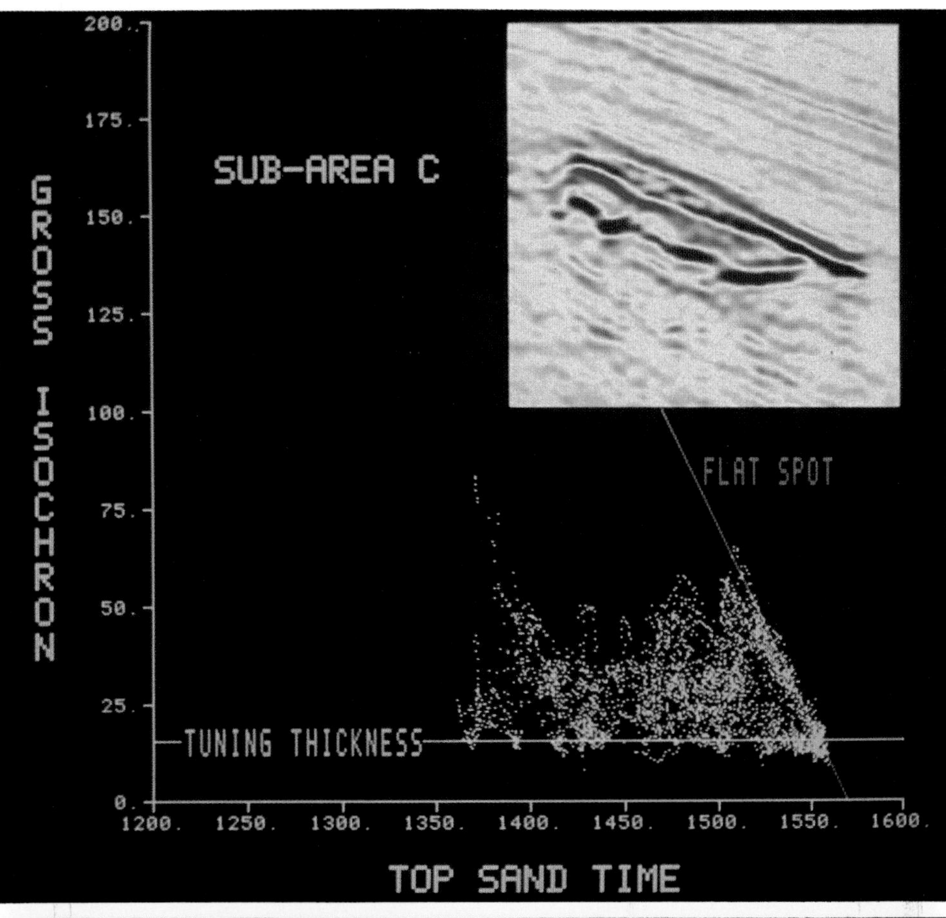

Fig. 7-11. Interactive crossplot of gross isochron against top sand time for sub-area C with exemplary data insert.

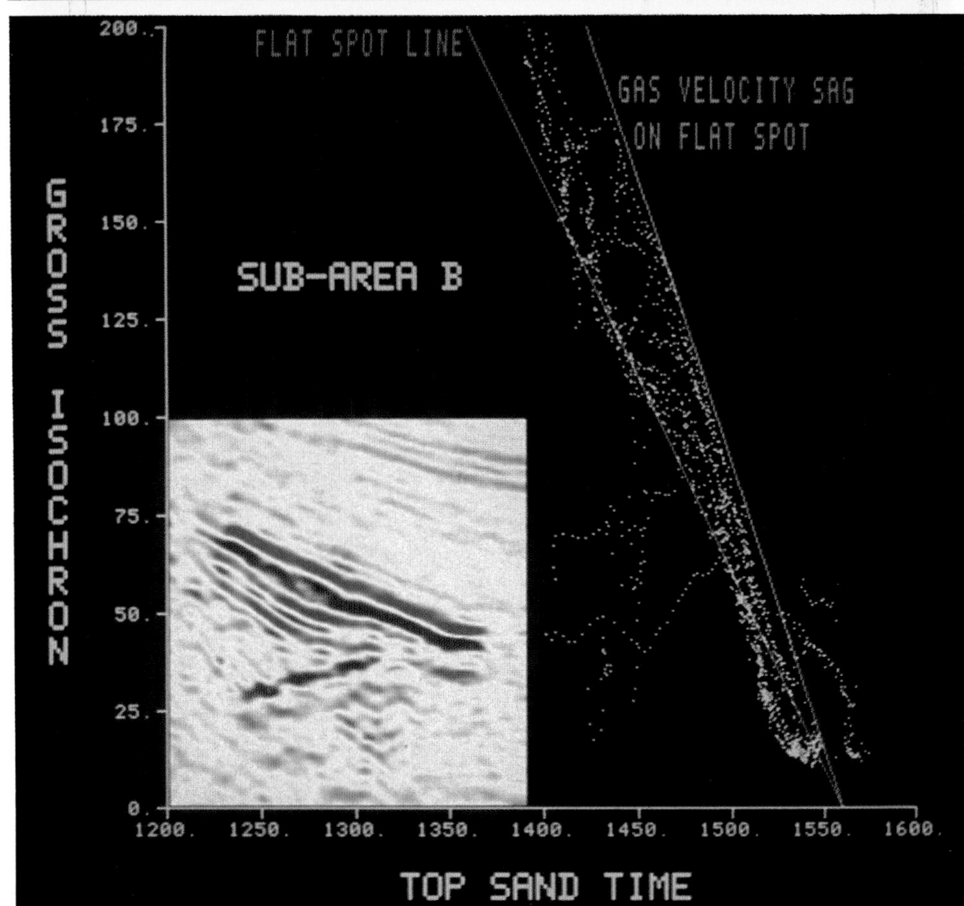

Fig. 7-12. Interactive crossplot of gross isochron against top sand time for sub-area B with exemplary data insert.

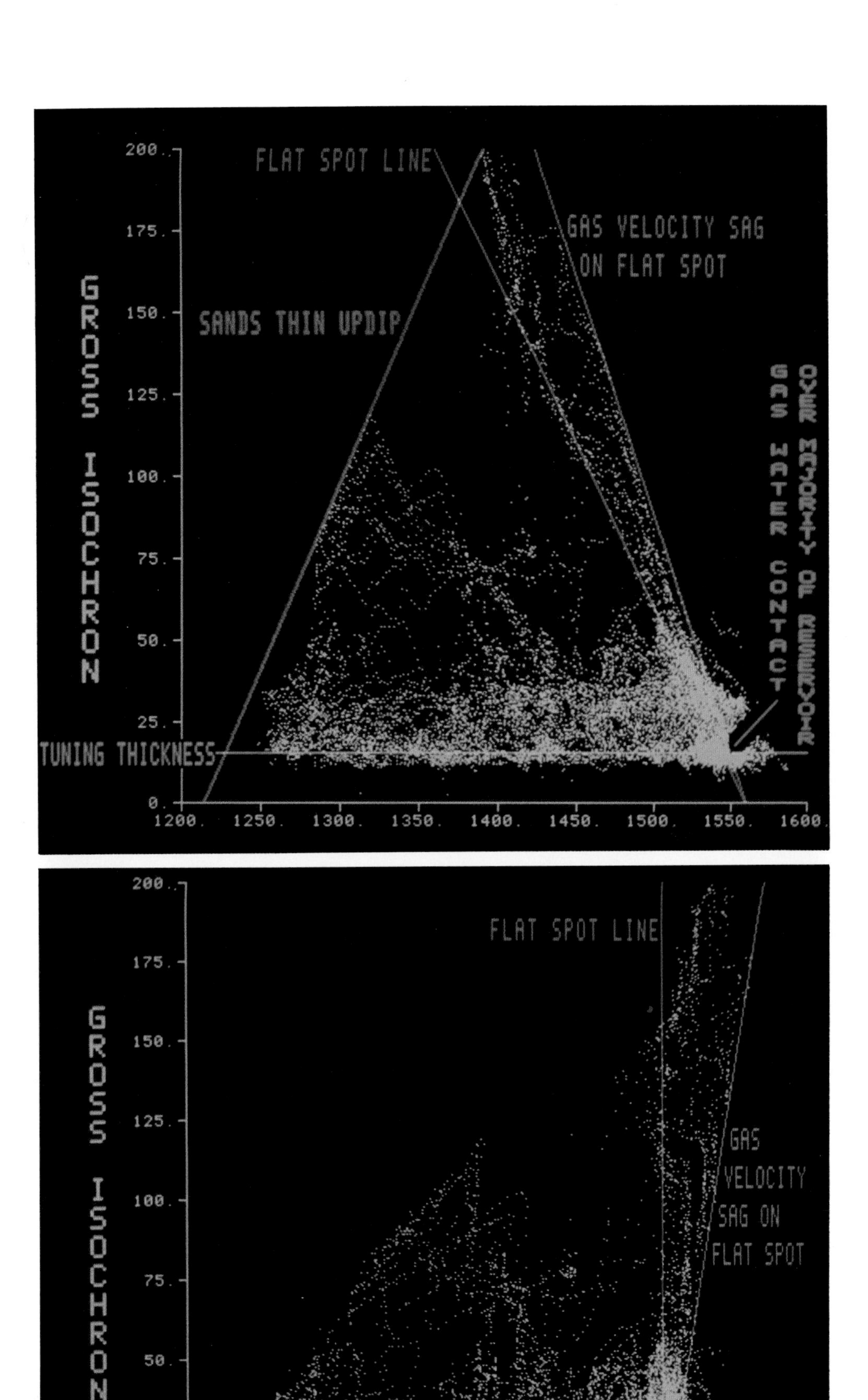

Fig. 7-13. Interactive crossplot of Figure 7-10 showing interpretation of lineations.

Fig. 7-14. Interactive crossplot of gross isochron against base sand time showing interpretation of some of the lineations.

Fig. 7-15. Line 478 with automatic tracks for top and base sand reflections. Same flattened on top sand track in an attempt to deduce the dip of the paleoslope. (Courtesy Chevron U.S.A. Inc.)

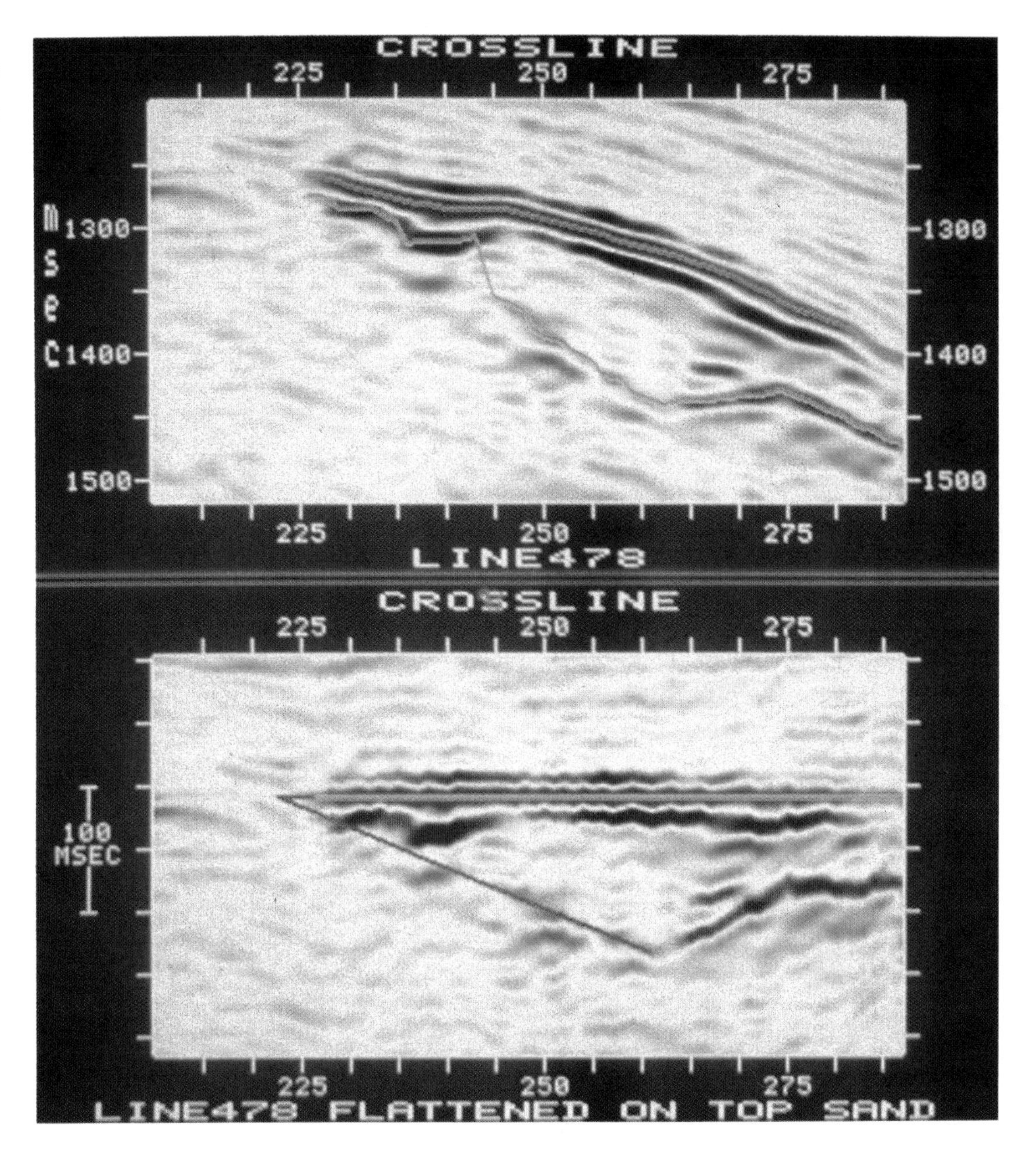

Meckel, L. D., Jr., and A. K. Nath, 1977, Geologic considerations for stratigraphic modeling and interpretation, *in* C. E. Payton, ed., Seismic stratigraphy–applications to hydrocarbon exploration: AAPG Memoir 26, p. 417-438.

Robertson, J. D., 1983, Carbonate porosity from S/P traveltime ratios: Proceedings, SEG 53rd Annual Meeting, p. 356-358.

Wyllie, M. R. J., A. R. Gregory and G. H. F. Gardner, 1958, An experimental investigation of factors affecting elastic wave velocities in porous media: Geophysics, v. 23, p. 459-493.

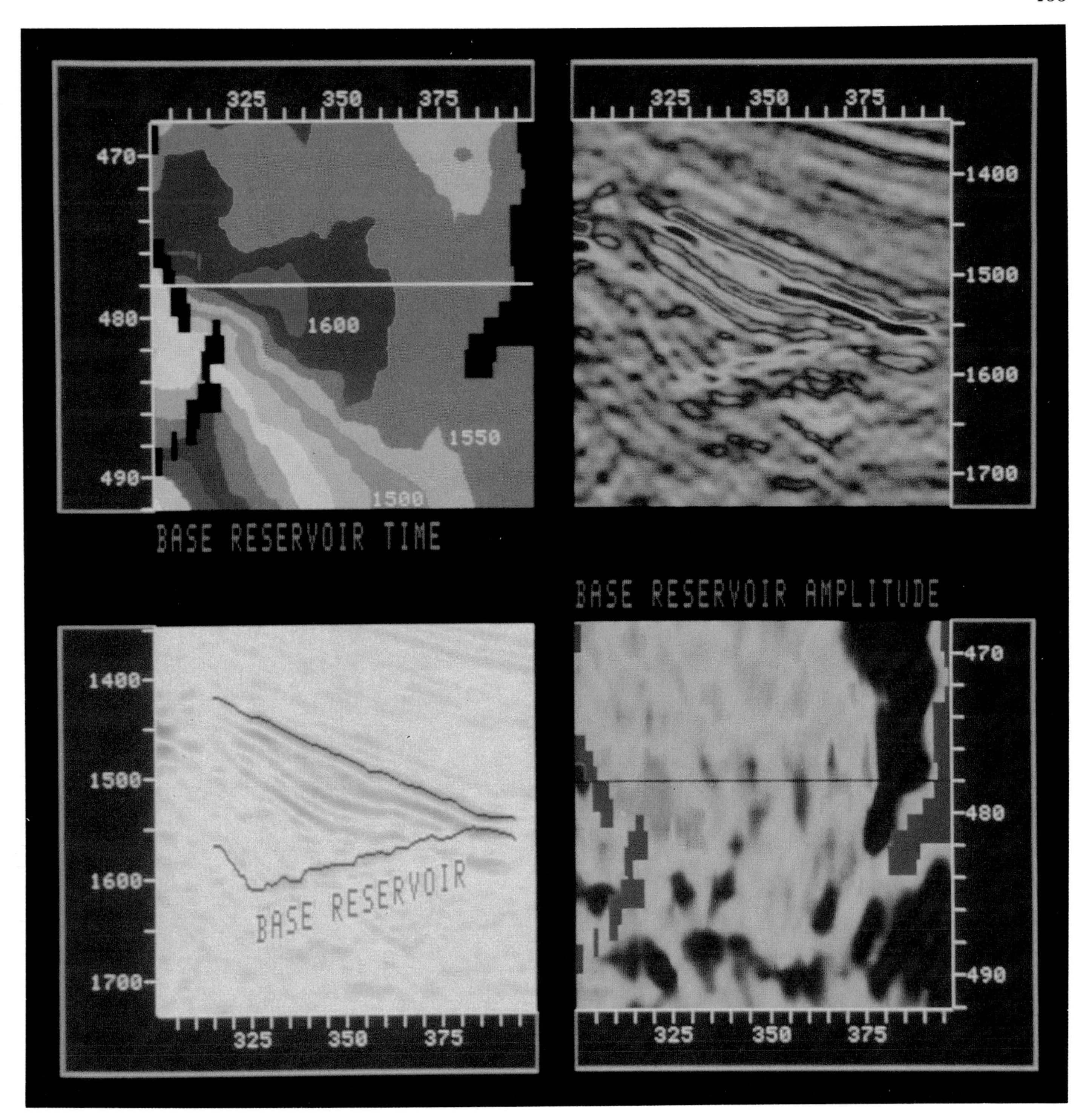

Fig. 7-16. (Upper Right) Vertical section through complex reservoir sand showing amplitude variation along fluid contact reflection.
(Upper Left) Time structure map on base reservoir reflection showing region of large gas velocity sag.
(Lower Left) Vertical section showing automatic tracks on top and base reservoir reflections.
(Lower Right) Horizon Seiscrop section for base reservoir reflection indicating internal reservoir layering by patterns in amplitude.
(Courtesy Chevron U.S.A. Inc.)

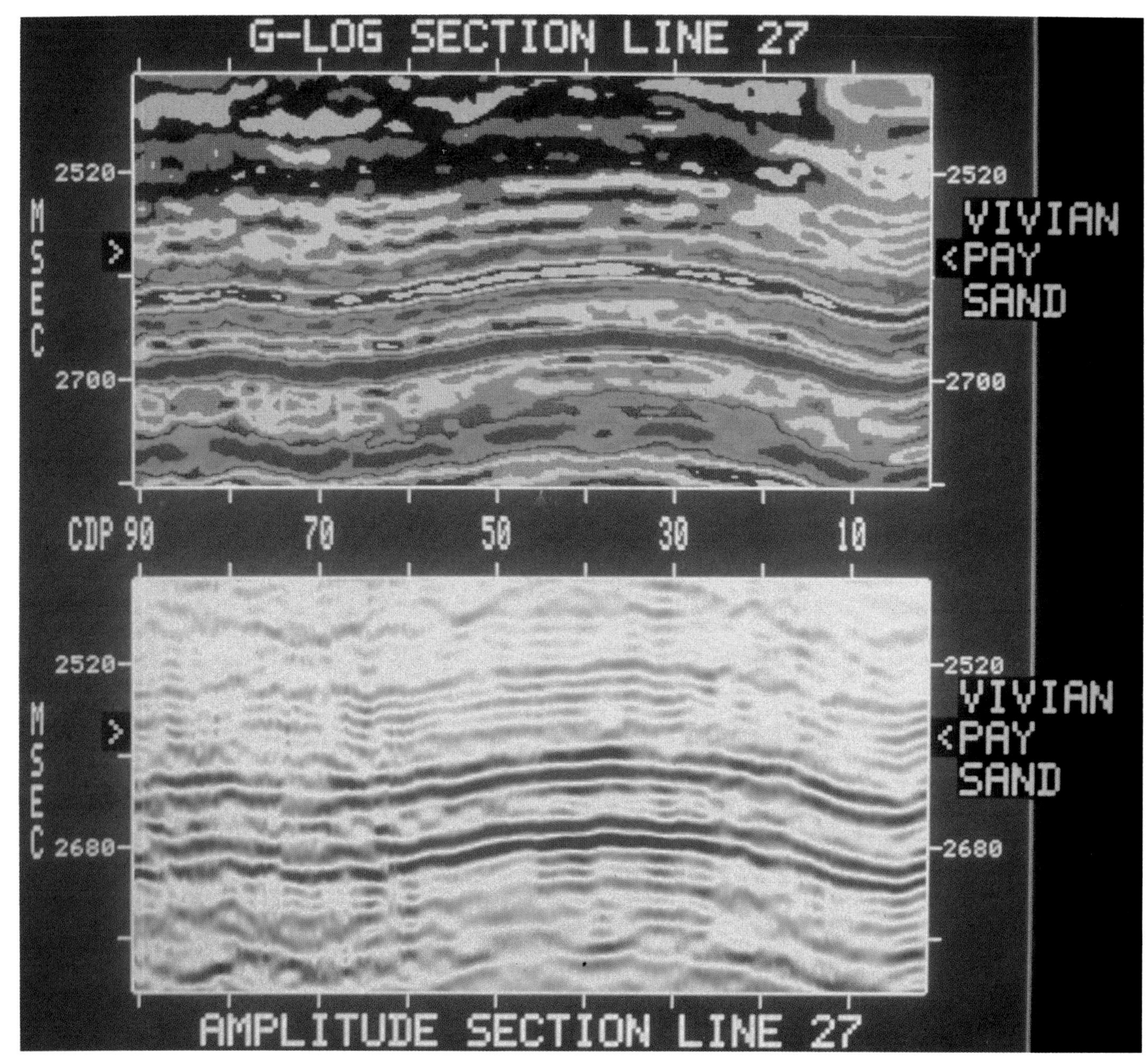

Fig. 7-17. G-LOG velocity section and color amplitude section from 3-D survey in Peru indicating Vivian Pay Sand. Velocity legend is in Figure 7-18. (Courtesy Occidental Exploration and Production Company.)

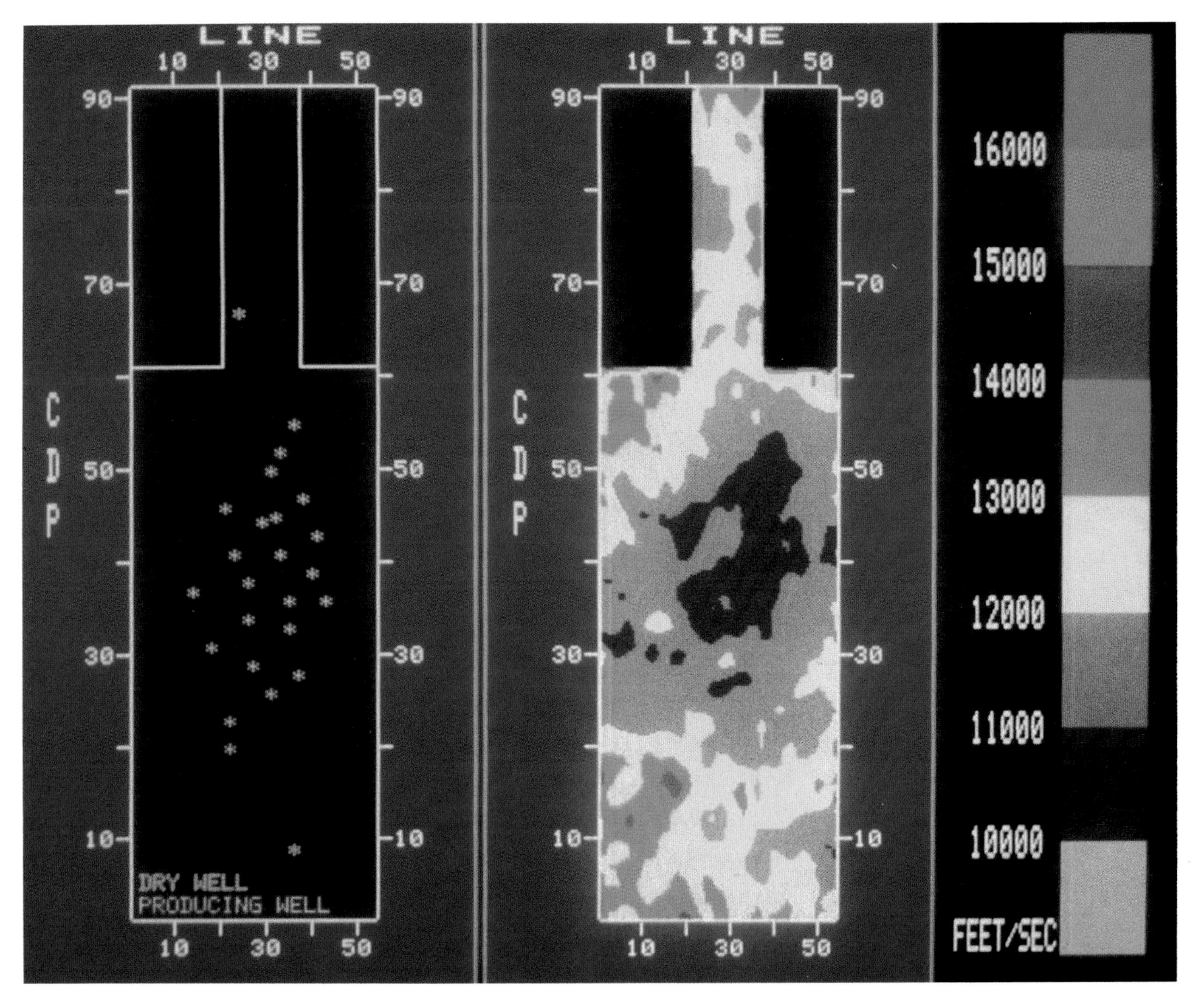

Fig. 7-18. Horizon Seiscrop velocity section through Vivian Pay Sand showing low velocity zone enclosing the area of producing wells. (Courtesy Occidental Exploration and Production Company.)

CASE HISTORIES OF THREE-DIMENSIONAL SEISMIC SURVEYS

This chapter presents four abbreviated case histories demonstrating the solution of subsurface problems with 3-D seismic surveys. The four have been selected on the basis of their diversity: land and marine environments, U.S. and overseas locations, structural and stratigraphic objectives, development and production arenas, authors from four major oil companies.

Case History 1

East Painter Reservoir 3-D Survey, Overthrust Belt, Wyoming

Donald G. Johnson, Chevron U.S.A. Inc.

The discovery of the East Painter Reservoir field in mid-1979 led to the initiation of the first major 3-D survey in the Wyoming Overthrust Belt. A 3-D survey was necessary because interpretation of conventional 2-D seismic data over the East Painter area did not provide a sufficiently reliable picture of the structure on the objective Triassic Nugget horizon to permit an aggressive development program. Field data for the 17 sq mi (44 sq km) East Painter 3-D survey were collected during the winter of 1979-80, and the final migrated sections were in hand by July 1980.

Interpretation of the final 3-D products resolved the previous structural ambiguities and showed the East Painter structure to be continuous and almost as large as the main Painter Reservoir feature. Information from the 3-D mapping allowed up to six development wells to be drilled at one time and helped to guide the locations of the last 13 development wells — all of them successful. The average cost per well was between \$4 and \$5 million. The cost of the 3-D survey was \$1.6 million, which turned out to be a good value.

Introduction

The complex structures of the Wyoming Overthrust Belt in the western United States are revealed with varying degrees of clarity by the conventional 2-D seismic reflection method. In some instances, however, additional structural definition is essential for exploration and production purposes, and the 3-D seismic method can make the difference in the resolution of the structural problem.

The East Painter Reservoir 3-D survey was prompted by results of the Chevron 11-5A well, a new field discovery in 1979 located approximately 1 mile east of the eastern fault-edge of the Painter Reservoir field. This well encountered the objective Triassic Nugget horizon dipping steeply to the northwest, which verified the existence of a frontal thrust structure to the Painter Reservoir feature. Interpretation of the Nugget horizon on conventional CDP seismic data suggested that the East Painter structure could be almost equal in size to the main Painter Reservoir field. However, data were very discontinuous, to the extent of nonresolution, over the central portion of the structure. The very poor data quality resulted from scattering and from destructive interference by out-of-plane energy. Because the 2-D seismic data did not provide a reliable interpretation, a 3-D survey was recommended to provide a better structural picture to facilitate development of the field.

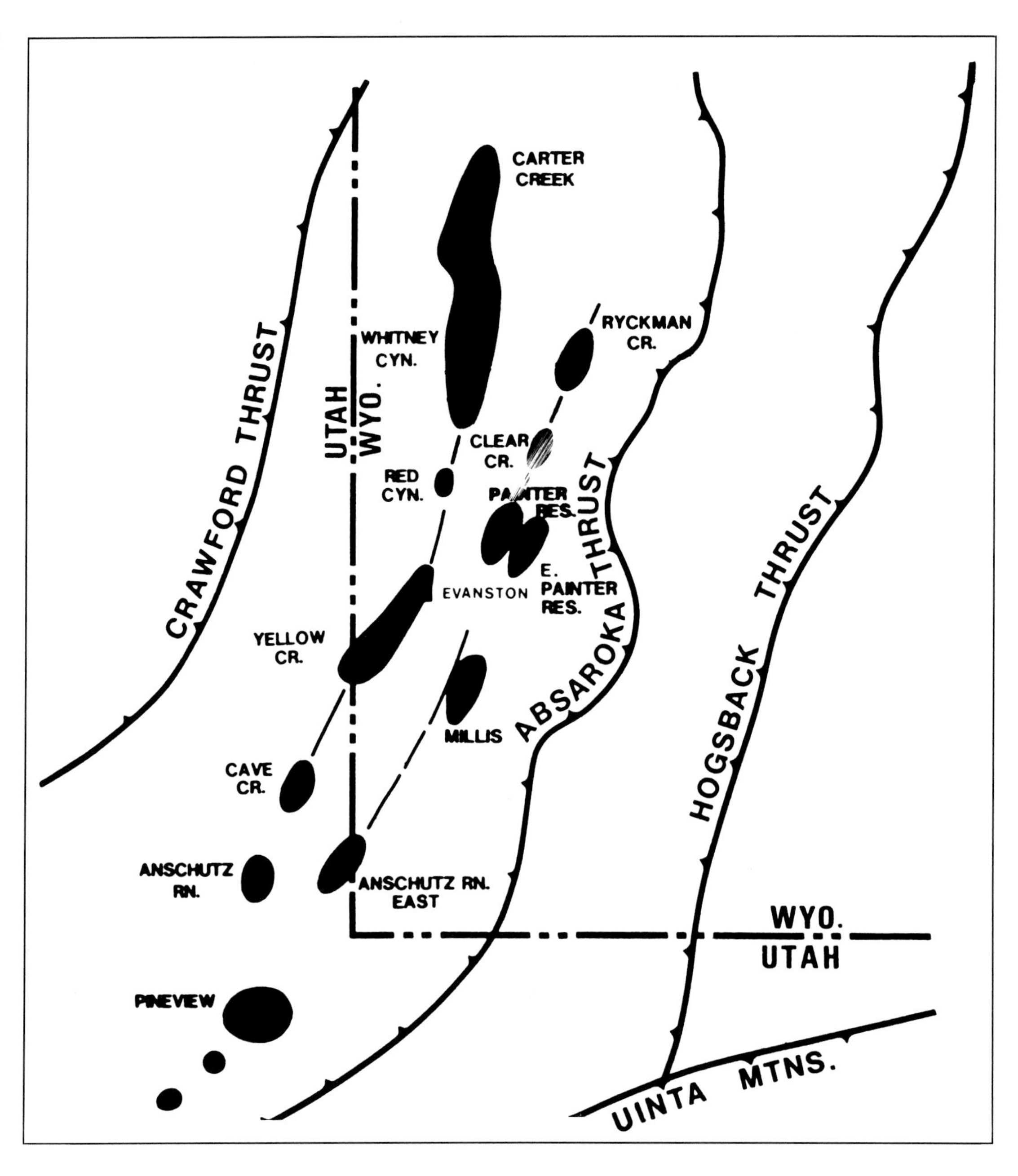

Fig. 8-1-1. Oil and gas fields of the Fossil Basin, Wyoming and Utah.

Geology

The major geologic objectives in the Fossil basin portion of the Wyoming Overthrust Belt are located in Jurassic, Triassic, Permian, Mississippian, and Ordovician sediments which have been folded and faulted into trap position on the Absaroka thrust plate. The Cretaceous sediments lying beneath the Absaroka thrust are the key source of hydrocarbons found in the Absaroka plate structures.

In the central Fossil basin, the frontal or easternmost structural trend on the Absaroka plate includes Ryckman Creek, Clear Creek, and Painter Reservoir fields (Figure 8-1-1). Their major production is oil, condensate, and sweet natural gas out of the Triassic Nugget sandstone. The Nugget horizon is cut off by the Absaroka thrust just east of the Ryckman and Clear Creek structures, but in the Painter Reservoir area, another fold-thrust trend is developed east of the Painter-Ryckman trend. It is this frontal trend that was disclosed by the East Painter Reservoir discovery (Figures 8-1-2 and 8-1-3).

Field Program

Partners in the East Painter 3-D survey were Amoco Production Co., Champlin Petroleum Co., and Chevron U.S.A., Inc. Chevron was the operator with a 50% interest in the survey. Geophysical Service Inc. (GSI) was contracted to conduct the field work and to process the data.

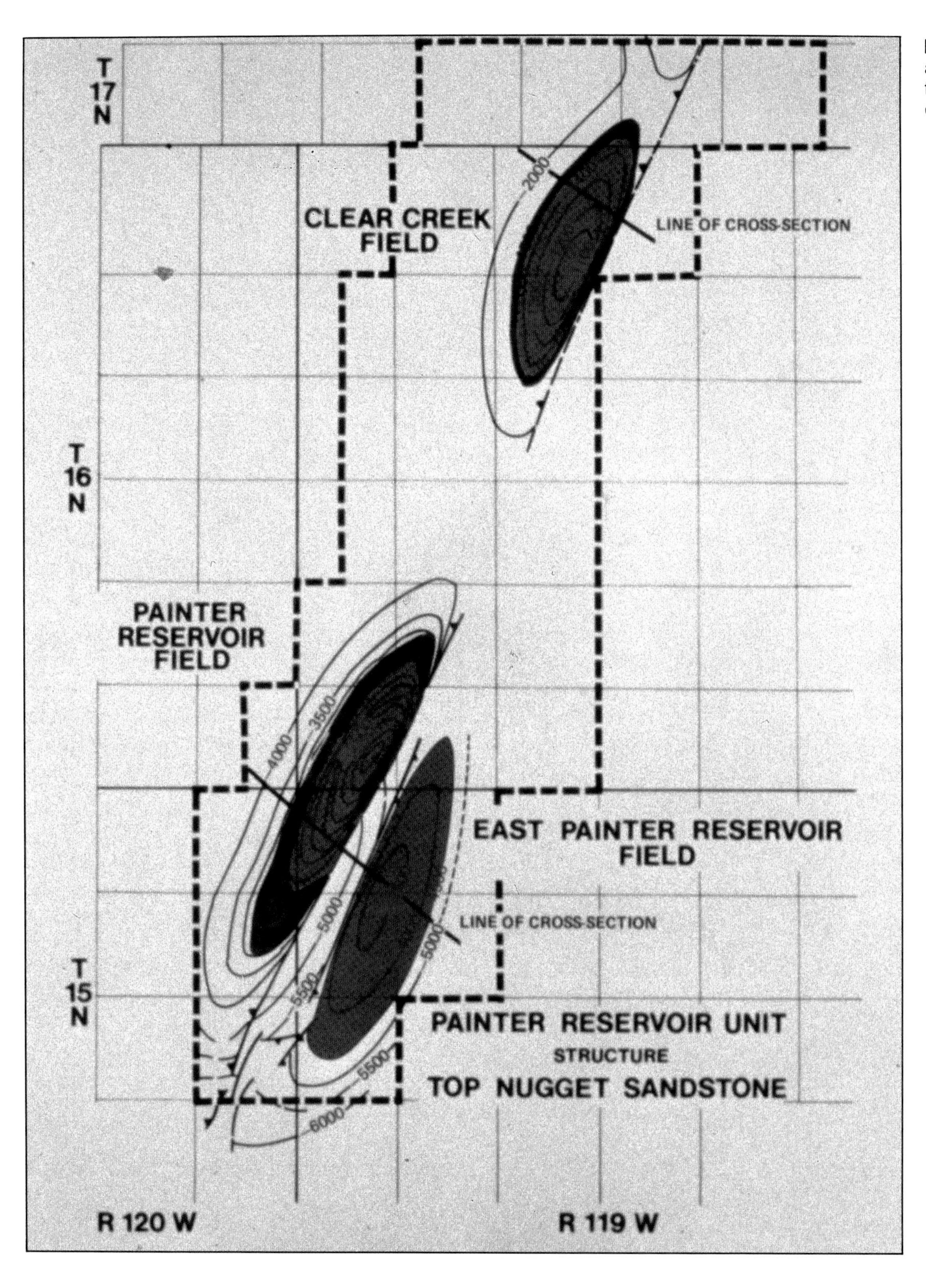

Fig. 8-1-2. Painter Reservoir and East Painter Reservoir fields, structural contour map on top of Nugget Sandstone.

The survey was carefully laid out so that the entire areal time expression of the East Painter feature could be recorded. This required 17 sq mi (44 sq km) of 3-D control. The CDP sampling was designed to be twice as fine in the dip direction (100 ft; 30.5 m) as in the strike direction (200 ft; 61 m) to prevent spatial aliasing of steeply dipping data. A 4-line "swath" shooting method was used with dynamite in shot holes as the energy source. Where shot holes could not be located because of rough topography or because of close proximity to drilling wells and pipelines, substitute shot locations were carefully determined by Chevron and GSI to ensure adequate 3-D

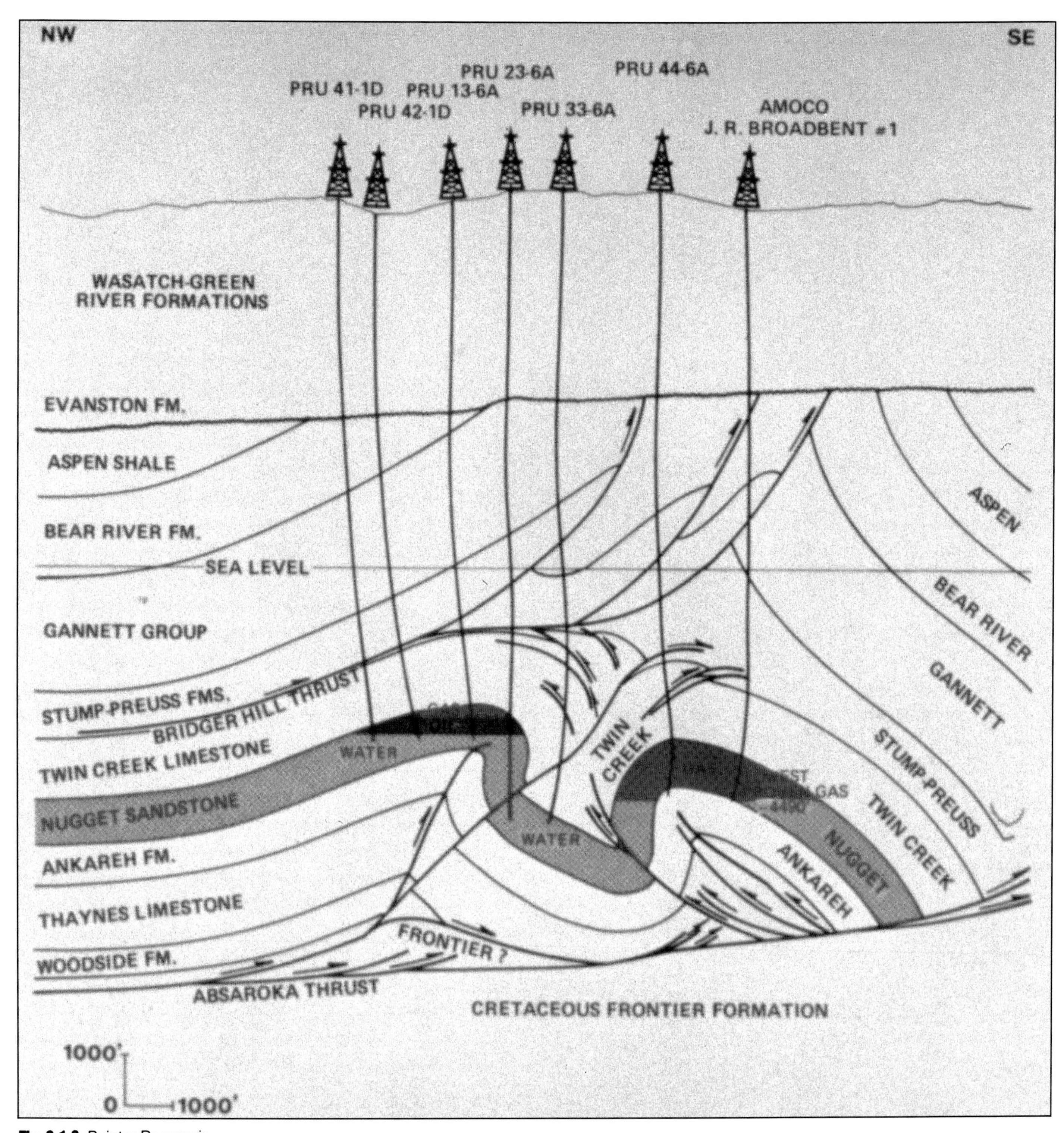

Fig. 8-1-3. Painter Reservoir and East Painter Reservoir fields, structural cross section.

coverage. Shooting began in September 1979 and was completed in March 1980. The migrated products were received in early July 1980.

Interpretation and Results

Final migrated data from the 3-D survey clearly resolved the structural configuration of the East Painter feature (Figures 8-1-2 and 8-1-4). The resulting interpretation showed the structure to be as large as previously mapped and the questionable central portion of the structure to be continuous. To date, a total of 16 wells have been drilled on the East Painter Reservoir structure. Thirteen of these were spudded after the 3-D survey was completed and their locations were guided by the 3-D mapping used in conjunction with the incoming subsurface control from the

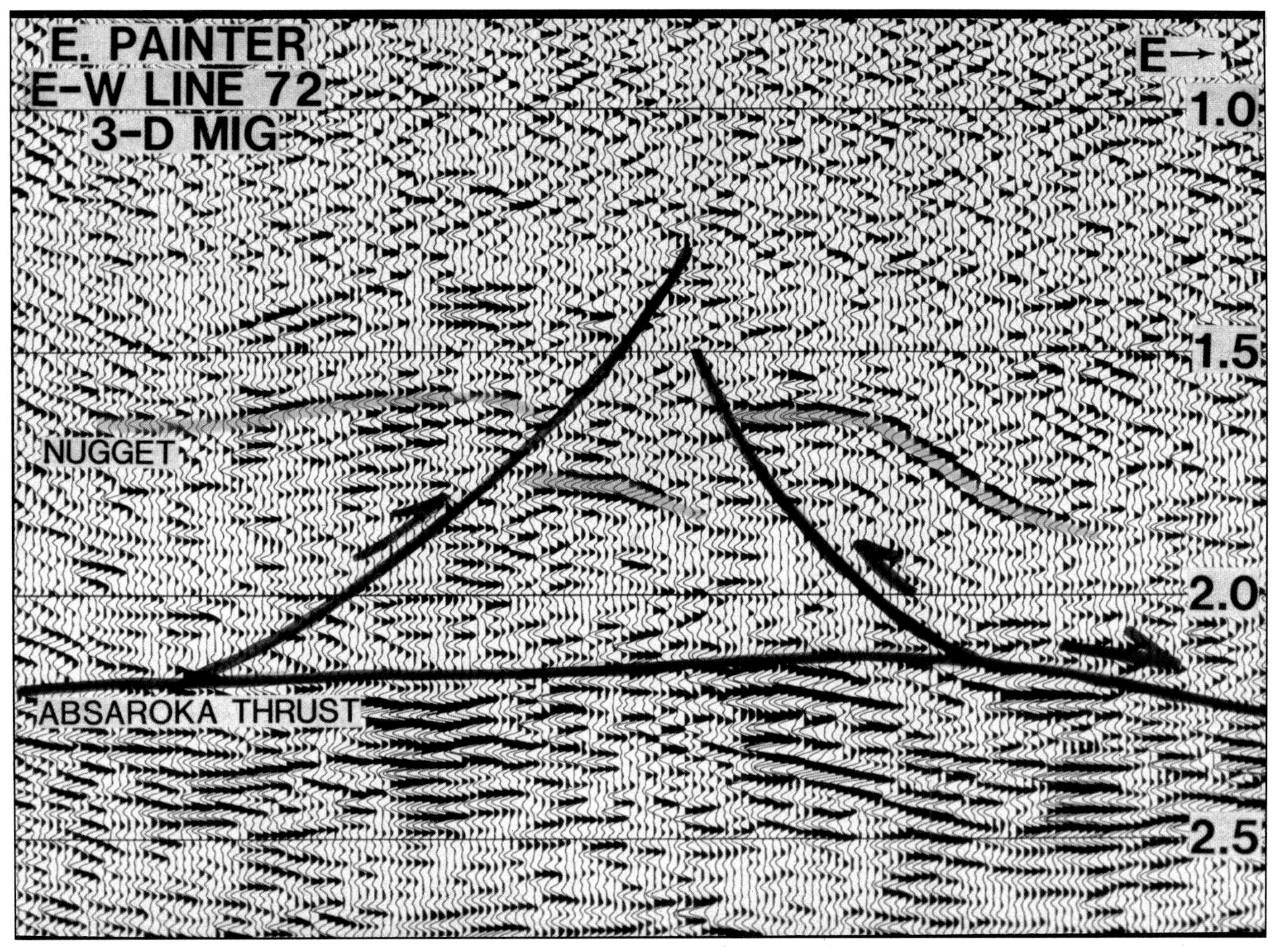

Fig. 8-1-4. East Painter 3-D migrated line 72 showing interpretation of Nugget horizon.

development drilling. All of the wells have been successful without any structural surprises on the Nugget horizon. The 3-D mapping allowed up to six development wells to be drilled at one time which greatly accelerated development of the field. The wells were drilled to an average depth of 12,500 ft (3,800 m) with an average cost per well of $4 to $5 million. The cost of the East Painter 3-D survey was $1.6 million — a good value!

Conclusion

Following the success of the East Painter 3-D survey, four additional thrust belt surveys were conducted during the next three years, three of which were larger than 45 sq mi (116 sq km). Without question, the 3-D seismic method is now an accepted and established exploration and development tool in the Overthrust Belt of the western U.S. As a final but necessary comment, it should be noted that the success of a 3-D survey is not automatic. Careful planning and the application of both geologic and geophysical expertise are essential to ensuring optimum results.

Statistics

Date of shooting	October 1979-March 1980
Area of coverage	17.4 sq miles
Fold	600-700%
Number of shots	1207
Collection cost	$1,214,500
Processing cost	$ 332,300
Total cost	$1,546,800
Cost per sq mile	$ 88,900

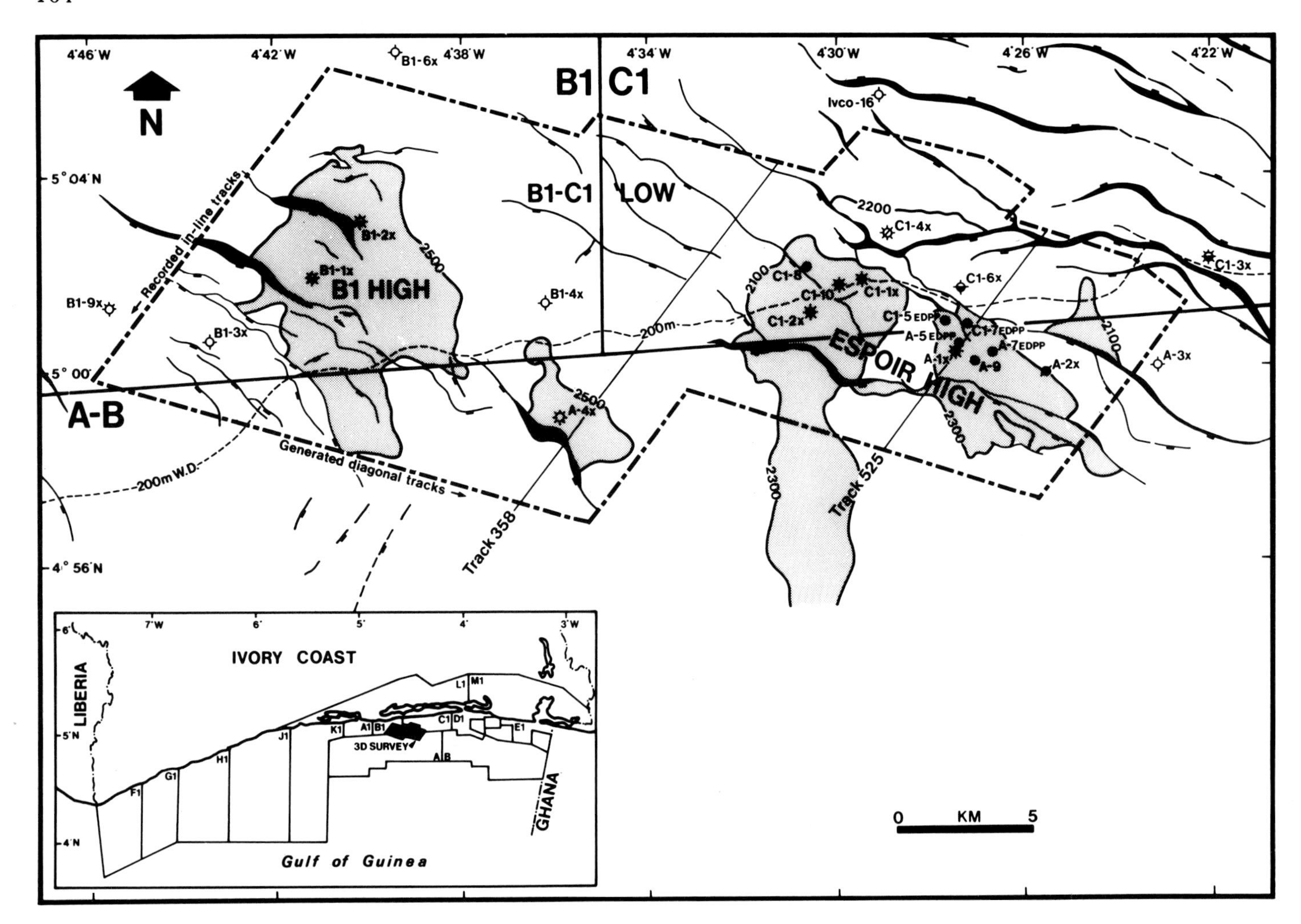

Fig. 8-2-1. Location map of 1981-82 Ivory Coast 3-D seismic survey showing structure of Albian unconformity. Contours in meters.

Case History 2

Three-Dimensional Seismic Interpretation: Espoir Field Area, Offshore Ivory Coast

L. R. Grillot, P. W. Anderton, Phillips Petroleum Co. Europe-Africa
T. M. Haselton, Consultants of Tri-D, Denmark
J. F. Dermargne, Phillips Petroleum Co. UK Ltd, England

The Espoir field, located approximately 13 km (8 mi) offshore Ivory Coast, was discovered in 1980 by a joint venture comprising Phillips Petroleum Co., AGIP, SEDCO Energy, and PETROCI. Following the discovery, a 3-D seismic survey was recorded by GSI in 1981-82 to provide detailed seismic coverage of Espoir field and adjacent features. The seismic program consisted of 7,700 line-km of data acquired in a single survey area located on the edge of the continental shelf and extending into deep water. In comparison with previous 2-D seismic surveys the 3-D data provided several improvements in interpretation and mapping including: (a) sharper definition of structural features; (b) reliable correlations of horizons and fault traces between closely-spaced tracks; (c) preparation of detailed time contour maps from time-slice sections; and (d) an improved velocity model for depth conversion. The improved mapping aided in the identification of additional well locations; the results of these wells compared favorably with the interpretation made prior to drilling.

Introduction

The discovery well, A-1X, was drilled in approximately 1,700 ft (518 m) of water to test a structural high at the Albian unconformity level (Figure 8-2-1). The well encountered hydrocarbon-bearing, reservoir-quality sands beneath this unconformity surface and an

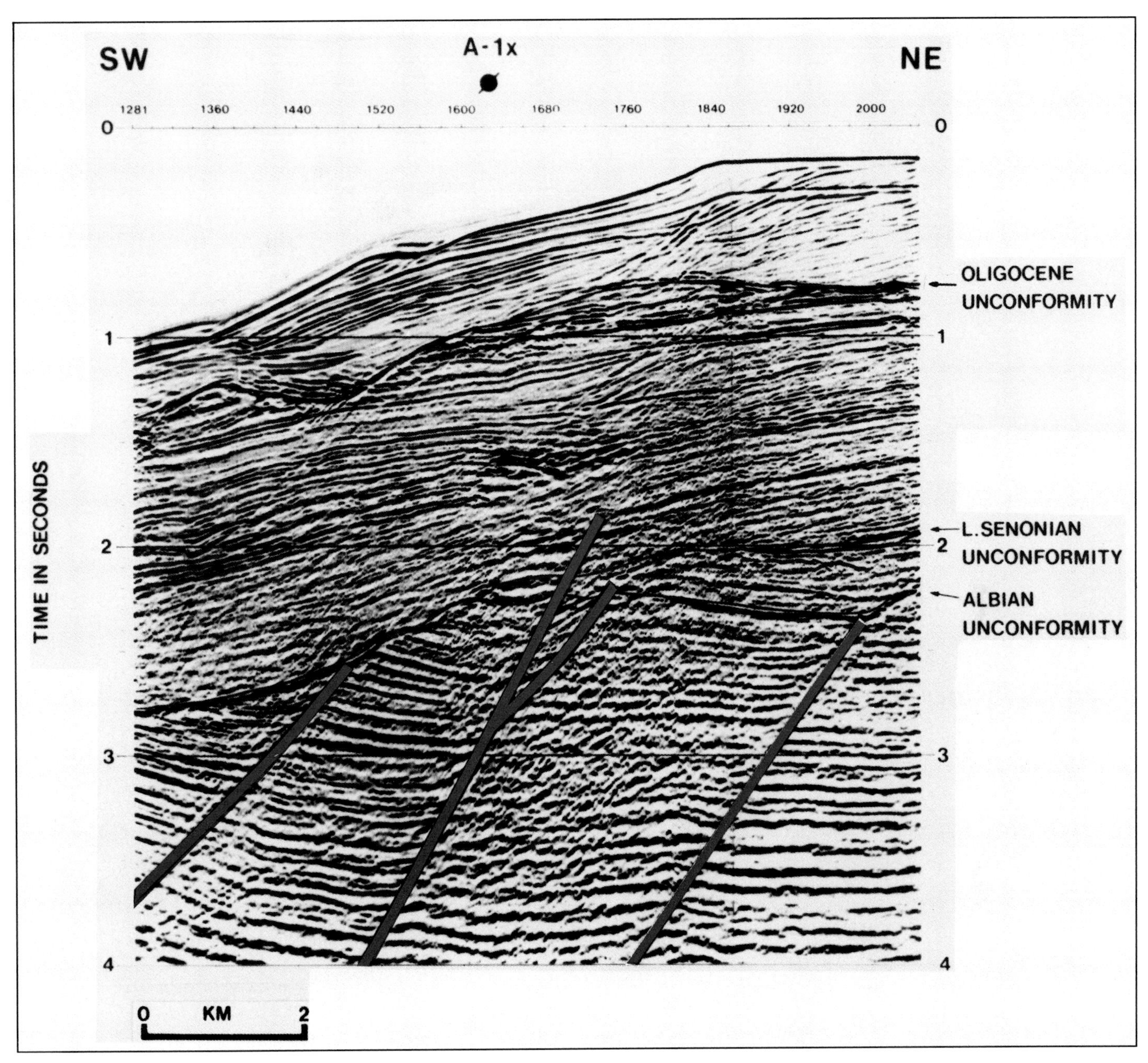

Fig. 8-2-2. In-line 525 crossing A-1X well location, Espoir field, and showing clear definition of rotated fault blocks beneath Albian unconformity.

appraisal well, A-2X, confirmed the presence of a significant accumulation in the Espoir area. Also, additional exploration work in the adjacent B1 block revealed other features of interest associated with the Albian unconformity. On this basis, the joint venture decided to undertake a 3-D seismic program which had as major objectives the detailed mapping of the Albian structure, as well as definition of the complex faulting which appeared to be present beneath the unconformity.

The rhomboid shape of the survey area (Figure 8-2-1) was devised to include both Espoir field and adjacent structures in a single survey and to orient the recording direction perpendicular to the major faults. The survey consisted of 525 northeast-southwest-trending lines recorded during a four-month period from October 1981 to February 1982. Data were recorded using a conventional 2,400-m cable and GSI's 4,000-cu inch air gun source. Recorded line lengths ranged from 8 to 15 km (5 to 9 mi).

Results and Interpretation

The resulting in-line sections (e.g. Figure 8-2-2) clearly demonstrate the primary mapping surface (Albian unconformity) and the tilted fault blocks typical of the structural style in the area.

Fig. 8-2-3. Time-slice at 2,848 ms across Espoir field showing traces of major faults beneath the Albian unconformity.

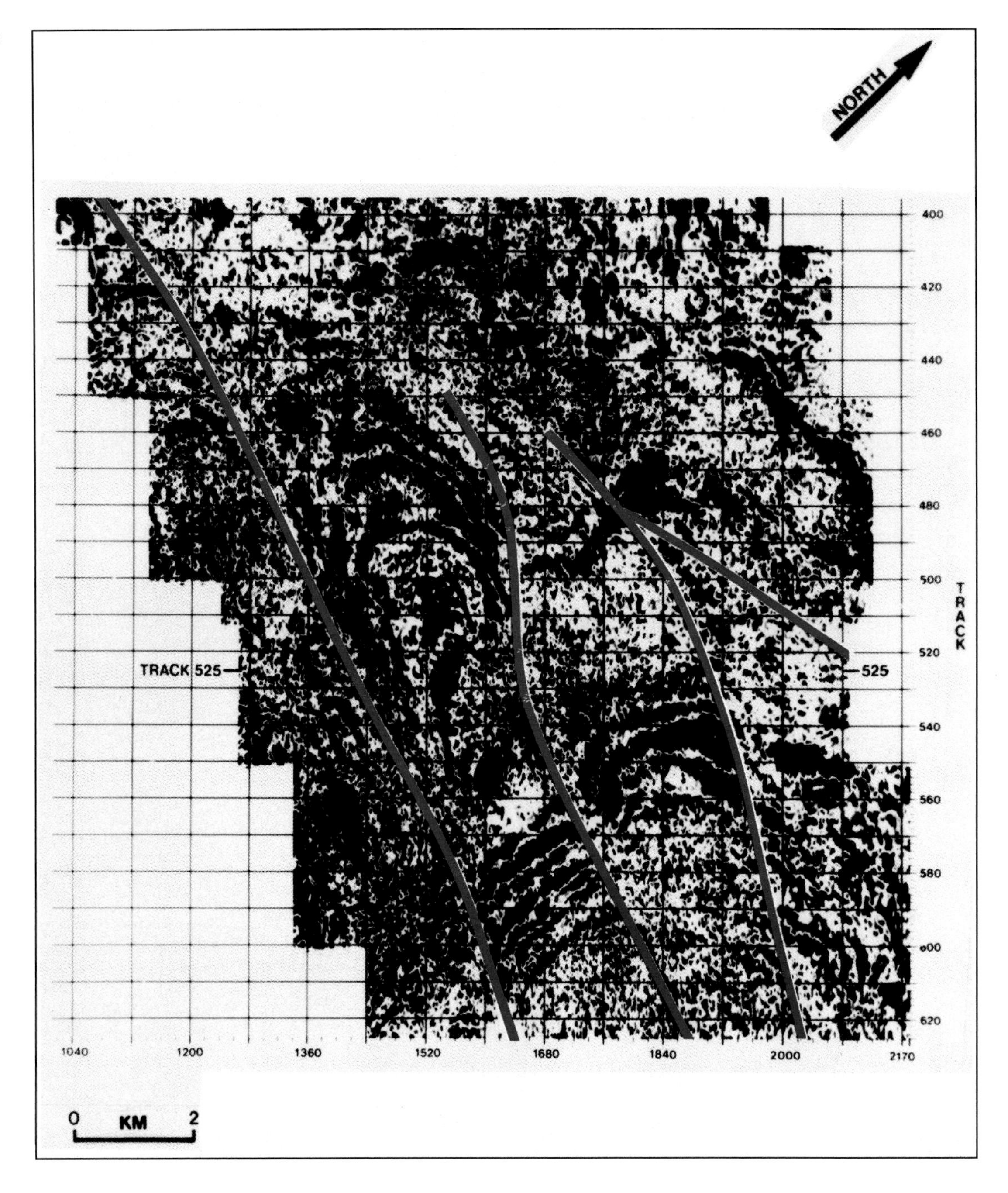

Figure 8-2-2 also demonstrates the sloping water bottom which gives a distorted structural picture on seismic time sections, both with regard to the closure at the Albian unconformity level and to pre-unconformity dips. This phenomenon was of particular concern during the velocity analysis and depth conversion stages of the mapping.

The bulk of the mapping was based on combined interpretation of vertical sections and horizontal time slices. Time slices were most useful where the reflections were distinct and not closely converging. In these areas, fault trends could be identified on time slices but generally the traces could not be mapped with the required precision. Figure 8-2-3 shows a time slice taken well beneath the Albian unconformity, which demonstrates these points. The red bands mark the traces of major faults; the individual seismic character of each fault block can be identified. However, the top part of the figure shows a zone where data quality is poorer and fault traces cannot be adequately mapped. In these areas, conventional interpretation of the closely-spaced (60 m; 200 ft) vertical sections was necessary to define fault traces, to correlate weak or complex reflections, and to map smaller depositional units.

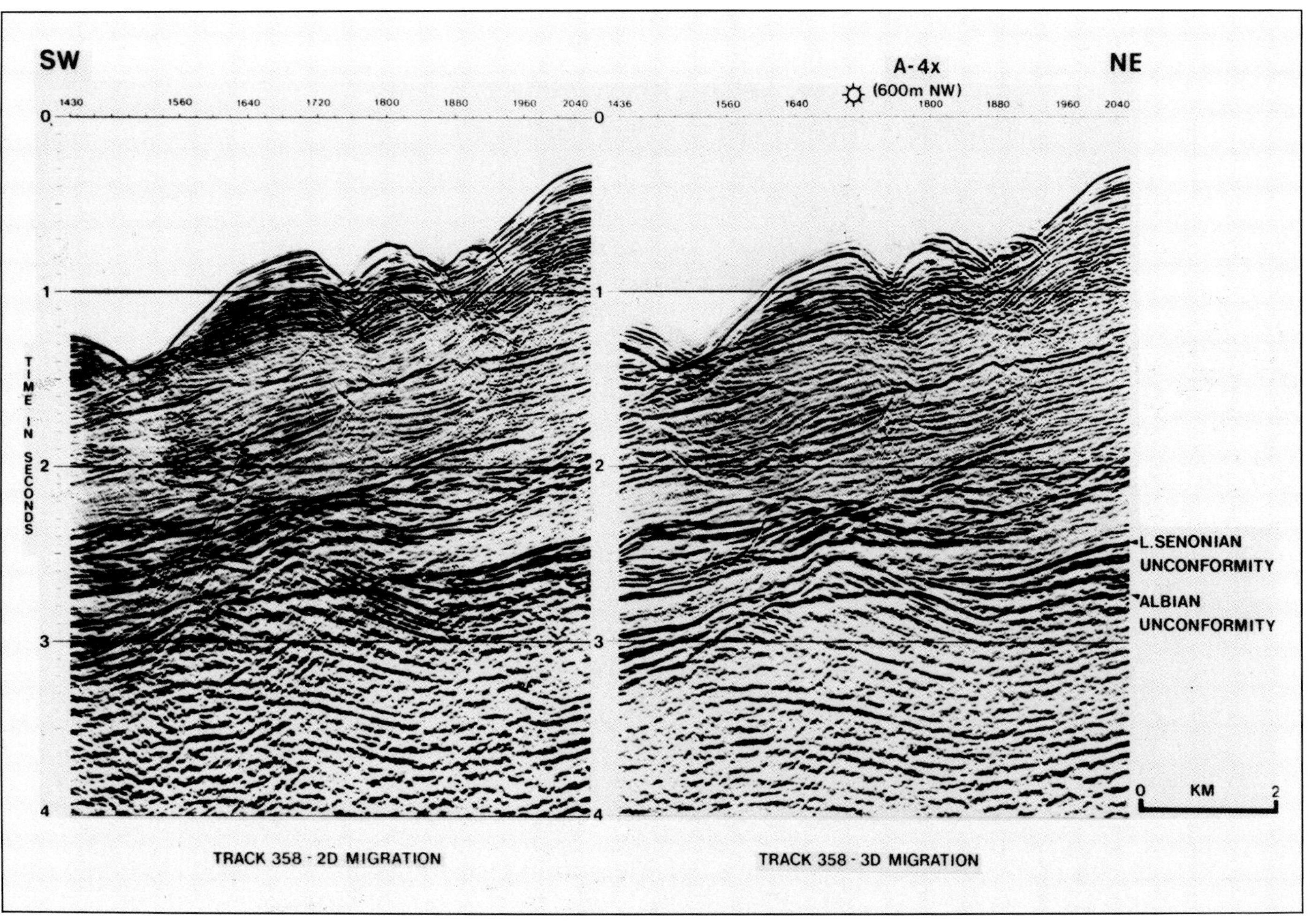

Fig. 8-2-4. Comparison of 2-D migration and 3-D migration sections across structure drilled by well A-4X showing improved definition of erosional high on Albian unconformity and fluid contact (flat spot).

In general, the 3-D data showed improved definition of the Albian unconformity surface across the entire survey area. This resulted in significant mapping revisions to the top of the reservoir interval and to changes in the extent of mapped closure over major structures in the area. Compared with previous efforts, the new maps showed increased closure at the Albian level over east Espoir together with a southward shift of the structural crest, especially in the vicinity of well A-2X.

A particularly interesting change in mapping occurred on the feature tested by well A-4X where the erosional high on the Albian unconformity is well resolved by the 3-D data. Figure 8-2-4 shows two versions of line 358 which crosses this feature near A-4X. The section on the left shows data at the intermediate 2-D migration stage. Although there is evidence of an anomaly in the center of the figure at about 2.7 seconds, the feature itself is not clear. The section on the right shows the same data after 3-D migration. The improvement in detail is noticeable and steeply-dipping intra-Albian reflections can be seen cutting through a flat spot which is close to a fluid contact defined in well A-4X. The slight tilt of the flat spot is due to the sloping water bottom. The 3-D mapping confirmed the structural isolation of the A-4X feature from the larger structure to the west.

Conclusion

In addition to the improvements in interpretation already discussed, benefits included better definition of pre-unconformity reflections, which resulted in improved mapping of intra-Albian horizons and better correlation across major faults. The ability to generate sections across well locations and individual features contributed to a better understanding and interpretation of the area. In the final stages of work, the improved velocity model derived from the closely-spaced velocity analyses aided the preparation of depth maps at the reservoir levels; this contributed to development of Espoir field and identification of further appraisal locations in the area. Overall, the 3-D survey has been a positive contribution to the evaluation of the Espoir area.

Case History 3

Field Appraisal With Three-Dimensional Seismic Surveys Offshore Trinidad

Robert M. Galbraith, Texaco Inc., Latin America/West Africa Division
Alistair R. Brown, Geophysical Service Inc.

A consortium operated by Texaco Trinidad Inc. commenced exploration in the South East Coast Consortium block offshore Trinidad in 1973. After four years of intensive exploration, a gas/condensate discovery was announced in early 1977 on the Pelican prospect. Later that year, in anticipation of the possible future need to site drilling/production platforms, a three-dimensional (3-D) seismic survey was recorded over the prospect. This survey resulted in improvements in seismic record quality, multiple attenuation, and fault resolution. A coordinated geologic-geophysical interpretation based on the 3-D seismic survey, a re-evaluation of log correlations, and the use of seismic logs differed significantly from earlier interpretations. Because of this, it is anticipated that the development of the field will need to be initiated in a different fault block from that previously envisioned.

A second 3-D survey contiguous to the Pelican survey was recorded in 1978 over the Ibis prospect. Results show significant data enhancement in the deeper part of the section and improved fault resolution relative to previous two-dimensional (2-D) control. The 3-D interpretation has revealed a much more complex fault pattern than originally mapped. Separate fault blocks will have to be individually evaluated, thus greatly increasing exploration risk.

Introduction

The republic of Trinidad and Tobago lies approximately 8 mi (13 km) off the northeast coast of Venezuela on the continental shelf of South America. The South East Coast Consortium was formed in 1973 to evaluate an offshore license obtained from the Government of Trinidad and Tobago in that year. The Consortium comprises Texaco Trinidad Inc. (operator), Trinidad and Tobago Oil Company Ltd., and Trinidad-Tesoro Petroleum Company Ltd.

The license area lies approximately 30 mi (48 km) off the southeast coast of Trinidad in the Galeota basin. This basin covers approximately 5,000 sq mi (13,000 sq km) in which thick Pleistocene to upper Miocene deltaic sandstones contain hydrocarbons in traps formed in gravity-induced structures. Closures consist of large diapiric anticlinal ridges and rollover features developed downthrown to major growth faults. To date, four major oil fields and four major gas fields have been discovered in the basin and recoverable reserves have been estimated at1 billion bbls of oil and 13+ trillion cu ft of gas.

Exploratory drilling in the Consortium block was carried out between 1975 and 1977 with a total of nine wells drilled on four separate structures. Of this total, three were drilled on the Pelican prospect with a gas/condensate discovery declared in 1977. However, even after four years of intensive exploration, including the recording of 1,400 mi (2,250 km) of 2-D seismic data, the Consortium was still unable to determine a location for a development platform. In seeking a solution, the Consortium engaged GSI to conduct a 3-D seismic survey over the Pelican structure in 1977. Following this, the Ibis 3-D survey was recorded in 1978.

All data were recorded with 24-fold geometry along lines oriented southwest-northeast, the predominant dip direction over the block. The lines were 100 m (330 ft) apart, and the subsurface interval along each line was 33 m (108 ft). The currents in the area were commonly 6 to 8 knots at right angles to the shooting direction, so the cable drift was high. Continuously recorded streamer tracking data provided the location of each depth point for each shot. A common-depth-point (CDP) set was then defined as those traces whose source-receiver midpoints fell within a bin 67x100 m. This limited the lateral subsurface smear to an acceptable level with a consequent improvement in the stack response.

One of the reservoirs in the Pelican area occurs at the top of the Miocene. The dip at this level between the Pelican-1 well and the northwestern boundary of the 3-D survey area was mapped to be 2,000 ft (610 m) on the pre-existing 2-D data. After the primary reflections had been correctly identified using the 3-D data, less than 1,000 ft (305 m) of dip were mapped on the north flank. This decrease in dip increased the interpreted hydrocarbon-bearing area under closure by approximately 20%, thus significantly affecting reserve estimates and development economics.

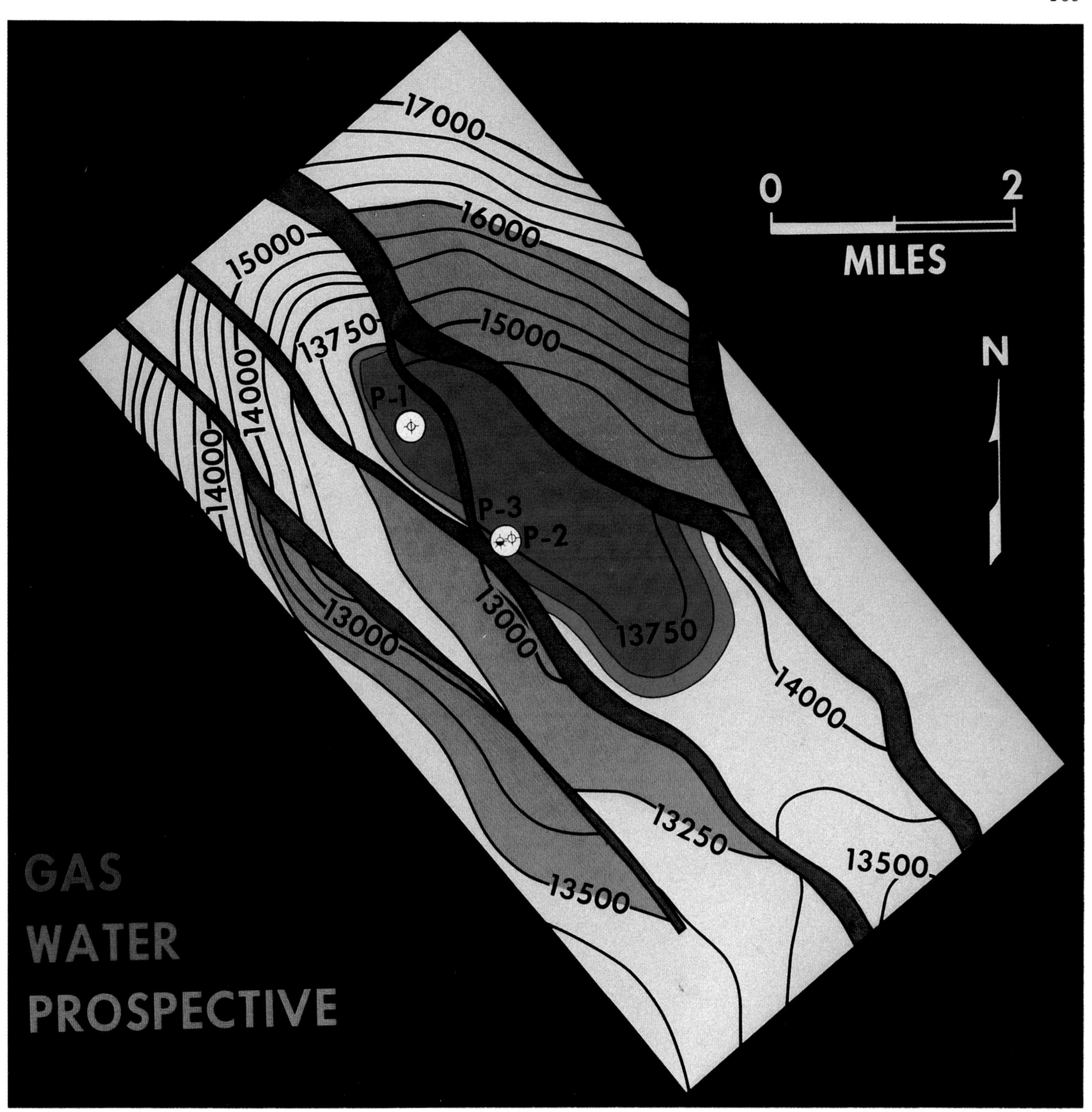

Fig. 8-3-1. Map of Pelican-3 sand, offshore Trinidad, interpreted from 2-D data. Contour interval 250 ft (76 m).

Results and Interpretation

The prime reservoir in the area is the Pelican-3 sand. Figure 8-3-1 shows the interpreted map at this level before the 3-D survey. Figures 8-3-2 and 8-3-3 show two interpretations made from the 3-D survey data. While a similar difference in the northwest dip exists at this level as was mapped at top Miocene, the principal difference between pre- and post-3-D interpretations concerns the faulting.

Initial interpretation of the logs from Pelican-1 and Pelican-3 wells indicated different water levels in the Pelican-3 sand. This was explained by a cross-fault separating the two wells (Figure 8-3-1). The 3-D data precluded the possibility of this cross-fault. Instead, the growth fault has

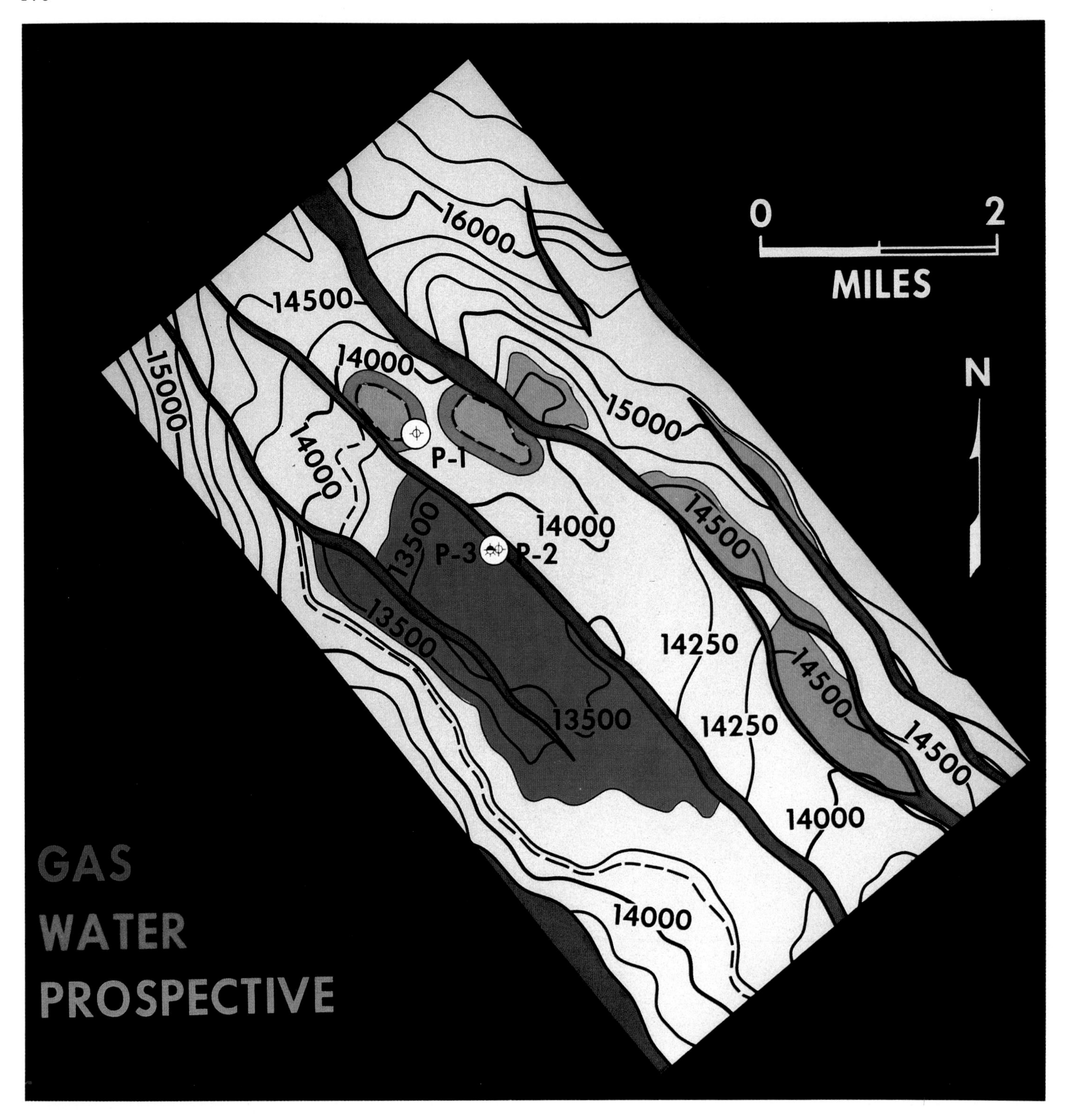

Fig. 8-3-2. Map of Pelican-3 sand interpreted from 3-D data with southeastern structural closure but not honoring water level in well. Contour interval 250 ft (76 m).

been interpreted farther northeast, thus separating the two wells at the Pelican-3 sand. The impact of this on the interpreted position of the reserves is shown in Figure 8-3-2. The recommendation based on the 3-D interpretation was therefore to initiate development drilling in a different fault block from the one proposed prior to the acquisition of 3-D control. This change in interpretation has probably saved the South East Coast Consortium the expense of at least one dry hole and possibly the cost of mislocation of a development platform.

The water level in the Pelican-3 sand in the Pelican-3 well is near 13,800 ft (4,210 m). The contour at this level is shown by a dashed line in Figure 8-3-2. This is 200 ft (60 m) deeper than

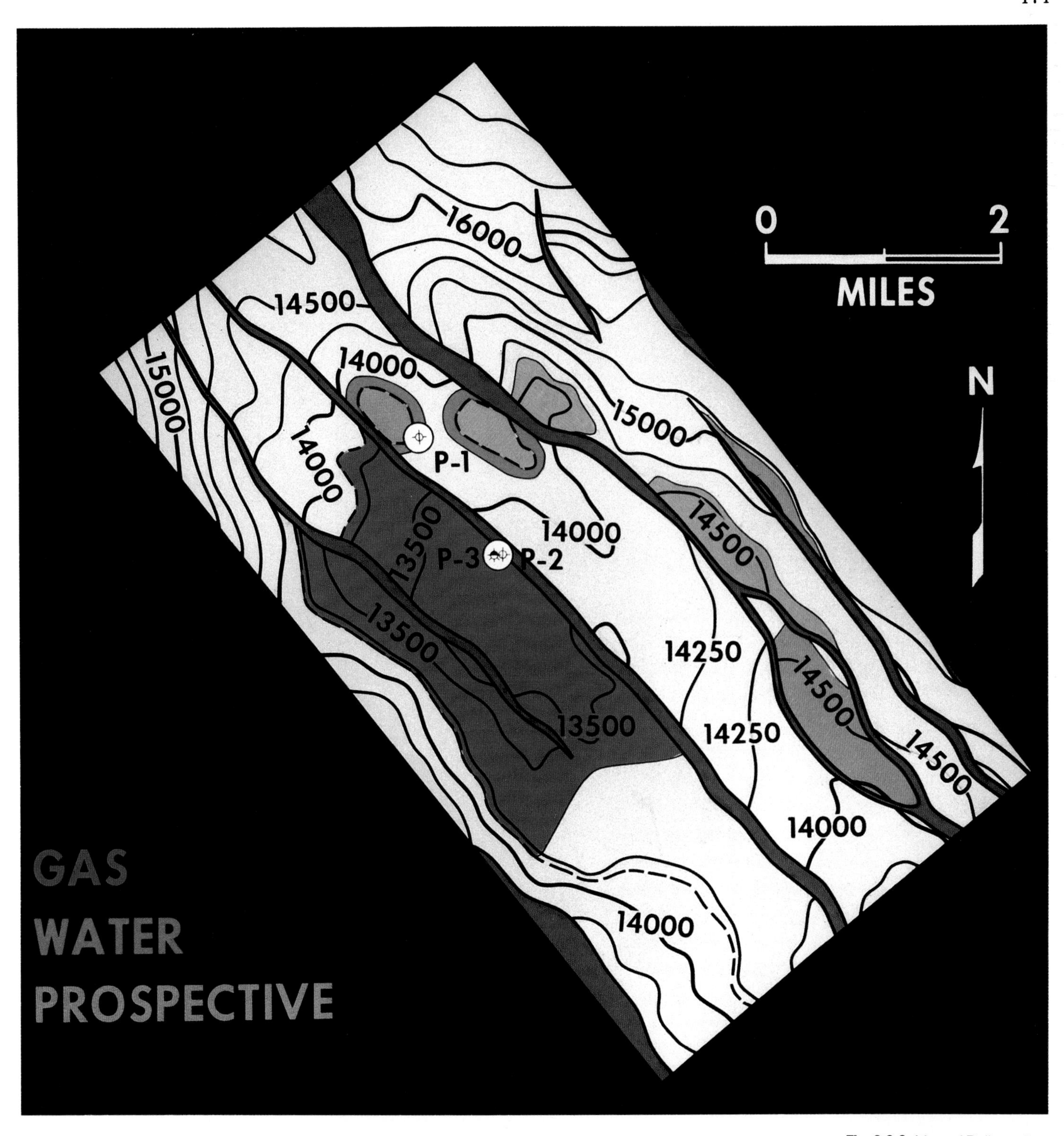

Fig. 8-3-3. Map of Pelican-3 sand interpreted from 3-D data with southeastern stratigraphic boundary and honoring water level in well. Contour interval 250 ft (76 m).

the structural spill point of 13,600 ft (4,150 m) which, on the basis of structural closure alone, would control the downdip extent of the gas. An alternative interpretation which honors the water level in the well is shown in Figure 8-3-3. This invokes a stratigraphic reservoir boundary on the southeast.

The seismic section along crossline 87, northwest-southeast through Pelican-3 well, shows a very marked character change at the Pelican-3 reservoir level southeast of the well. This probably indicates the position of the stratigraphic boundary. This character change is evident on seven crosslines which intersect the boundary, and also on several Seiscrop sections, from

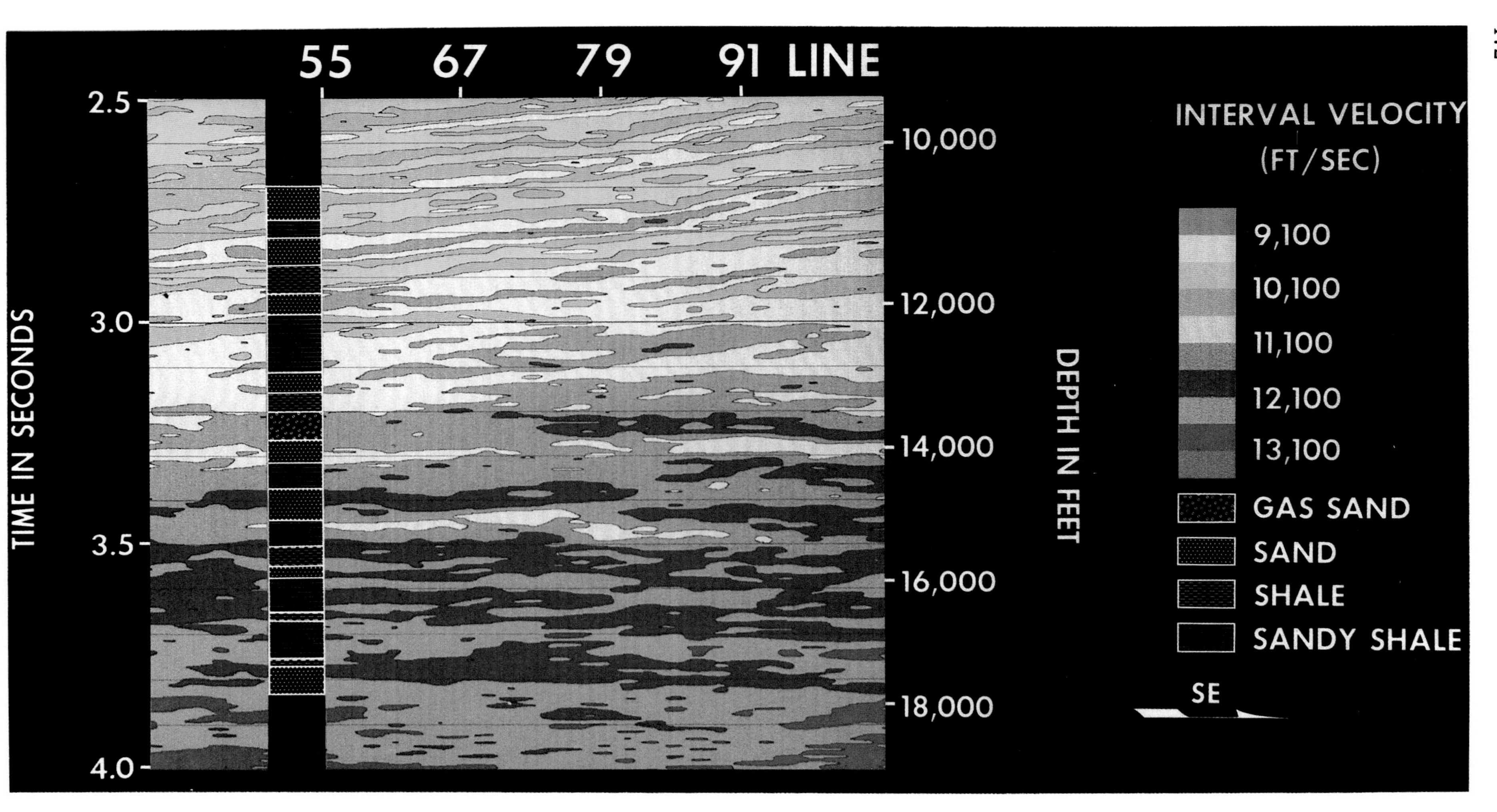

Fig. 8-3-4. G-LOG velocity section along crossline 87 through Pelican-3 well showing lateral velocity transition across inferred southeastern reservoir boundary.

which its position was mapped (Figure 8-3-3).

The G-LOG* process of seismic inversion was applied to crossline 87 through Pelican-3 well in an attempt to study the nature and validity of the stratigraphic boundary. The resulting G-LOG section in color is shown in Figure 8-3-4. Generally, the higher velocities correspond to the sands and the lower velocities to the shales. Cyclical sand-shale deposition is evident above 3.0 seconds.

Conclusions

The simplified lithology in the well shows the Pelican-3 gas sand between 3.20 and 3.26 seconds (Figure 8-3-4). There is no low velocity expression of this interval on the G-LOG section. However, away from the well to the southeast, the correlative interval shows an abrupt lateral increase in velocity; this is interpreted as the stratigraphic reservoir boundary. Close examination of the transition suggests layering which is also observed in the well; in the upper portion of the reservoir the transition occurs at line 70, in the next layer at line 79, and in the lower half of the reservoir at line 73. The magnitude of the velocity contrast across the boundary is approximately 600 ft/sec (180 m/sec). It is concluded that this lateral change from low to high velocity indicates the change from a porous gas-filled sand to a tight sand, in which the pores are filled with cement which is probably clay.

Data quality has been improved. Processing took into account cable drift, a major problem offshore Trinidad, thus limiting subsurface smear during stack. Some deep primary events have been observed for the first time. Because of increased data density, fault definition is excellent. Structural interpretations are more reliable with removal of energy from outside the plane of the section. The flexibility which permits an interpreter to generate lines in any direction is a significant benefit. The probable containment of the principal Pelican reserves by a stratigraphic reservoir boundary to the southeast has been substantially validated after a detailed study of its nature.

The 3-D results have caused major changes in the Pelican field development plans. The interpreted area under closure has been increased. The possibilities of drilling an initial dry hole and mislocating a development platform have been reduced due to improved reliability of the coordinated geologic-geophysical interpretation based on the 3-D seismic survey and a re-evaluation of log correlations. This has had a positive effect on development economics.

The 3-D seismic method has proved to be a useful tool for field appraisal in this area offshore Trinidad and will be considered over other prospects prior to commitment to expensive offshore development programs.

*Trademark of Geophysical Service Inc.

Fig. 8-4-1. Location map of shots and receivers for enhanced oil recovery 3-D seismic survey, north Texas.

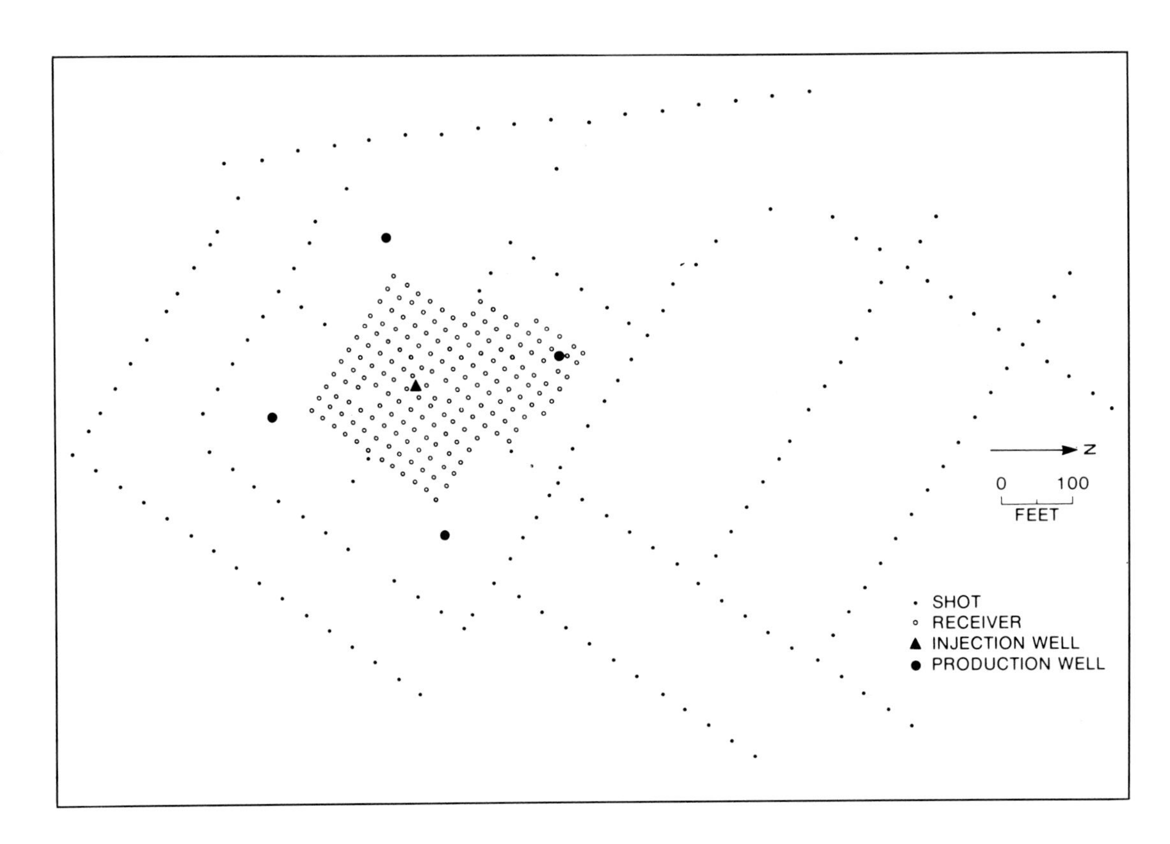

Case History 4

Three-Dimensional Seismic Monitoring of an Enhanced Oil Recovery Project

R. J. Greaves, T. J. Fulp, Arco Oil and Gas
P. L. Head, Arco Exploration

The 3-D seismic survey technique has been used to monitor the progress of an enhanced oil recovery project in which production is stimulated by in-situ combustion driven by injected gas. A baseline 3-D data volume was recorded before the initiation of the combustion program. After combustion had been allowed to proceed for some time, the 3-D survey was repeated. Since the basis for tracking the effects of the combustion process was comparison, great care was taken to duplicate field geometry, recording parameters, and data processing. VSP data were also recorded to locate precisely the target sand reflection time and character.

Before the analysis of the 3-D data, synthetic traces were generated from well log data modified in several ways to simulate the effects of the combustion process. The target sand is characterized seismically by an impedance contrast caused by low density. The predicted changes in reflection character are primarily due to changes in density caused by increased gas saturation. Complex trace attributes were computed to examine amplitude and other waveform changes. Comparison of pre-burn to post-burn data shows differences that can be explained by increased gas saturation.

Introduction

The quantitative measurement of the success of enhanced oil recovery methods is a very important but difficult task for petroleum engineers. One measure of success is the volume of the producing formation contacted by the enhancement process. The purpose of this project was to test the ability of reflection seismology to map the propagation of a particular type of enhanced oil recovery — in-situ combustion. A seismic program was carried out to collect duplicate sets of 3-D seismic data before and after combustion had occurred so that the areal extent of the burn process could be mapped and some estimate of the vertical thickness could be made.

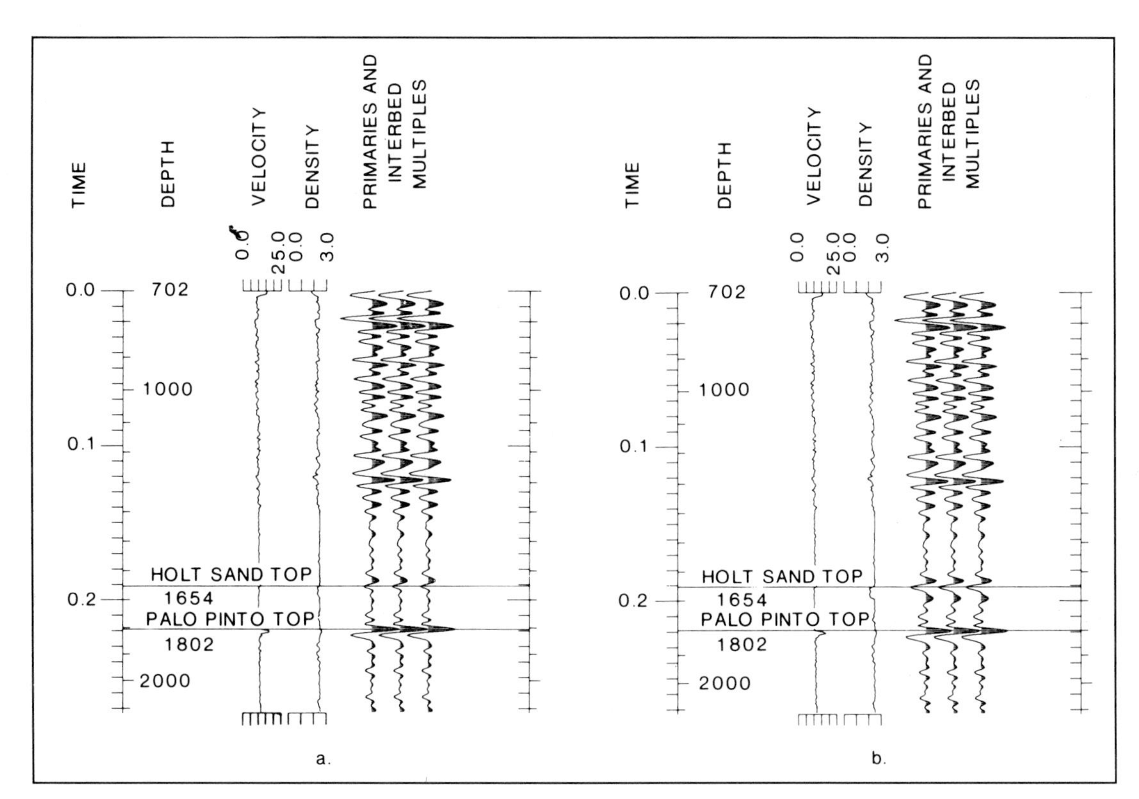

Fig. 8-4-2. Modeling the effect of combustion on reflection amplitude: (a) pre-burn velocity and density and the corresponding synthetic trace; (b) velocity and density with a 5 and 10% reduction, respectively, and the corresponding synthetic trace.

The engineering process used a 5-spot well pattern with four production wells, spaced 300 ft (91 m) apart, surrounding a combustion gas injection well. The producing formation is the Holt sand of the Pennsylvanian Canyon group. It is 40 ft (12 m) thick, sandwiched within a shale section, and approximately 1,650 ft (500 m) deep. Equations have been derived for calculating temporal resolution and tuning thickness in terms of wavelet frequency characteristics. Since the velocity of the sand from well log data is about 10,000 ft/sec (3,050 m/sec), a predominant frequency of 125 Hz would allow a tuning thickness of 20 ft (6 m). For areas where burn thickness is less than the tuning thickness, burn thickness might be estimated from the interference amplitude, but calibration would be difficult. These estimates of possible resolution are based on a simple model and do not take into account the effects of noise. To maximize potential resolution, data collection techniques were chosen to collect the highest possible frequencies with minimum noise levels over a broad frequency range.

Acquisition

A 3-D patch survey geometry was laid out for data collection. This type of shooting geometry allowed high-fold CDP coverage over the target area in spite of limited access at the site due to production facilities, roads, and power lines. Another advantage is that no movement of receivers was required during the project, so the geophones were permanently buried. This helped reduce receiver noise and made duplication of the survey a much easier task. The shot and receiver patterns are shown in Figure 8-4-1. A modification to the simple 3-D patch geometry was made by adding shot and receiver positions to the north. This accounted for migration of reflections due to a slight northward dip of the sand and surrounding beds.

The data were recorded using a 192-channel GUS BUS system at a 1 ms sample rate with 50 Hz low-cut and 320 Hz anti-alias high-cut filter. A single 40 Hz geophone was used for each receiver group and buried at 20 ft (6 m); the group spacing was 20 ft (6 m). The source was 3 lb of explosives buried at 75 ft (23 m) at each shotpoint. These parameters were chosen to minimize surface wave noise but retain a high-frequency spectrum.

Modeling

The ability of reflected seismic signals to detect the combustion zone depends on the effect of the combustion process on the acoustic properties of the Holt sand. Rock physics laboratory tests performed on core samples from the formation indicated that a total replacement of the pore

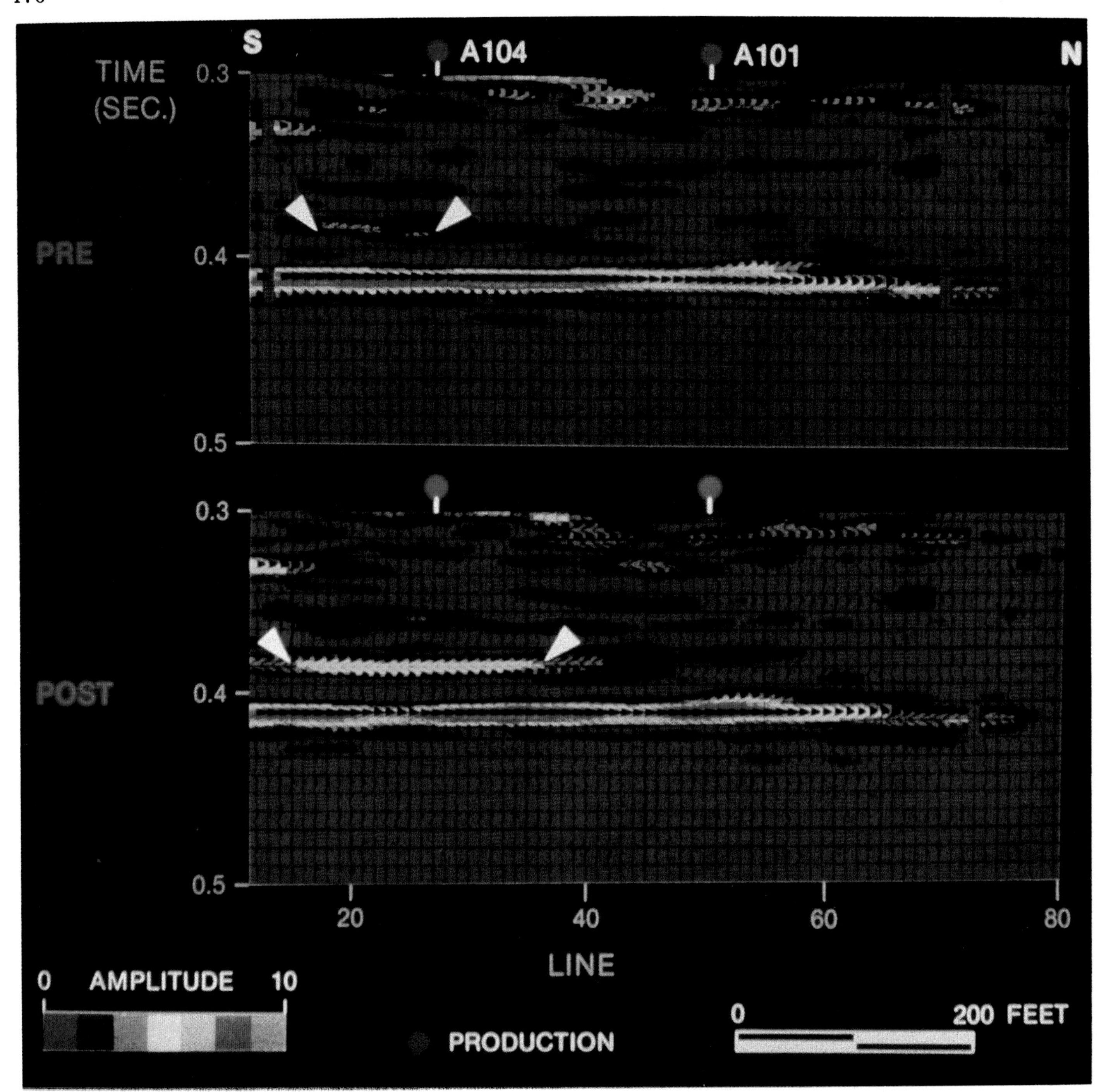

Fig. 8-4-3. Envelope amplitude, line 14, showing top of Holt Sand reflection, indicated by arrows, from pre-burn (above) and post-burn (below) data volumes.

fluid by gas in the burned-out zone would result in about a 5% decrease in velocity, with a corresponding 10% decrease in density.

Figure 8-4-2a shows the sonic velocity and density logs recorded in a well penetrating the Holt sand. The outstanding acoustic feature of the Holt sand is its low density relative to the surrounding shale. Acoustic impedances calculated from these logs were used to generate synthetic traces. Then for modeling purposes, the estimated density and velocity decreases were applied to the logs and the synthetic traces regenerated (Figure 8-4-2b). The effect of the burn is a significant increase in reflection amplitude (that is, a bright spot) created by increased gas content. Increased reflection amplitude from the sand, then, is one feature to look for in locating the burned volume. It can also be projected that in the shadow of the bright spot a decrease in the reflection amplitude of the Palo Pinto limestone below the sand might be observed. This dim spot would be generated by a decrease in energy transmitted through the sand due to the burn effects.

Interpretation and Conclusion

The data were processed using standard CDP methods but with special care to preserve true relative amplitude. Figure 8-4-3 shows an example of a profile sliced out of the pre-burn and post-burn 3-D migrated data volumes. Careful comparison of these profiles was made, looking especially for increased reflection amplitude at the Holt sand but also for any significant differences in the data character of the sand and limestone reflections. Three zones with significant differences in amplitude or character are observed in the post-burn data (lower section, Figure 8-4-3). These are a slight increase in amplitude of the Holt reflection, a small change in the side lobe interference character, and a significant decrease in the amplitude of the deeper Palo Pinto reflection.

The Holt amplitude increase when mapped through the 3-D data volume covered an area larger than expected for the burn propagation and may be caused partially by gas injected into the formation previous to combustion. The areal extent of the change in interference character and of amplitude decrease is of the dimensions expected for the burn volume and is initially interpreted to be an indication of maximum combustion.

Detailed analysis and interpretation of the data are aided by use of a color graphics CRT. Seismic attributes have been calculated, and color displays show substantial differences in envelope amplitude and instantaneous frequency. Time slices of the data with structure removed (horizon slices) are used to map trace amplitude, attributes, and their differences automatically. The effectiveness of the seismic data in mapping the combustion zone was tested by comparison to the burn volume deduced from postburn core holes.

INTERPRETATION EXERCISE

Background Information

The object of this exercise is to map the structure and extent of a turbidite sand, known from well control to be visible on the 3-D seismic data as a high amplitude peak. You are provided with two vertical sections (Figures A-1 and A-2) and eight horizontal sections (Figures A-3 through A-10). The coordinates of a point at which you can identify the turbidite reflections are:

Line 539
Crossline 600
Time 1,600 ms

Structural Component of Exercise

Follow the crest of the identified blue event on each horizontal section to yield a time structure map on this horizon with a contour interval of 20 ms. Be careful to follow structural continuity regardless of lateral changes in amplitude. Faults will be seen as lateral displacement, not simply as amplitude changes. As a guide, continuity without amplitude superimposed is visible directly along the adjacent zero crossings.

Stratigraphic Component of Exercise

Using the structure map to identify the high amplitudes associated with the turbidite, outline the dark blue areas for this horizon on each horizontal section supplied. Connect these outlined areas interpretively to yield a stratigraphic map of the extent and possible flow direction of the turbidite. Because you are supplied with horizontal sections at only 20 ms intervals, there will be some gaps in coverage in the direction of dip. More interpolation and smoothing will thus be needed in the dip direction than in the strike direction.

Procedure

Take a piece of transparent paper and register it on the annotation frame of the horizontal sections. Use the vertical sections as a guide to structural continuity only. Complete the structural component before attempting the stratigraphic component.

Solution

One interpreter's map of the extent and structure of the turbidite is shown in Figure A-11. This map is based only on the data supplied for the exercise. The horizon slice and superimposed structure generated interactively and based on all the data are shown in Figure A-12.

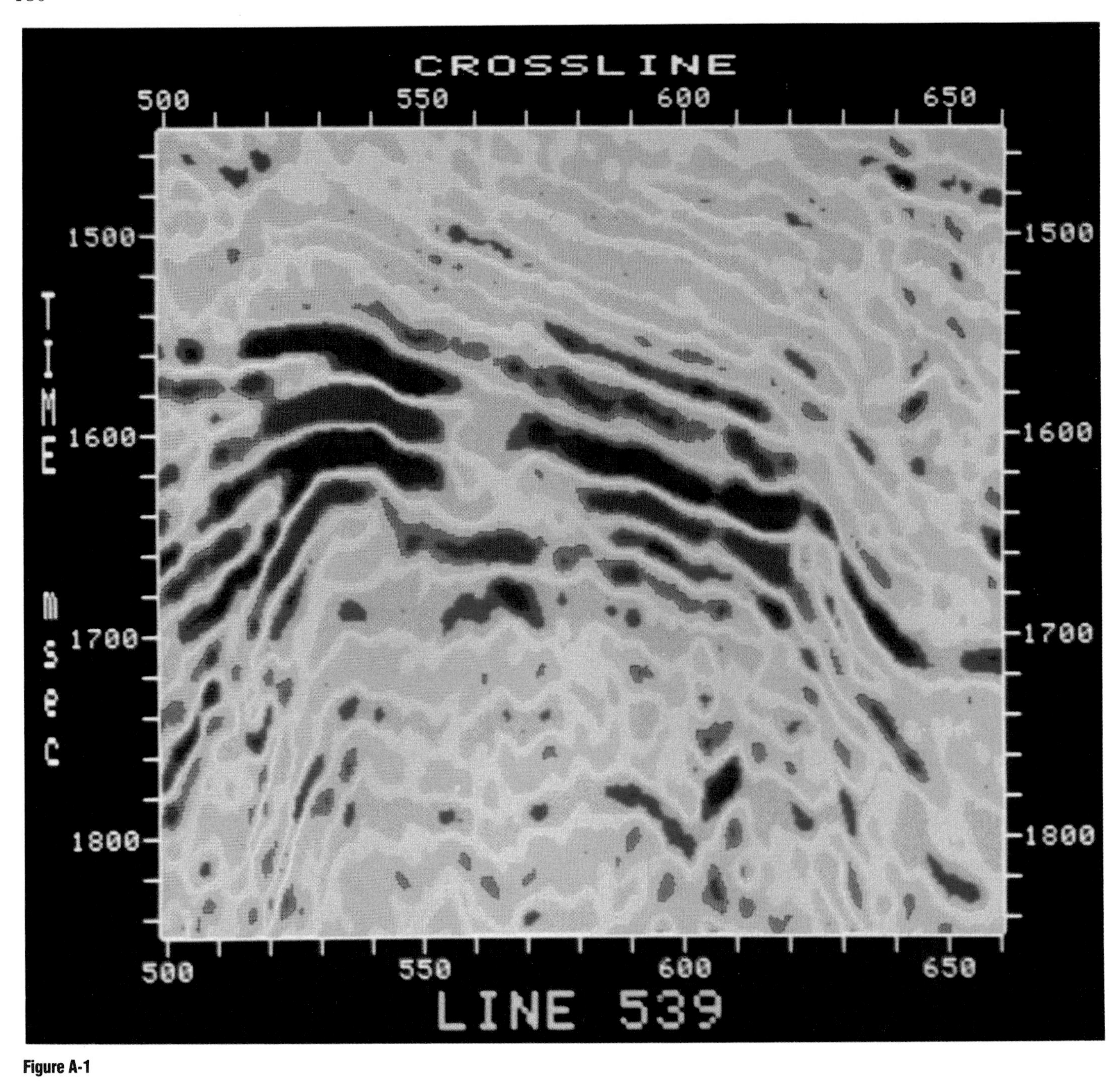
CROSSLINE
500
550
600
650
1500
1600
1700
1800
TIME
msec
500
550
600
650
LINE 539

Figure A-1

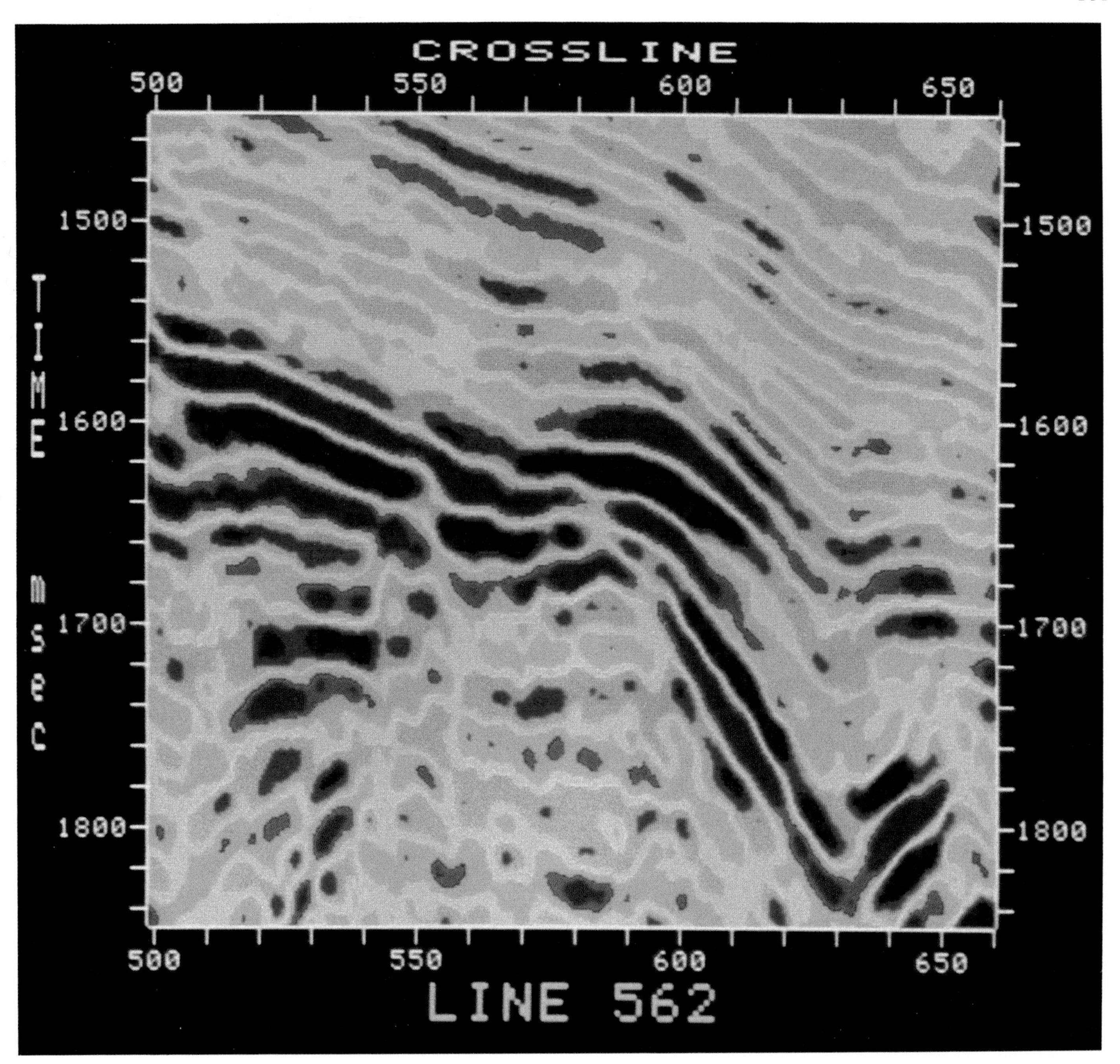
CROSSLINE
500
550
600
650
TIME
msec
1500
1600
1700
1800
LINE 562

Figure A-2

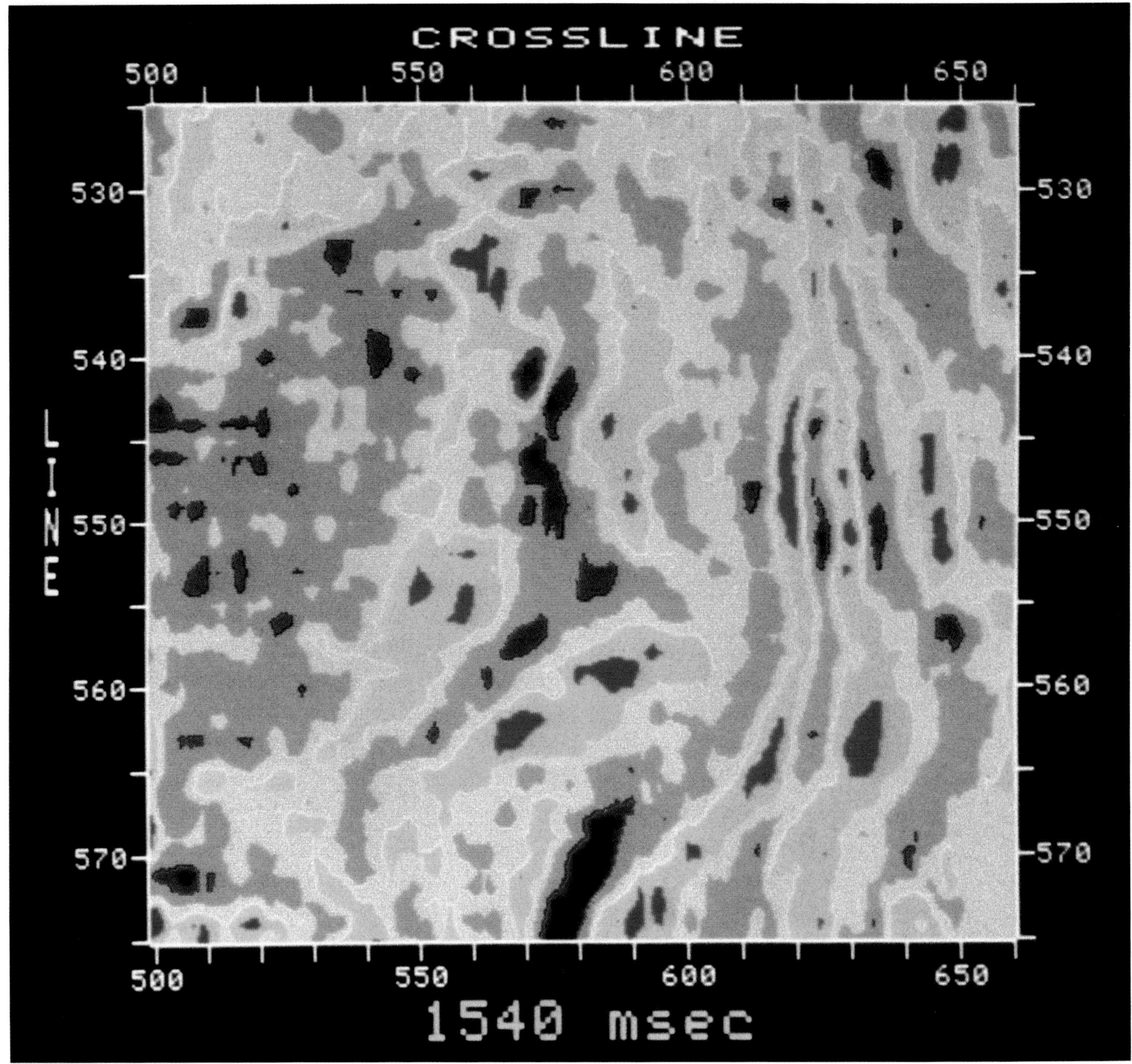

Figure A-3

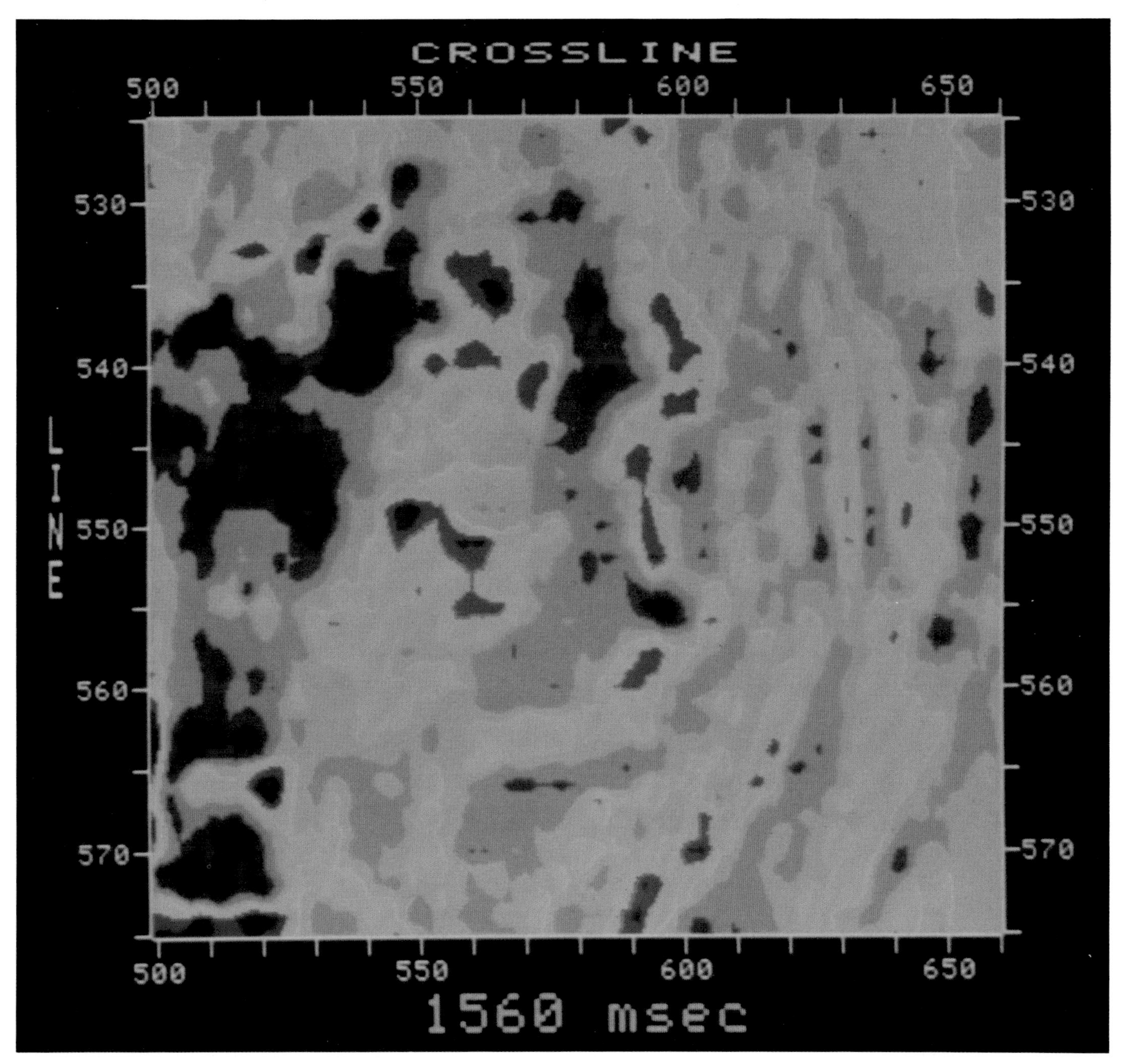

Figure A-4

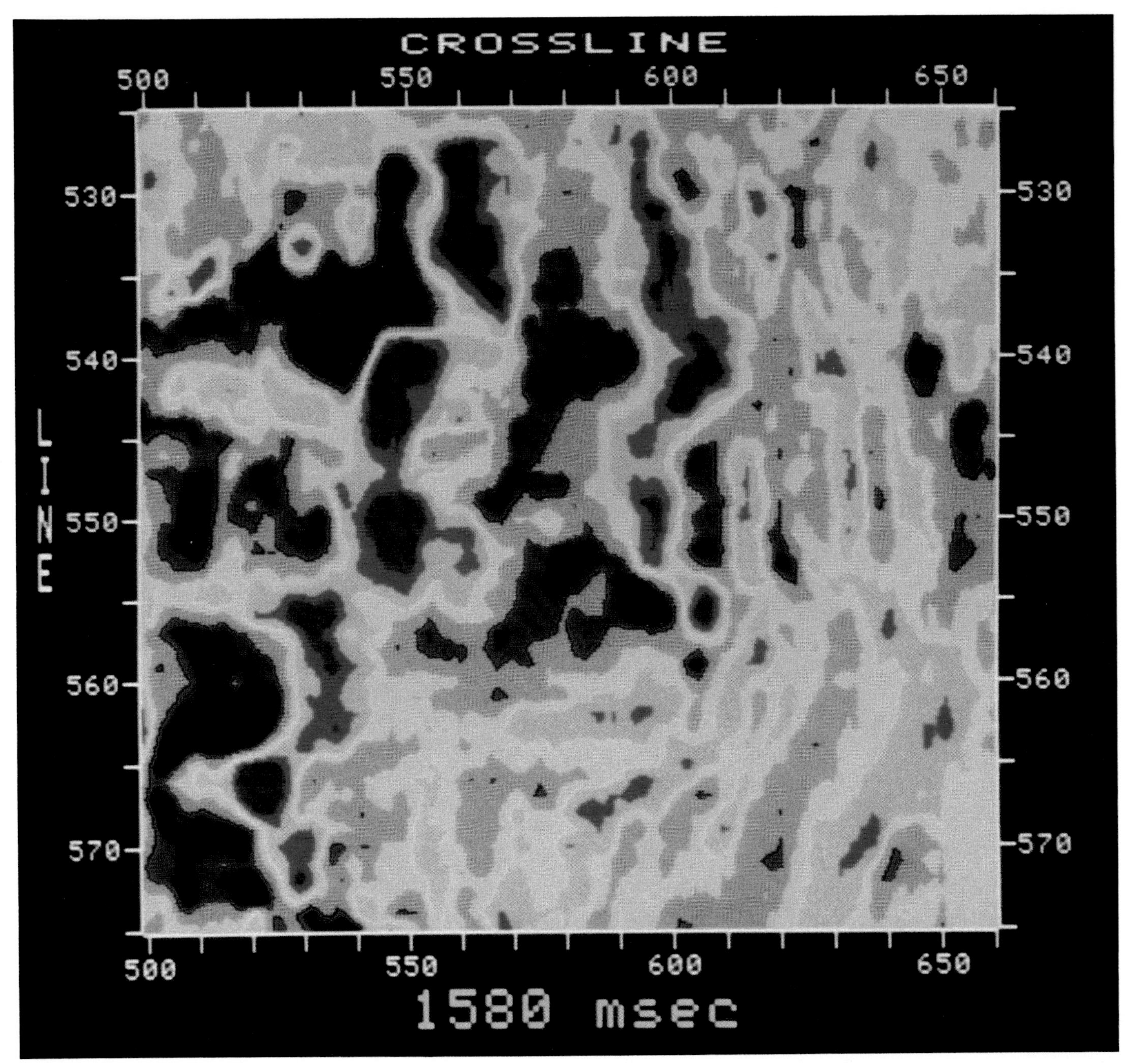

Figure A-5

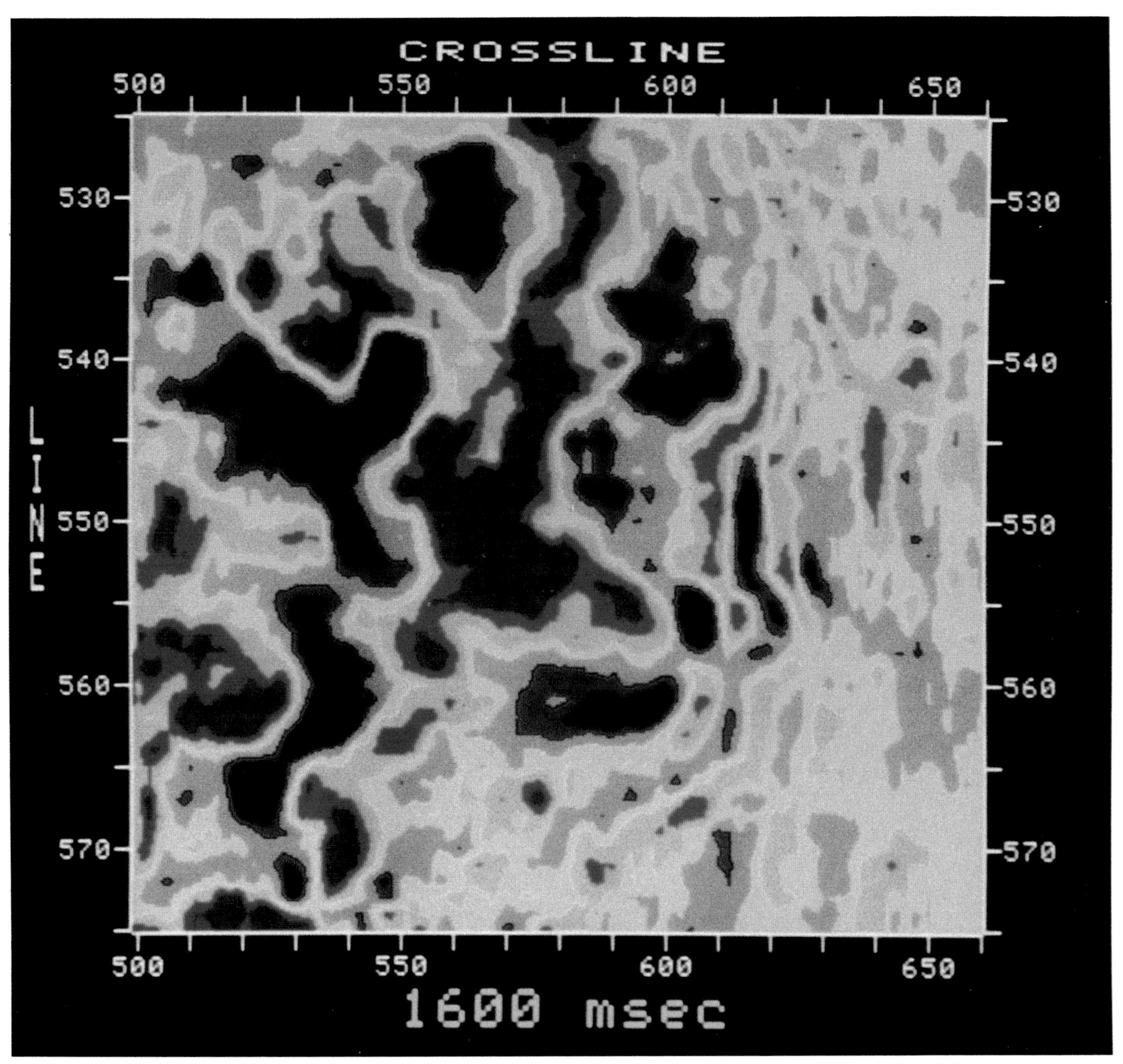

Figure A-6

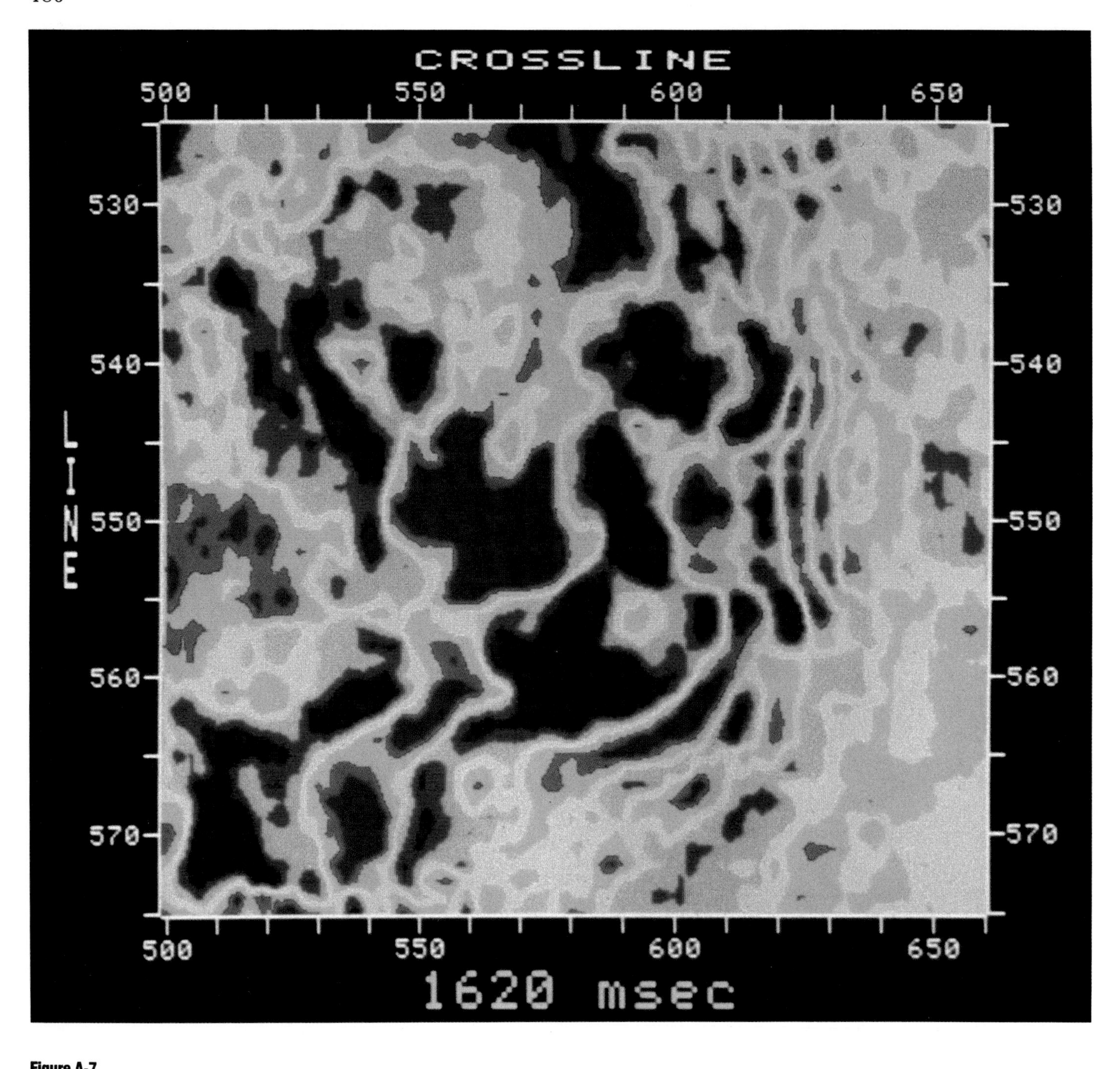

Figure A-7

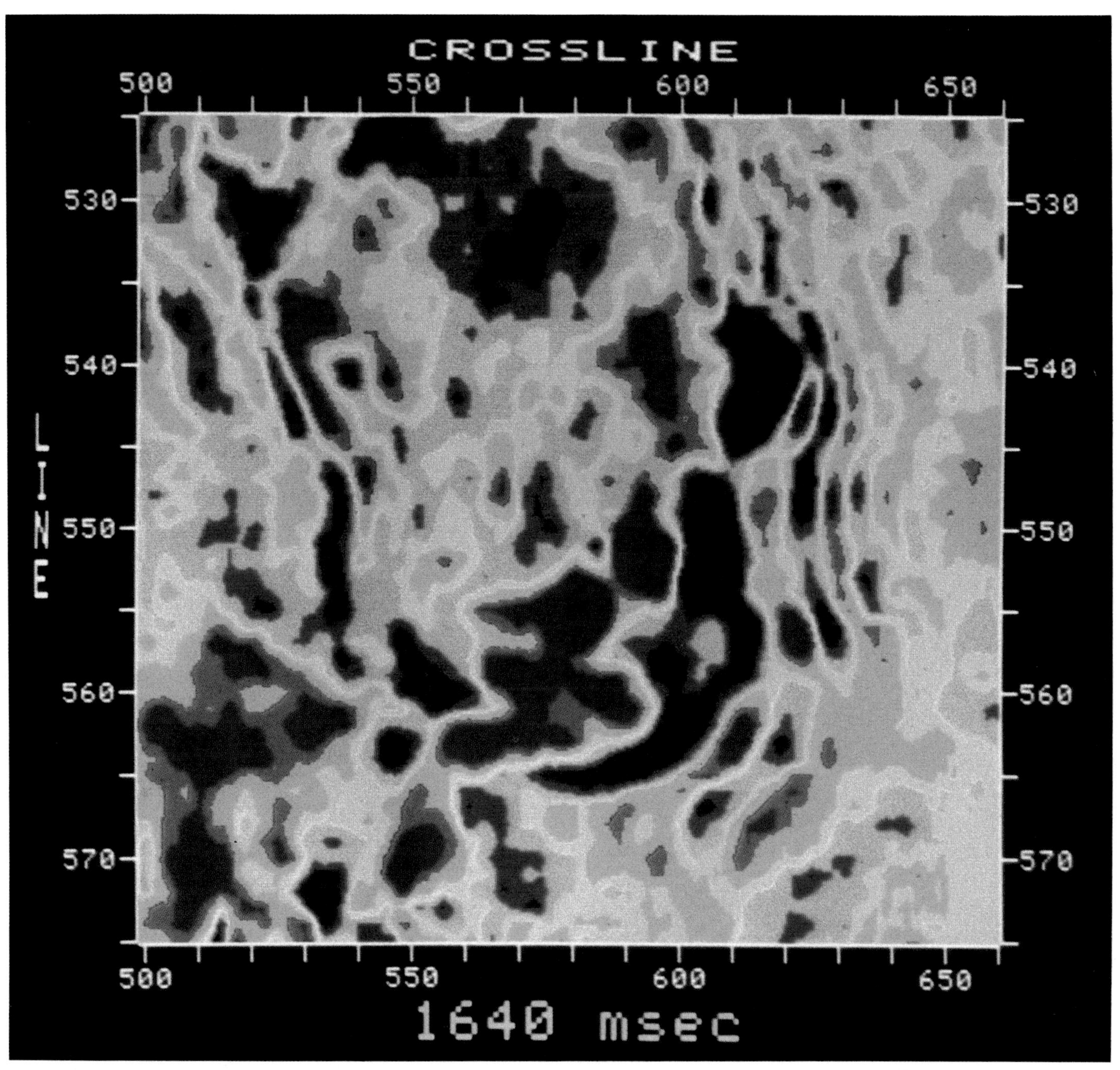
CROSSLINE
500
550
600
650
530
540
550
560
570
LINE
500
550
600
650
1640 msec

Figure A-8

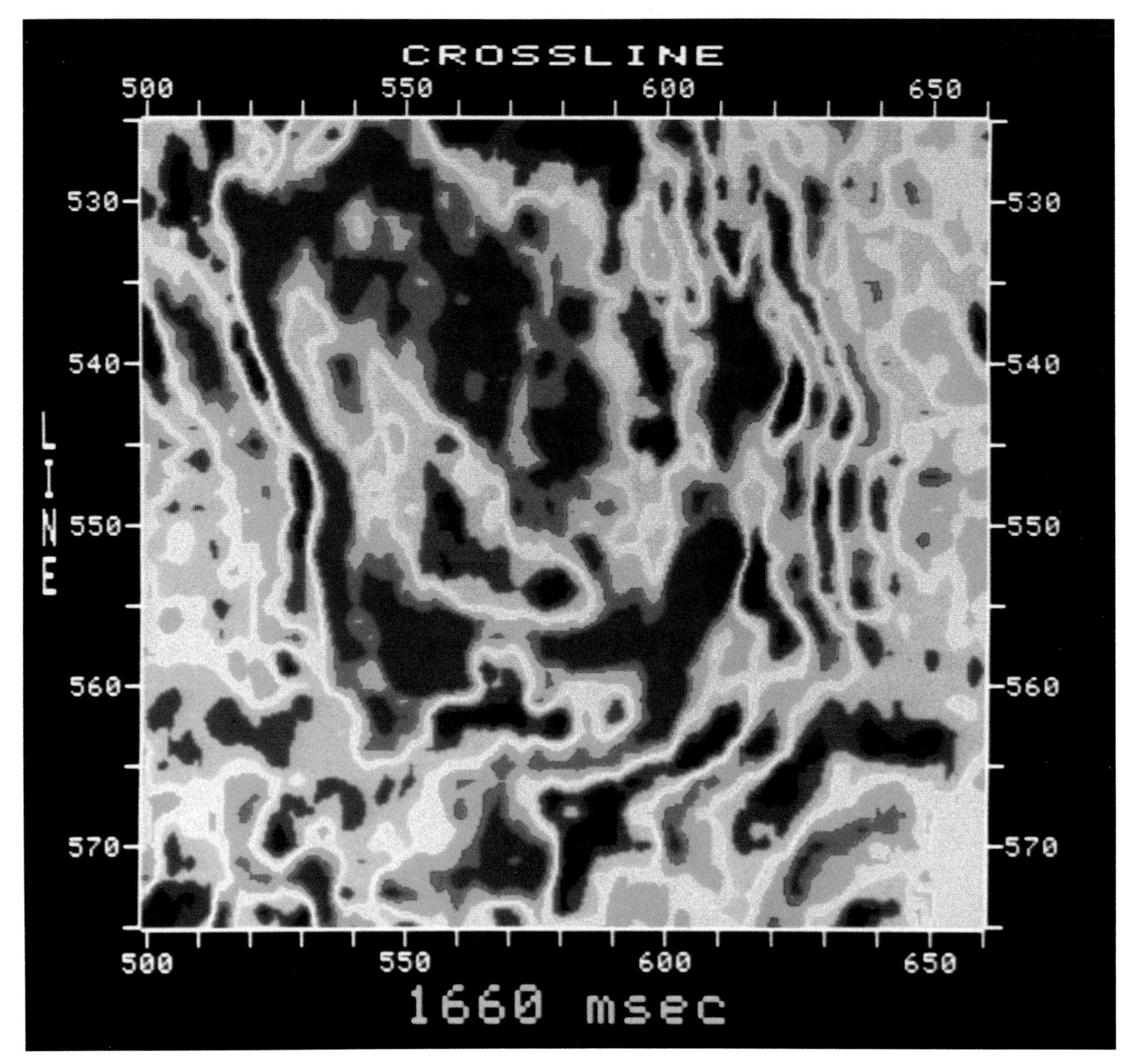
CROSSLINE
500
550
600
650
530
540
550
560
570
LINE
530
540
550
560
570
500
550
600
650
1660 msec

Figure A-9

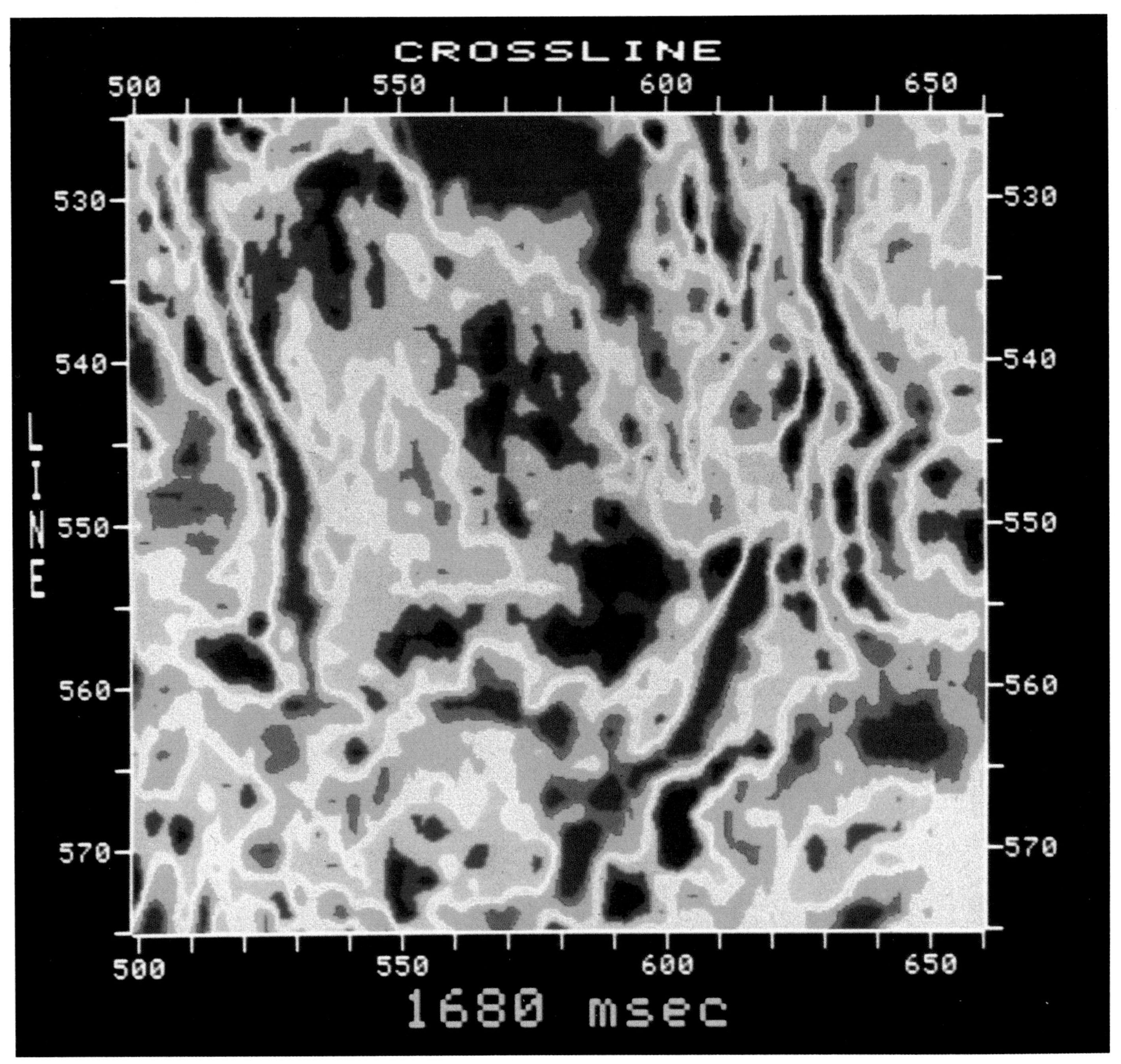
CROSSLINE
500
550
600
650
530
540
550
560
570
LINE
530
540
550
560
570
500
550
600
650
1680 msec

Figure A-10

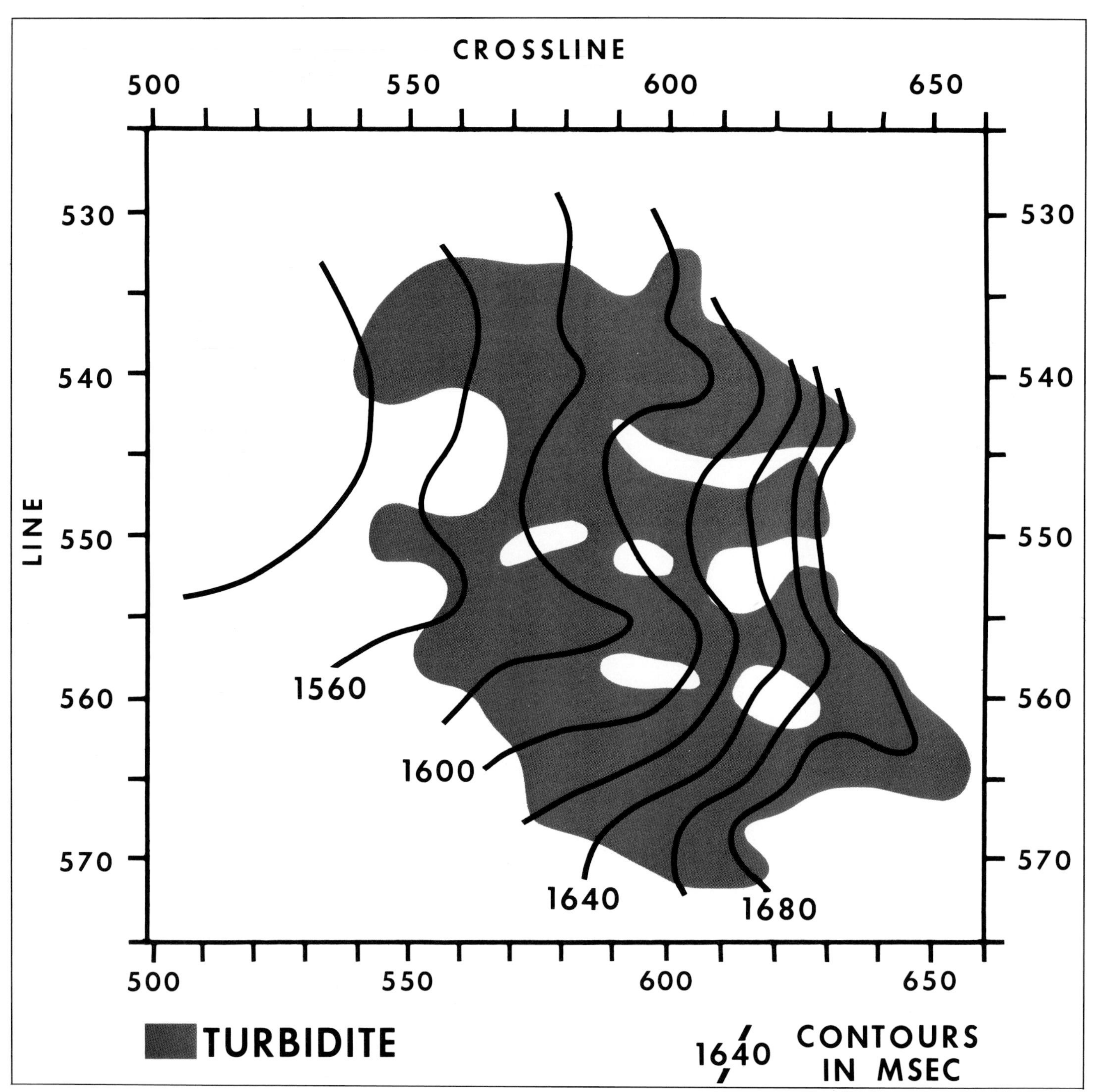
CROSSLINE
500
550
600
650
530
540
550
560
570
LINE
1560
1600
1640
1680
TURBIDITE
1640
CONTOURS
IN MSEC

Figure A-11

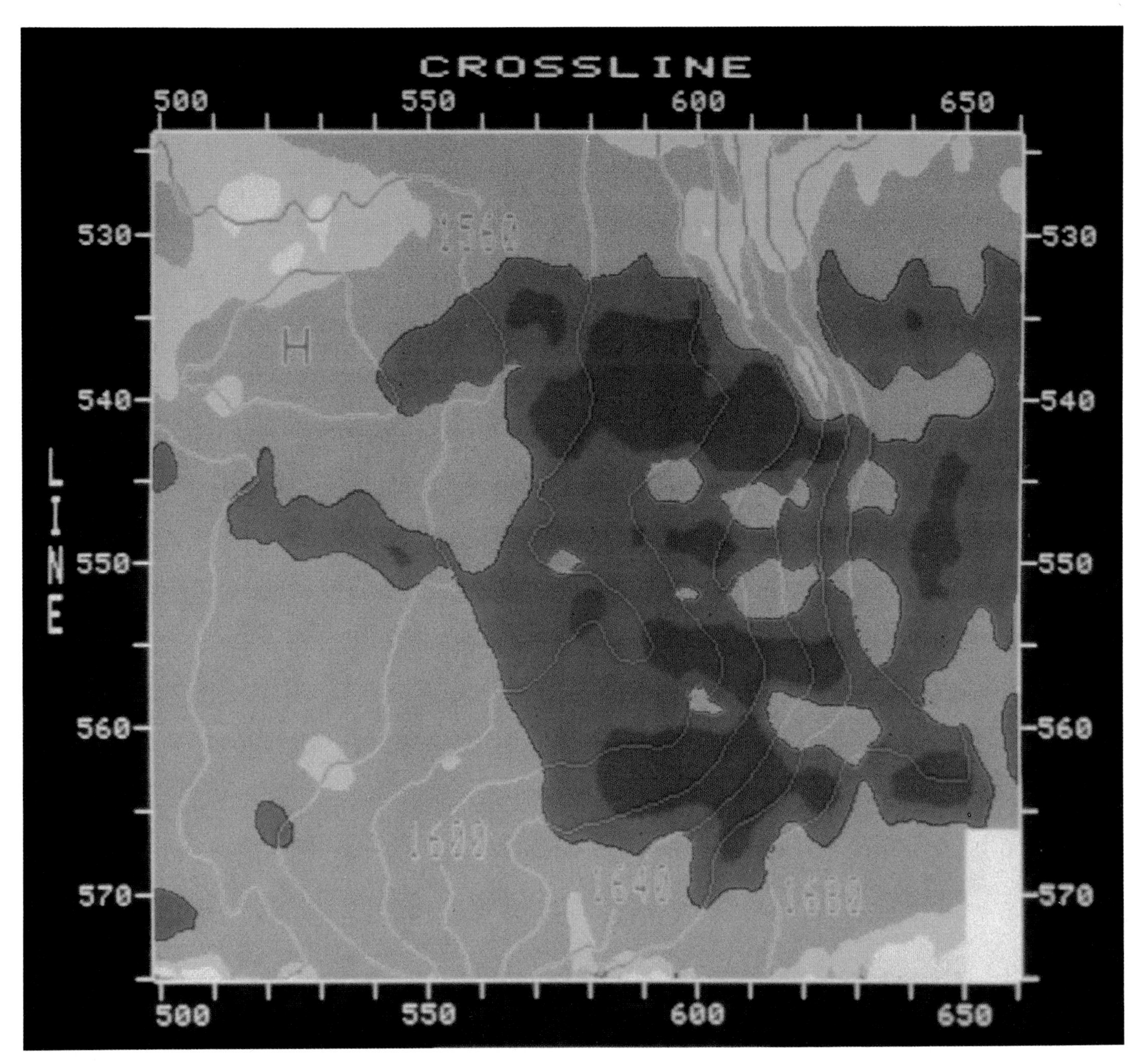
CROSSLINE
500
550
600
650
LINE
530
540
550
560
570
1560
H
1600
1640
1680

Figure A-12

Index

A reference is indexed according to its important, or "key" words.
Three columns are to the left of a keyword entry. The first column, a letter entry, represents the AAPG book series from which the reference originated. In this case, ME stands for AAPG Memoir Series. Every five years, AAPG will merge all its indexes together, and the letters ME will differentiate this reference from those of the Studies in Geology Series (ST) or from the AAPG Bulletin (B).
The following number is the series number. In this case, 42 represents a reference from AAPG Memoir 42. The third column lists the page number of this volume on which the reference can be found.

† = titles * = authors